D. Eisma

Suspended Matter in the Aquatic Environment

With 135 Figures

Springer-Verlag
Berlin Heidelberg New York
London Paris Tokyo
Hong Kong Barcelona
Budapest

Professor Dr. DOEKE EISMA
Netherlands Institute
for Sea Research (NIOZ)
P.O. Box 59
1790 AB Den Burg
Texel, The Netherlands

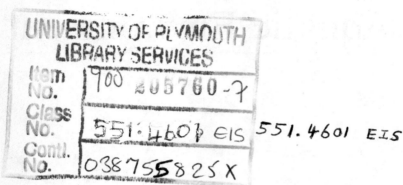
ISBN 3-540-55825-X Springer-Verlag Berlin Heidelberg New York
ISBN 0-387-55825-X Springer-Verlag New York Berlin Heidelberg

Library of Congress Cataloging-in-Publication Data. Eisma, D., Dr. Suspended matter in the
aquatic environment / Doeke Eisma. p. cm. Includes bibliographical references and index. ISBN
3-540-55825-X. — ISBN 0-387-55825-X (U.S.) 1. Sediment, Suspended. I. Title. GB 1399.6.E37
1992 551.46'01 — dc20 92-27025

Typesetting: Macmillan India Ltd., Bangalore 25

32/3145/SPS-5 4 3 2 1 0 – Printed on acid-free paper

Preface

The purpose of this book is to give an introduction to the most important aspects of suspended matter in the aquatic environment: its origin and composition, the concentration distribution, transport and deposition, and the most important physical-chemical-biological process that affects suspended matter: flocculation. In Chapter 1 the development of suspended matter observation and study throughout history is given, with the coming of a more modern approach during the 19th century and the first half of the 20th century, and the development of the present science of suspended matter after 1945. The sources of suspended matter in rivers, lakes, estuaries, and the sea are discussed in Chapter 2, which includes the supply of detrital particles as well as the formation of new particles in the water (organic matter, carbonate, opal). The concentration distribution of suspended matter in rivers, lakes, estuaries, tidal areas, lagoons, and in the sea is discussed in Chapter 3, to which is added a discussion on the sampling of suspended matter and on methods to determine its concentration. Particle composition is treated in Chapter 4, to which is added a section (4.6) on the compositional analysis of suspended particles. Also included is a discussion on particle surface characteristics and the adsorption of elements and compounds onto particles. In Chapter 5, the transport of suspended matter by currents and waves is discussed, which includes the initiation of particle motion (scour, erosion, resuspension) and particle settling, as well as transport and deposition in stratified waters and autosuspension. This is preceded by a short introduction on turbulence (Sect. 5.1), which is the principal agent in bringing and keeping suspended particles in suspension. Particle size and flocculation are discussed in Chapter 6, with a section (6.4) on measuring suspended particle size. A more integrated description of the behavior and fate of suspended matter in natural aquatic environments (rivers, lakes, estuaries, intertidal areas, fjords, and the sea) is given in the last Chapter (7).

By necessity this book has certain limits. The physical, chemical, biological, and sedimentological background is not treated extensively but is referred to, or considered to be known. Also the complex details of the various processes involved have been left out: an extensive reference list is provided. In all this the emphasis is on processes where they affect the particulate matter in suspension. The reverse – the influence that suspended particles have on chemical and biological processes in particular – is treated in only a general way. The reader who wants to understand more is referred to the various publications cited in the text where these processes form the central theme.

During writing and publishing I received much help from G.C. Cadée, R.W. Duck, K. Dyer, L. Maas, R.H. Meade and W. Salomons who read the manuscript, or parts of it, critically. Mrs. J. Hart and Mrs. J. Schröder I thank very much for typing the manuscript and Mr. B. Aggenbach and Mr. B. Verschuur for preparing the figures and photographs.

Texel, 1992   D. Eisma

Contents

1 Introduction

By convention, particulate matter in suspension is defined as the material that is retained on a 0.4 to 0.5 μm pore size filter. Smaller material is considered to be "dissolved" but actually it may be colloidal or particulate: particles as small as 0.02 μm have been observed in natural waters (Gordon 1970; Harris 1977; Eisma et al. 1980), although it is not certain that such particles are not artifacts produced out of larger particles during sampling. As very little is known about such particles in natural waters, particles smaller than 0.4 to 0.5 μm, colloids, large or small molecules, and ions, although chemically important and of possible importance for the flocculation of suspended matter, remain outside the scope of this book.

The upward size limit of particulate matter in suspension is not fixed. Relatively large and heavy particles like sand grains and gravel sink rapidly to the bottom, but very large low-density structures can remain in suspension for quite a long time. Because of gravity, all suspended material with a density greater than the surrounding water will eventually sink to the bottom unless there is a force that keeps it suspended. Normally this force is the drag provided by the turbulent motion in the water: whether the material remains in suspension or not is closely related to the intensity of the turbulence, as well as to the size, density, and shape of the particles involved. In strong turbulence even heavy and large sand grains can go into suspension and remain suspended for some time, as happens in rapidly flowing streams or in the heavy surf along beaches. This can also happen when sand is suspended in high concentrations of fine mud: the settling of the sand grains is hindered by the large number of fine particles. Particles containing gas (air) or low-density organic material can be buoyant and move predominantly upwards through the water.

Deposition of suspended matter usually results in a fine-grained deposit ("mud") with a grain size less than ca. 100 μm. When the suspended material contains large amounts of relatively large hollow biogenic particles such as foraminifera tests or diatom frustules, an "ooze" is formed that contains particles much larger than 100 μm. In lakes and reservoirs behind dams, deposition of suspended material supplied by a fast-flowing stream also can result in a relatively coarse-grained deposit.

The fine-grained material in suspension has a large specific surface (surface-area per unit weight), which is larger when the particles are finer. Because organic matter is usually fine-grained, and forms coatings on the particles, while the fine mineral particles in suspension are mostly clays, the fine-grained suspended material is highly surface-active. This implies that dissolved or colloidal material is easily adsorbed onto the particle surfaces. A partitioning takes place between material that is adsorbed and material that remains in solution, which depends on the concentrations and specific properties of both the dissolved substances and the particles in suspension. During the past ca. 125 years this property has had – and still has – large consequences for the dispersal of polluting substances such as trace metals and a number of synthetic organic compounds. In this way, pollutants also can become concentrated and buried in sediments. Pollutants that are discharged in fine-grained particulate form (organic waste, fly-ash, gypsum), also end up in mud deposits, in so far as they are not dissolved (gypsum) or mineralized (organic matter) on the way before being deposited.

In other fields, too, suspended matter plays an essential role. Particulate organic matter often has a high nutritional value and thus can be a major source of food for many aquatic organisms, but mineralization of accumulated particulate organic material can result in very low dissolved oxygen values and even in a high mortality of aquatic fauna when anoxia occurs. Both processes have consequences for the abundance of aquatic life, including fish and large mammals, and are of great interest to fisheries. Sedimentary geologists have an interest in the formation of recent fine-grained deposits from suspension in order to understand how, and under what conditions, fine-grained mudstones or shales in the geological record have been formed. The organic matter in such sediments has been a basis for the formation of oil and natural gas. An early interest from engineers and soil scientists came from the accumulation of mud in coastal areas, which leads to siltation of harbors and shipping channels, but also to the formation of new land (marshes, flats) that can be reclaimed.

Suspended matter is present in all natural waters of the world. It may be a very small amount, as in the crystal-clear waters in caves and in some parts of the ocean, but microscopic inspection up to now has always indicated the presence of at least some suspended particulate material. Because of this ubiquitous presence, and because of the physical and chemical properties of the particulate material itself, the suspended matter forms an integral part of the worldwide geochemical, biological, and geological cycles in the aquatic environment.

1.1 Suspended Matter Observations in History

The interest in suspended matter and the realization of its importance are quite recent. In Antiquity, the interest was mainly restricted to the effects of large rivers carrying (suspended) sediment into the coastal sea, causing extensive siltation in embayments and filling up the shallow straits between some islands and the shore (Berger 1903; Forbes 1963). The Nile and the Don were the classic examples (Aristoteles Meteor I.XIV; Herodotus II, 10–12) but also for other rivers there were descriptions of river mouths choked with mud (Po, Anio rivers), seaward shifts of a delta coast (Meander and Pyramus rivers in present Turkey), islands becoming attached to the mainland (Pharos island in Egypt) and formation of offshore mud deposits (Po river; Danube; Thucydides II.102; Polybius IV.41; Plinius III.1, II.1.5; Strabo V.1.5, X.2.19, XII.2.4-19). Soil erosion was a recognized phenomenon in ancient Greece (Plato Critias III), and Pausanias (VIII.24.11) in the 2nd century A.D. related the heavy siltation in the valley and the mouth of the Meander (Menderes river) to the intense cultivation of the river basin.

During the Middle Ages, the old observations were preserved but little was added to them except by the Arab travelers, who gave descriptions of south Asian rivers and coasts, of the upper Nile, and of the Aral Sea with adjacent waters (Miquel 1975, 1980). In China, however, a description of 137 rivers was already made in the 1st century B.C. which in the 6th century A.D. was extended to more than 5000 rivers by Li Tao-Yuan (Shui Ching Chu; Needham 1959, 1971). Shen Kua described in 1070 the muddy silt-bearing rivers in northern China which erode the mountains and carry the mud eastwards, where it is deposited. Shan O, in the 11th century, wrote a book on water conservancy based on observations in the area around Taihu Lake near the lower Chang Jiang and a study on the Yellow River valley was published in the 12th century by Fu Yin. As water engineering flourished in China, practical knowledge on the behavior of suspended matter was developed early. This included experience with extensive canalization, dike construction, and silt precipitation from irrigation water, which rendered former alkaline soils fertile. Along the Yellow River dikes were made according to the observation that where the stream is wide, flow is slow, sediment is deposited, and the river bed rises, whereas where the river is narrow, flow is rapid, sediment is transported away, and the river deepens (Pan Jixun, 15th century A.D.). Water level records existed in China already ca. 2000 years ago. The first known measurement of suspended matter content in a river was made by Chang Jung in ca. 1 B.C. in the Ching river, located in the loess area and a tributary of the Yellow River (60% was recorded, probably based on decantation, as the amount of suspended matter is expressed in dry volume per total sample of river water; modern measurements give up to 67%; Needham 1971).

In Europe after the Middle Ages, observations as made during Antiquity were continued by Leonardo da Vinci (Notebooks 1508–1519, 1948), who also looked into the conditions of mud deposition: at low current velocities (particularly where a river broadens) and in the quiet areas between plant roots along the borders of a river. With the scientific revolution in the 17th and 18th centuries, the approach gradually changed and emphasis came on systematic observation of nature coupled to a theory to explain the observed phenomena (Ellenberger 1988). The theory could only be accepted on the basis of consistence and proof – in the natural sciences usually observations that can be predicted and repeated. John Ray (1673) did not yet come further than to explain lowland mud deposits – and particularly those extending to 100 ft depth below Amsterdam – as supplied by rivers, but gradually more observations were made so that Lyell could make an extensive review of them in his *Principles of Geology* (Part I) which appeared in 1830. Knowledge on rivers, lakes, and the coastal sea had increased enormously by that time, and accurate descriptions ranged from the Mississippi and the Great Lakes to the large European lakes and rivers, the tideless deltas of the Baltic and the Mediterranean, large sediment-loaded rivers like the Ganges-Brahmaputra, Yellow River, and Amazon, and the coastal areas of western Europe. Lyell noted the lack of satisfying measurements of the sediment load of rivers: only a few were available, the oldest one known to him being an estimate of the suspended load of the Yellow River ($5 \, \mathrm{g \cdot kg^{-1}}$) made by Staunton (1797).

1.2 Development of a More Quantitative Approach

The lack of quantitative data changed during the 19th century and series of observations became available for many rivers, for some rivers like the Danube and the Rhine extending for almost 30 years (Penck 1894; Santema 1953). Measurements were made in relation to current velocity, total discharge, and water temperature. The latter was done because the relation between the settling velocity of particles in water and the viscosity of the water (which is related to its temperature) had been worked out by Stokes (1845). The methods of determination, however, were not very precise, nor were they uniform or calibrated, so that the suspended matter concentrations obtained during that period can only be used in a very general way. A common method was to let a known volume of water stand quietly for some time and then measure the thickness of the mud that was deposited. As this depends on the degree of consolidation of the mud and its water content, often the concentrations estimated in this way were much too high. Reliable figures, however, were obtained already in 1865 for the Rhine by drying and weighing the settled mud (van der Toorn 1868).

During the 19th century particularly biologists developed an interest in the turbidity of the water: the penetration of light into the water influences plankton growth and regulates the phototropic movements of zooplankton. Hearne (1705) already lowered a white disc in Lake Vättern in Sweden and several other devices (white dishes and screens) were used from the end of the 17th century to estimate the transparency of the water (Thoulet 1905; Hutchinson 1957). Secchi (1865) standardized this by lowering a white disc of 30 cm diameter into the Mediterranean and measuring the depth at which it became invisible. Although this method is not very accurate, it gives a reliable and quick estimate of turbidity (or transparency) of the water. It has been used now for more than 125 years.

Other measurements in the sea followed the general development of oceanography since the end of the 19th century. Ocean sediments were studied systematically for the first time during the Challenger Expedition in 1873–1876

(Murray 1891). The Secchi disc had been used in 1873 in Lake Geneva by Forel (1892–1901) and was for the first time on a large marine research cruise lowered from the *Michael Sars* in 1910 (Murray and Hjort 1912). Thoulet (1905) made calibration curves to estimate the suspended matter concentration from the depth of Secchi disc visibility, which was followed by in-situ photometry of turbidity with photo-electric cells (Knudsen 1922; Shelford and Gail 1922) and measuring water samples in a spectrophotometer (Kalle 1935, 1937). From there it was only a short step to develop a beam transmissometer that made it possible to determine turbidity (also horizontally) at any water depth (Petterson 1934).

Kolkwitz (1912) had defined all particulate matter in suspension in lakes as "seston", which was extended to particulate matter in seawater by Krey (1949). Seston was separated into plankton (living), "detritus" (nonliving remains of organisms) and mineral particles. Thoulet (1922), in his book on oceanography, gave much attention to sediment supply and its dispersal in the sea. He recognized that supply from the continents and from oceanic plankton growth were the principal sources of particulate matter in the ocean, besides a smaller contribution from aeolian transport, icebergs, and volcanism. The importance of particle transport processes was emphasized and he studied the relation between particle supply at the ocean surface and deposition on the ocean floor. At that time the "neutral line" (Cornaglia 1891) or "mud" line (Murray and Hjort 1912; Thoulet 1912) was seen as an important sedimentological feature marking the outer limit of seaward sediment dispersal from the coast. It has been related by Cornaglia (1891) to wave action and was considered an important parameter for the location of a harbor. Thoulet (1922) showed that there are as many "neutral lines" as there are particles of different size, density, and shape.

1.3 Suspended Matter Transport Studies

While hydraulics were already studied in Antiquity and had resulted in practical applications (notably by Archimedes in the 3rd century B.C.; Rouse and Ince 1957), the first (theoretical) approach to particle transport mechanics was by Newton (1687), who gave a formula for the drag of a particle, which is the tractive force exerted on the particle by flowing water. When this force acts on a particle on the bottom and is larger than the friction force that resists the tendency of the particle to move, the particle is picked up. This was first expressed in a formula by Brahms (1757), based on Newtonian mechanics, and later independently by Airy (1834), whose name is still attached to it. Measurements made by Du Buat (1786) on the pick-up velocity of various types of (single) particles agreed well with this formula. Since particles are usually of variable size and weight and the fine ones tend to fill the voids between the larger ones, other scour criteria than those dealing with single particles became of interest so that mixed sediments could be dealt with. Still in use is the one developed by Chézy (ca. 1770, 1921; Mouret 1921), who related the tractive force per unit of surface (or drag) to the flow velocity, in order to be able to compare flow conditions in different streams. During the 19th century it became accepted that also the velocity gradients in the flow are of importance: a dynamic equilibrium (i.e., a constant suspended matter concentration) would exist when the lifting force, related to the vertical change in the tractive force, is balanced by the specific gravity of the particle. This theory, however, could not explain the large concentration variations that were observed and, like modern theory, was hampered by the fact that fine-grained material is cohesive. In practice, these difficulties were overcome by introducing semi-empirical formulae with coefficients whose numerical value was determined from field or laboratory measurements.

The basic relations for fluid motion, which mark the beginning of present-day fluid dynamics, had been formulated by Euler in 1755, and the basic equations of motion (the Navier–Stokes equations) were derived by Stokes in 1845 largely based on the earlier work of Navier in 1827. These were relating to viscous flow, but gradually turbulence became recognized as being fundamental to all transport phenomena in natural flow. Du Buat in 1786 (Lugt 1983) had already given the first evidence that a fluid adheres to an immersed solid surface (the no-slip condition). This is important for understanding the development of turbulent whirls (vortices) as this tangential surface force exerts a torque on the fluid above it, imparting angular momentum to it. Boussinesq, in 1877, formulated an exchange coefficient for turbulent flow to replace the molecular viscosity coefficient used in laminar flow theory. The work of Von Helmholtz (1858), Lord Raleigh (1893) and Lord Kelvin (Sir W. Thomson 1869) made clear that natural flow is almost always unstable, which results in the turbulence that can be observed everywhere in flowing water, as well as in air. Criteria defining the boundary conditions marking the transition of laminar flow to turbulent flow – or the reverse – were given by Reynolds (1883, 1894). Prandtl (1905), who is generally regarded as the founder of present-day fluid mechanics, developed the concept of a thin laminar boundary layer present at any surface in contact with flowing water, with turbulent flow at some distance away from it. This focused the attention to the flow conditions in the immediate vicinity of a particle. Around 1900, also a more empirical approach to suspended sediment transport originated (Leliavsky 1955), which related the capacity for suspended matter transport to the depth of a channel (Kennedy 1895) or to the sixth power of the current velocity (Lacey 1930). The concept of a suspended matter concentration distribution being the result of turbulent diffusion of suspended particles through a flow (by eddy diffusivity) was first worked out by Rouse (1937), while theory based on basic flow dynamics was first expressed by Velikanov (1944). These theories, particularly the diffusivity concept, form the basis of the present approach to suspended sediment transport, but the quantification of suspended sediment transport continues to be elusive, as the in-situ behavior of suspended matter remains poorly known.

Filtration of river and lake waters to determine suspended matter concentrations and to collect material for chemical analysis was carried out at the beginning of the century with asbestos or paper filters. Also filtration of seawater for chemical analysis was already done early in this century (e.g., Atkins 1926) but was applied for the first time by Gry in 1942 to determine suspended matter concentrations (in the Danish Wadden Sea with standard laboratory filters). Only after the second World War was systematic determination of the suspended matter concentration and composition by filtration started in the sea on a large scale: by Krey (1949, 1950, 1952, 1953) in the Baltic and the Southern North Sea, and by Armstrong and Atkins (1951) in the English Channel. This was followed by the development of standard filters. Molecular (membrane) filters of cellulose acetate were first used by Goetz and Tsuneishi (1951) for bacteriological analysis of water and their application to oceanographic research was worked out by Goldberg et al. (1952). Nuclepore filters, made of polycarbonate and with a more precisely defined pore size, came into use around 1975. Shortly after the introduction of the molecular filters, the present custom of separating "particulate" from "dissolved" material with a 0.4- or 0.5-μm pore size filter became established. The formation of organic particles (algae) in the water was initially studied by microscope and through measurements of the production in the water of dissolved oxygen, which is produced during primary production. This method had already been used in 1916 by Gaarder and Gran in Oslofjord (Report ICES 1927). After 1950, the more quantitative determination of the aquatic primary production using ^{14}C began to receive interest (Steeman Nielsen 1954; Lieth 1975), although the correct equation for photosynthesis ($CO_2 + H_2O$ + light → plant matter + O_2 + chemical energy) had already been developed between 1771 and 1845 (Rabinowitch 1971), and Von Liebig had formulated the importance of nutrients for plant growth in 1840.

After 1945, as a result of the development of nuclear physics, natural stable and radioactive isotopes (the U and Th series, ^{14}C) as well as artificial ones supplied by bomb explosions and industrial discharges were increasingly used for dating carbon-containing materials (shells, wood) and fine-grained sediments, and as tracers for suspended sediment transport and deposition (Broecker and Peng 1982). To collect quantitatively the amount of suspended matter settling through the water down to the bottom, sediment traps were developed. They were first used in lakes in 1873 (Heim 1900), and in 1975 for the first time in the ocean (Phleger and Soutar 1971). The realization that also in the deep ocean, currents may be strong enough to maintain high concentrations of suspended matter and transport large quantities, came after Wüst in 1955, on the basis of data collected in 1925–1927, had demonstrated the possible existence of such currents in the Atlantic Ocean. Size analysis of the suspended particles was for a long time limited to microscope observations and the use of Stokes' Law for the settling of particles in a fluid. This was improved by the development of the Coulter counter, which measures particle size electronically and was first developed to count blood cells (Coulter 1956). The first in-situ observations of suspended matter in the water from underwater observation chambers were made by Beebe (1934) and Suzuki and Kato (1953). Recently, it became generally recognized that suspended matter is mainly present in the form of fragile flocs and in-situ systems were developed for measuring their size, shape, and settling velocity (Bale and Morris 1987; Eisma et al. 1990).

With accurate filtration and in-situ observation, combined with precise chemical, mineralogical, microscopic, and particle size analysis, transport modeling, the development of sediment traps, and the use of isotopes, the modern research on suspended material that forms the basis of this book took shape. Publications on suspended matter, turbidity, or fine-grained bottom sediments now contain few references to work done prior to 1948 and only rarely to publications from before 1920. Even for rivers, for which early reliable quantitative data exist, the data, with a few exceptions, cover only 50 to 100 years, but usually a much shorter period. This means that the science of suspended matter developed mainly after the activities of man (canalization and dam construction in rivers, irrigation of river plains, coastal engineering and land reclamation, deforestation, agriculture, mining, road construction, and pollution) had already caused large regional changes in the supply, composition, transport, and dispersal of the material in suspension. These changes and their possible or known adverse effects, besides being of general scientific interest, have been, and still are, an important stimulant for suspended matter studies.

2 Sources of Suspended Matter

The amount of water in the sea is 1.37×10^9 km^3, which is ca. 94% of all water in the hydrosphere (van der Leeden 1975; Degens 1989). In rivers and lakes, 280 000 km^3 is present, which is ca. 0.019%. The remaining ca. 6% is contained in ice (ca. 1.6%), groundwater (ca. 4%), and the atmosphere (0.001%). The total amount of suspended matter in the sea is ca. 3×10^{16} g (at an average concentration of ca. 0.02 mg.l^{-1}). The total amount of suspended matter in rivers and lakes is less well known, but can be estimated to be in rivers ca. 0.425×10^{12} g (at an average concentration of 340 mg.l^{-1}), as will be discussed in more detail below.

Virtually all suspended matter is supplied either by terrestrial erosion (runoff, mass flows, ice flows, eolian supply) or through the production of organic matter, biogenic carbonate, and biogenic opal. Volcanism supplies an insignificant amount, but can be regionally and temporarily an important source. Authigenic (inorganic) mineral formation and cosmic material are quantitatively negligible. Authigenic minerals can regionally or locally be of importance in relation to geochemical processes; the presence of particles of cosmic origin can be important for the reconstruction of cosmic events.

2.1 Suspended Matter in Rivers

Milliman and Meade (1983) estimated that rivers carry ca. 13.5×10^{15} g.y^{-1} to the sea in suspension together with $1-2 \times 10^{15}$ g.y^{-1} of bed load (Milliman and Meade 1983). This includes some sand as well as silt and clay and is equivalent to what is called "wash load". The 25 rivers with the largest suspended load ($> 40 \times 10^6$ t.y^{-1}) are indicated in Table 2.1. Most of the river supply – ca.70–80% – reaches the sea in south and east Asia, where most of the large rivers with a high sediment load are located – the Ganges-Brahmaputra, the Huang He or Yellow river, the Irrawaddy, and the Chang Jiang being the largest – and from the large oceanic islands of the western pacific. The Amazon is the only large river with a high suspended load outside this area (Fig. 2.1). Recent work by Milliman and Syvitsky (in press) showed that when considering mainly large rivers, the contribution from the more numerous smaller ones is underestimated. Material eroded in small river basins has a better chance to reach the sea than material eroded in large river basins with flood plains and

Table 2.1. Average sediment discharge to the ocean of the 25 rivers with the largest sediment load. (Meade in press)

	Average sediment discharge[a] (10^6 t.y^{-1})	Average water discharge (km^3.y^{-1})	Average concentration (mg.l^{-1})
Amazon	1000–1300[1000–1300]	6300	190
Yellow River (Huang He)	1100 (100) [1200]	49	22040
Ganges-Brahmaputra	900–1200	970	1720
Chang Jiang	480	900	531
Irrawaddy	260 [260]	430	619
Magdalena	220	240	928
Mississippi	210 (400) [500]	580	362
Godavari	170	92	1140
Orinoco	150 (150) [150]	1100	136
Red River (Hung Ho)	160	120	1301
Mekong	160	470	340
Purari/Fly	110	150	1040
Salween	~ 100	300	300
Mackenzie	100 (100) [100]	310	327
Parana/Uruguay	100	470	195
Zhu Jiang (Pearl)	80	300	228
Copper	70 (70) [70]	39	1770
Choshui	66	6	11000
Yukon	60 (60) [60]	195	308
Amur	52	325	160
Indus	50 [250]	240	208
Zaire	43	1250	34
Liao He	41	6	6833
Niger	40	190	210
Danube	40 [70]	210	190
Rivers that formerly discharged large sediment loads			
Nile	0 [125]	0 (was 39)	
Colorado	< 1 [125]	1 (was 20)	
Other rivers that discharge large volumes of water			
Zambesi	20	220	90
Ob	16	385	42
Yenesei	13	560	23
Lena	12	510	24
Columbia	8 [15]	250	32
St. Lawrence	3	450	7

[a]( ) Presumed natural level; [ ] year 1890.

deltas. They estimated that prior to dam construction about 20×10^{15} g.y^{-1} was transported to the sea. About half of this was supplied by rivers with basins smaller than 10 000 km^2.

For most rivers, the data on which the estimates for average yearly suspended load are based are inadequate. The suspended matter concentration

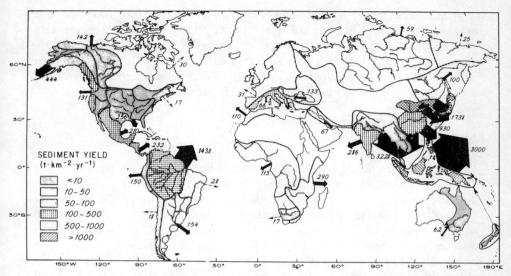

Fig. 2.1. Annual discharge of suspended sediment to the oceans from various drainage basins in the world. *Width of arrows* corresponds to relative discharge, *numbers* indicate average annual input in 10^6 tons. *White areas* essentially no discharge to the ocean. (Milliman and Meade 1983)

varies across a river channel, and mixing of suspended material supplied from tributaries may not be complete so that gradients can persist over large distances. There may also be hydraulic effects within a channel, resulting in an inhomogenous distribution of suspended material. Supply varies with time, in some rivers reaching zero during the dry season when there is no water flow and being at maximum during periods of flood. Added to this are variations over the years, changes induced by man and exceptional or catastrophic events resulting in very high suspended loads. To estimate sediment transport, usually a relation is established between the suspended load and water discharge or water level at a fixed station, but this relation itself is not constant: suspended matter concentrations often are higher during rising water level than during falling water because, during the rising stage, material is picked up that was deposited during the preceding falling stage. But also the reverse can happen, as in the Chang Jiang, where the rice fields are drained during the falling stage, and much suspended material is then supplied to the river (Meade 1988). Where measurements include an entire cross-section and have been carried out for several decades, a good estimate of suspended sediment supply, can be made. Of 21 rivers listed in Table 2.1, which together supply ca. 49% of the total river supply, the data for only 14 rivers are considered adequate to make a reliable estimate of average suspended load; the data for the remaining seven rivers, including the Ganges-Brahmaputra, the Amazon, and the Irrawaddy, are considered inadequate (Milliman and Meade 1983).

The reliable river sediment supply data cover at most 50 to 100 years, but usually a much shorter period. Rivers like the Rhine, for which reliable data were obtained already in 1858, are an exception. The present data therefore reflect a mixture of natural and man-influenced supply, but in spite of human influence, the river suspended load generally reflects the properties of rocks and soils in the basin, as well as topography and climate.

2.1.1 Storage of Sediment

The total amount of suspended matter carried by rivers is not the amount that reaches the ocean. Part of the material is brought to inland seas, particularly to the Caspian and Aral Seas, where some large rivers discharge (the Volga, the Amur, and Syr Darya). Based on the area involved and the supply from these rivers (Holeman 1968; Milliman and Meade 1983; UNESCO 1972, 1979), the total yearly supply in the inland drainage areas (the endoreic areas) (Fig. 2.2) can be estimated to be ca. 150×10^6 t . y^{-1}. But also in the areas that drain into the oceans (the exoreic areas), part of the material eroded from watersheds and reaching the rivers by runoff and mass flows remains behind. It is mostly deposited in valleys, basins, and lowlands, where often inland deltas or sediment fans are formed. Particularly the large river basins have a high trapping capacity (Schumm and Hadley 1961). This has been enhanced by man's activities: dam construction has led to accumulation of suspended river sediment in the storage basins behind the dams, and sediment transport and deposition have been influenced by canalization and diversion of streams to provide water for

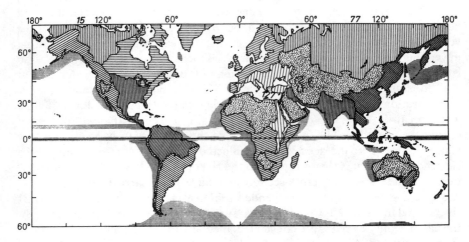

Fig. 2.2. Main drainage basins on land and areas of upwelling in the ocean. *Dotted areas* inland drainage areas. *Lined patterns* are oriented in the general direction of drainage. (Emery and Milliman 1978)

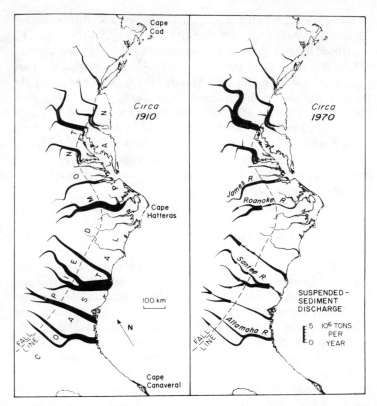

Fig. 2.3. Average annual loads of suspended sediment carried by rivers of Atlantic drainage of the United States during years near 1910 (*left*) and 1970 (*right*). (Meade and Trimble 1974)

agriculture (Fig. 2.3). No general worldwide estimates are available, but Holeman (1980) indicates that in the U.S. soil erosion produces in the order of 5 $\times 10^9$ t . y^{-1} of sediment, while from the same area only 0.4×10^9 t . y^{-1}, or 8% reaches the ocean through rivers. Data for the Huang He indicate that only 24% of the sediment load reaches the ocean and that the remainder is deposited in the lowlands of the lower reaches and in the delta area (Milliman and Meade 1983).

A distinction has to be made between storage on a seasonal scale, which does not influence the yearly average sediment load when this is measured adequately, and storage for a longer period, in the order of decades to thousands of years. Man-induced erosion following deforestation, crop-farming, surface mining, road construction, or urbanization has led to a large increase of river sediment load: to four to five times in the eastern U.S. (Meade 1969) and to more than ten times in the Sacramento River basin, California (Gilbert 1917). Much of the material produced in this way is still stored in valleys, lowlands, and deltas, part of it already for more than 100 years. Also, the sediment stored behind dams or on irrigated farmlands is being stored for at least decades, in some cases for more than a century. Meybeck (1988) estimates that at present ca. 1.5 $\times 9$ t . y^{-1} is being stored behind dams. The total river sediment loads estimated by Meybeck (1988: 15.4×10^6 t . y^{-1}) and Gordeev (1983: 18.5×10^6 t . y^{-1}) are higher than the estimate by Milliman and Meade (1983), who estimated the sediment discharged to the oceans, and include the amounts stored in this way. Over a longer time period than ca. 100 years, increased erosion and sediment storage is important in areas where crop-farming has been practised for a long time (Mesopotamia, Eastern Mediterranean, China). The Yellow River became muddy when the wooded steppe that is the natural vegetation was cut for farming on a large scale, but became clearer again when it was turned into grassland between 69 and 600 A.D. (Ren and Shi 1986).

2.1.2 Particulate Organic Matter in Rivers

Besides mineral material, rivers transport particulate organic matter in suspension. Estimates on the total amount that is supplied to the oceans vary from 280 $\times 10^6$ t C . y^{-1} (Kempe 1985) to 780×10^6 t C . y^{-1} (Mantoura and Woodward 1983). Meybeck (1988) arrived at $300–380 \times 10^6$ t C . y^{-1}, which is probably low because the highland rivers in southeast Asia were not included. Degens and Ittekkot (1985) arrived at $420–570 \times 10^6$ t C . y^{-1}, Schlesinger and Melack (1981) at $370–410 \times 10^6$ t . y^{-1}. Using a factor of 2 for converting C into organic matter, and taking the total supply of suspended matter of 13.5×10^6 t . y^{-1} as given by Milliman and Meade (1983), the average organic matter content of the suspended matter is ca. 4.5%. There is a general inverse correlation between organic matter content (in %) and suspended matter concentration, so that in some rivers with high suspended matter concentration (> 500 mg . l^{-1}) the organic matter content of the suspended matter is less than 1% (Degens and Ittekkot 1985). This inverse relation points to the possibility that in turbid rivers light conditions are such that primary production is strongly reduced. Stronger erosion, however, results in increased supply of mineral particles, and thus also may lead to a reduction of the organic content. Organic matter in rivers comes from various sources: primary and secondary production in the river itself, as well as from lakes, soils, and human waste discharges. Lieth (1975) gives for the primary production in rivers and lakes a total of 1×10^9 organic matter t . y^{-1}, while de Vooys 1979 arrived at 0.58×10^9 t C . y^{-1} or 1.2×10^9 t organic matter per year. Rivers also discharge an average of 5.5 mg . l^{-1} of dissolved organic carbon (DOC), which is ca. 1.2 times the discharge of particulate organic carbon (POC). In total therefore, ca. 1×10^9 t C . y^{-1} (TOC) or ca. 2×10^9 t . y^{-1} of organic matter is supplied by rivers. Comparing this with the estimates for total primary production in rivers and lakes, it is clear that the freshwater primary production alone cannot account for all the organic matter that is transported: there is also production of organic matter in lakes that is not transported seaward by rivers and part of the primary production is consumed or mineralized before it reaches the sea. Part of the particulate organic matter therefore must have been produced outside the rivers and is derived from soils and shore vegetation.

This can also be demonstrated in another way. Taking the total surface area of rivers as 0.8×10^6 km^2 (the total area of inland waters is 2×10^6 km^2) and of lakes ca. 1.2×10^6 km^2; de Vooys 1979) and dividing the total supply of organic matter by rivers to the sea (ca. 400×10^6 t C . y^{-1}) by this figure, the average primary production in rivers would be ca. 500 g C . m^{-2} if all this organic matter had been produced in the river water. This value is somewhat higher than the maximum production in a natural eutrophic lake (Likens 1975) and therefore is certainly too high as an average for all rivers. Likens (1975) gives a total range of < 1–650 g C . m^{-2} . y^{-1} for temperate rivers and 1–$1000(?)$ for tropical rivers, while for the large subarctic rivers, analogous to the subarctic lakes, a low production must be assumed (< 50 g C . m^{-2} . y^{-1}). This also is an indication that part of the organic matter transported by the rivers, must come from other sources. An average concentration of 4.5–5.5% of organic matter in the suspended matter in rivers may seem unimportant in relation to the large mass of inorganic particles, but it is important for the flocculation of the suspended particles, for the interaction with the dissolved material in the river (including pollutants) and as a direct or indirect food source for aquatic organisms.

Plankton organisms that produce calcite are very rare in rivers – all biogenic calcite in rivers formed in situ, is produced by benthos – but diatoms producing opal can be abundant. No overall data on the opal production in rivers exist, and data are scarce anyway, but concentrations of up to ca. 1–2 mg . l^{-1} have been reported (Wang and Evans 1967: Illinois river; Admiraal et al. 1990: Rhine river). Part of the opal produced in rivers is deposited in lakes. Because of the paucity of data and the strongly seasonal character of diatom growth in fresh water at higher latitudes, the writer has not attempted to calculate an overall estimate of opal production in rivers.

2.2 Suspended Matter in Lakes

The suspended matter in lakes comes from rivers and streams that flow into them, from local run-off and mass flow, from primary production in the lakes themselves, from the atmosphere, and from waste discharges. The largest lake is the Caspian Sea (371 000 km^2), followed by Lake Superior, Lake Victoria, Lake Baikal, the Aral Sea, Lake Huron, and Lake Michigan, which are all between 55 000 km^2 and 84 000 km^2 (Times Atlas of the World 1990). The largest lake by volume is Lake Baikal (23 000 km^3). Lakes or ponds of the smallest sizes are measured in m^2. The largest lakes are of structural or tectonic origin or were excavated by land-ice during the Pleistocene. Long, deep natural lakes generally are of such origin; many deep glacial valleys have been turned into man-made reservoirs. Most lakes are formed in shallow depressions. In spite of the size of some lakes, and the fact that rivers may flow in and out of them, lakes essentially are closed systems with regard to suspended matter. Since the drainage area is usually large in relation to the lake area, river supply dominates over the other suspended sediment sources. The sediment loading and the sedimentation rates are usually about one order of magnitude higher than in the ocean, which has a very large area compared to its drainage area (Table 2.2, where the data for the ocean are added for comparison). An important feature in lakes is the relation between surface area and water depth. Shallow lakes can be mixed from surface to bottom and are generally sensitive to wind stress, whereas in deep lakes a density stratification can develop with warmer water at the surface, separated from colder and more stagnant deep water by a thermocline. Seasonal cooling, as occurs in temperate areas, results in overturning of the lake whereby cool surface water sinks to the bottom. This can also happen when winds become stronger and the mixed layer reaches to greater waterdepths. The deepest lake is Lake Baikal (ca. 1610 m). Lake Victoria is only 79 m deep and some large lakes like Tchaad and Lake Eyre dry out completely during the dry season. Also important is the size of a lake and its orientation towards the direction of the principal winds, which limits the development of surface waves.

2.2.1 River Supply and Primary Production

The river sediment supply to lakes varies by a factor of more than 1000. The supply from the Colorado river to Lake Mead was ca. $145 \times 10^6 \, t \cdot y^{-1}$ (Howard 1960 in Sly 1978), which is about equal to the total sediment supply to the Caspian and Aral Seas. Small streams in the Arctic or in limestone areas supply very small amounts. The total suspended matter supply to lakes can be estimated from the suspended matter supply of the entering rivers (as given by Holeman 1968; UNESCO 1974, 1978; Strakhov 1967; Sly 1978; Milliman and Meade 1983). Where river supply is not known, extrapolations can be made

Table 2.2. The ratio of surface area: land drainage area for different lakes. (Sly 1978)

	Lake surface area (km^2)	Ratio to land drainage area
Lake Eyre (Australia)	9300	1 :140
Lake Champlain (USA)	1130	1: 17
Lake Geneva (Switzerland)	600	1: 14
Lake Titicaca (Peru/Bolivia)	7600	1: 8
Lake Ontaria (USA/Canada)	19 000	1: 3.4
Lake Erie (USA/Canada)	25 800	1: 2.9
Lake Victoria (Central Africa)	68 800	1: 2.8
Lake Tahoe (USA)	500	1: 1.6
The ocean		1: 0.43

from the regional denudation rates. This gives for all lakes a total of ca. 1 $\times 10^9$ t . y^{-1}, which is of the same order as the amount yearly stored in reservoirs (Meybeck 1988). To this should be added an (unknown) amount of suspended matter from local runoff, shore erosion, and lake floor erosion, and from the atmosphere, as well as from waste disposal. Changes in agricultural practice and other changes in land use within a lake drainage area can have changed the supply considerably during the recent years, both in quantity and in composition (Duck and McManus 1984, 1989).

Estimates for total primary production in lakes, separate from rivers, are not available, but it can be assumed that most of the total freshwater primary production (ca. 1×10^9 t . y^{-1} of organic matter) is produced in lakes. The total surface area of lakes is ca. 1.2×10^6 km^2, and the surface area of rivers is about 60% of this. In lakes, the suspended matter settles more easily so that they are sediment sinks and the surface water is clearer. This in itself will enhance primary production but in rivers, the plankton, because of the turbulence, is regularly transported from the surface to deeper levels in the water where light conditions are less favorable, which, coupled to the greater turbidity, will reduce the primary production. Table 2.3 (from Likens 1975) indicates that at least in temperate rivers the primary production is generally less than in temperate lakes. In lakes, the primary production ranges from 5.1 g C . m^{-2} . y^{-1}, measured in an Antarctic lake below the ice, to more than 5000 g C . m^{-2} . y^{-1} in some Ethiopian lakes and more than 10.000 g C . m^{-2} . y^{-1} in a saline Australian lake (de Vooys 1979). Values higher than 2500 g C . m^{-2} . y^{-1} are exceptional in natural waters. Productive natural lakes have a primary production of ca. 1000 g C . m^{-2} . y^{-1}, but in spite of the high production in some lakes, the total primary production in lakes is only ca. 1.4% of the production in the ocean, because the oceans have a much larger surface area. Worldwide the incidence of solar energy has the largest effect on primary production (Table 2.3), but within the temperate zone nutrient levels are more important. Geomorphological differences have little influence on the productivity per unit area.

Table 2.3. Net primary productivity values for regional aquatic ecosystems. (Likens 1975)

Water system	mg C . m^{-2} . day^{-1}	g C . m^{-2} . y^{-1} [a]
Tropical lakes	100–7600	30–2500
Temperate lakes	5–3600	2–950[b]
Arctic lakes	1–170	< 1–35
Antarctic lakes	1–35	1–10
Alpine lakes	1–450	< 1–100
Temperate rivers	< 1–3000	< 1–650
Tropical rivers	< 1–?	1–1000?

[a] In most cases, averaged over estimated "growing season".
[b] Naturally eutrophic lakes may reach a maximum of 450.

2.2.2 Calcite Particles

Planktonic organisms that produce calcite are extremely rare in fresh water (Kelts and Hsü 1978): the calcite (or dolomite) in lakes is mainly detrital or derived from precipitation with, in bottom sediments, an admixture of carbonate produced by benthos. Precipitation of calcite in the water gives clouds of fine particles ("whitings") with particle sizes ranging from ca. 0.5 μm to more than 30 μm, but mainly less than 10 μm. Precipitation occurs when the water becomes supersaturated with respect to calcite. Calcite precipitation is defined by the relation $Ca^{2+} + CO_3^{2-} \rightarrow CaCO_3$. The dissolved carbonate includes besides dissolved CO^{2-} also dissolved CO_2 and HCO_3^-. Dissolved CO_2 reacts with water forming H_2CO_3 which quickly dissociates into H^+ and HCO_3^-. The principal mechanisms resulting in supersaturation are assimilation of CO_2 by

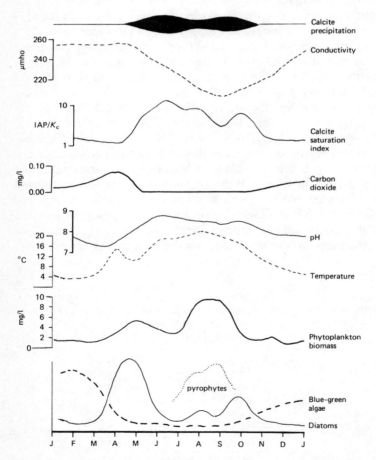

Fig. 2.4. Schematic correlation of several parameters observed during the seasonal cycle at 5 m depth in Lake Zürich, based on data for 1974 and 1975. (Kelts and Hsü 1978)

plants during photosynthesis, and temperature changes affecting the solubility of both CO_2 and calcite. Because the exchange of atmospheric CO_2 with water is rather slow, the CO_2 dissolved in the water can become depleted during photosynthesis so that the water becomes supersaturated with respect to calcite and Ca^{2+} is being precipitated together with CO_3^{2-}. Higher temperatures result in a decrease in the solubility of calcite and CO_2 (or HCO_3^-), which also leads to precipitation. Calcite precipitation is therefore often a seasonal process related to a temperature rise or a primary production peak (Fig. 2.4; after Kelts and Hsü 1978). Although also other carbonates can be produced, calcite most commonly is found; only when relative high concentrations of Mg^{2+} are present or the salinity of the water is high, are high-magnesium calcite or aragonite precipitated preferentially. Precipitation takes place through the formation of seed crystals, which form the nuclei for further crystallization. Large particles are formed when the supersaturation is not extreme so that there is a lack of seed crystals. The particles remain smaller when, at high supersaturation, large numbers of seed crystals are produced. The calcite particles may partially or wholly dissolve on settling down into deeper water where there is no supersaturation (e.g., because of a lower temperature).

2.2.3 Opal Particles

Particles of opal (amorphous or cryptocristalline SiO_2) are formed by diatoms in the form of external skeletons, which are the principal endogenic source of silica in lakes. The solubility of silica in fresh water is not well known and probably varies strongly from lake to lake. Excellent preservation of diatom frustules, indicating no dissolution at all, has been found in the bottom sediment of the east African lakes and shallow lakes and pools in The Netherlands (van Dam 1987), whereas partial dissolution is found in Lake Michigan (Jones and Bowser 1978). Diatom growth is based on photosynthesis and in mid- or high latitudes is seasonal. The seasonal changes in Lake Zürich (Fig. 2.4) show from February to April no diatom growth, no calcite precipitation, and low suspended matter concentrations. In April–May there is mass growth of diatoms followed by calcite precipitation due to CO_2 depletion during photosynthesis, which continues into October, starting with the formation of particles of 5–15 μm (up to 40 μm) and during the summer gradual formation of smaller particles down to less than 1 μm in diameter. Ca. 10% of this is dissolved again during settling in deeper water. The settling of particularly the fine particles is enhanced by the formation of aggregates.

2.2.4 Resuspended Particles

Where lakes are shallow, or where there is a strong circulation, resuspension may bring particles from the bottom into the water. In this way, particles containing high contents of iron, manganese, phosphorus, sulfide, or fluorite may be found in suspension. Iron and manganese are mainly deposited at the sediment/water interface as coatings on the particles or as nodules. Most of the ferromanganese material is very fine-grained and poorly crystallized. Phosphorus is mainly associated with the finer-sized particles and may be present as coatings or in the form of discrete particles of apatite or vivianite (Williams et al. 1971, 1976; Jones and Bowser 1978). Sulfides are formed under anaerobic conditions in quiet water, so that sulfides are not likely to be much resuspended. Fluorite is formed where lakes are situated in a volcanic region, as has been found in East African lakes.

2.3 Suspended Matter in Estuaries

An estuary is defined as a semi-enclosed coastal body of water which has a free connection with the open sea and within which seawater is measurably diluted with freshwater derived from land drainage (Pritchard 1967). Estuaries form the transition from rivers to the sea and present a separate environment, influenced by conditions in the river as well as in the coastal sea. The suspended sediment is supplied from various sources as summarized in Fig. 2.5. Depending on the local situation, one or more sources will dominate the supply. In the estuaries of large rivers with a high suspended load, the river supply dominates over all other sources. In many tidally mixed estuaries, river supply dominates in the inner part, and supply from the coastal sea in the outer part, but because of tidal mixing particles of marine origin may be transported well into the fresh-water

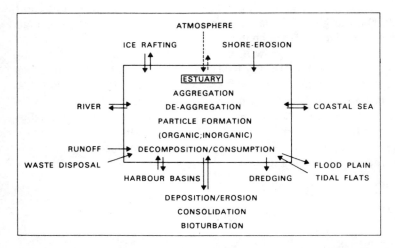

Fig. 2.5. Diagram of supply and removal of suspended matter in estuaries. (Eisma 1986)

tidal area, as occurs in estuaries along the east coast of the southern North Sea (Ems and Schelde rivers). This can be ascertained from the composition of the suspended material: suspended matter supplied from the sea contains particles produced by marine organisms (diatom frustules, sea-urchin spines, shell fragments), which serve as tracers. Resuspension of sediment from the bottom or from flood plains and tidal flats occurs regularly in estuaries because of the shallow depth. In this way also particles produced by benthic organisms – fragments of skeletons, iron/manganese-rich particles formed at the surface of bottom sediments, benthic diatoms,and fragments of plants – can be picked up. Because of the relatively small area of estuaries, supply from the atmosphere is usually negligible compared to supply from other sources. Only where dust or sand storms occur, can estuaries receive much sediment in this way (e.g., the Senegal river estuary). Supply from runoff, shore erosion, and ice rafting also depends very much on the local situation. Table 2.4 gives the relative amounts coming from different sources into the upper and middle Chesapeake Bay (after Biggs 1970): River supply does not reach to the middle part of the Bay, where shore erosion dominates. In the navigation channels of Delaware Bay, erosion from the estuary bottom dominates (Table 2.5), whereas in the Rhine estuary, which is entirely bordered by dikes, there is only supply from the river, from the coastal sea, and from waste discharges. In general, waste discharges, dredge spoils, and the construction of harbors and shipping channels strongly influence the suspended sediment supply where industrial centra and urban concentrations are located. Very large amounts of sediment can be reworked by dredging, and the estuary can be made to function as a large suspended sediment trap. In the Rhine mouth, where at present suspended matter supply from the river is ca. 1.7×10^6 t . y^{-1} (fine sand and mud), the annual influx from the coastal sea is ca. 13×10^6 t . y^{-1}, of which at least ca. 10.5×10^6 t . y^{-1} is dredged away and dumped in the nearby sea, and the remainder is transported seawards by the river outflow. Waste disposal from population centra can dominate over all other sediment sources: in Long Island Sound, which receives waste from the

Table 2.4. Sources of the organic and inorganic fraction of the suspended sediment in the upper and middle Chesapeake Bay, after Biggs (1970), in % of the total amount of sediment in suspension

	Organic (%)	Inorganic (%)	
Skeletal	2	0	
River supply	22	61	
Shore erosion	0	13	Upper Bay
Organic production	2	0	
From other parts of the estuary	0	0	
Skeletal	18	0	
Shore erosion	0	52	
Organic production	22	0	Middle Bay
From other parts of the estuary	5	3	

Table 2.5. Sources of sediment in navigation channels of Delaware Bay. (Data from Wicker 1973, in Meade 1982)

	$t \cdot y^{-1}$
River inflow	1.27×10^6
Erosion of the estuary bottom	2.07×10^6
Diatom production	1.35×10^6
Return from dredging	0.35×10^6
Sewers	0.12×10^6
Industrial pollutants	0.05×10^6
From the atmosphere	0.09×10^6
	5.30×10^6
Total deposited	6.20×10^6

New York-Boston area, waste discharges in 1960–1963 accounted for a suspended matter input of four times the natural river input, and of nearly nine times in 1964–1968 (Koppelman et al. 1976).

2.3.1 Supply of Biogenic Material

The contribution to estuarine sediments of organic matter, biogenic silica, and biogenic carbonate comes from estuarine organisms as well as from freshwater and the coastal sea, and is generally low compared to supply from other sources but can be locally and temporarily very high. Decomposition of the organic matter and dissolution of the silica usually reduce this to an amount that is much smaller than the total quantity of inorganic particulate matter that is present. Suspended matter in estuaries normally contains not more than 30% of biogenic material, estuarine bottom deposits even much less. For biogenic carbonate this is enhanced in tropical and subtropical areas because coral usually does not grow in or near to estuaries: the freshwater flowing out and the turbidity of the water prevent coral growth. Therefore a large source of biogenic carbonate sands and muds is usually absent in estuaries. Biogenic carbonate can accumulate, however, in the form of shell beds or calcareous sands.

Primary production is often limited in estuaries because of the relatively high turbidity of the water, which restricts the penetration of light into the water column and thus limits the depth of the euphotic zone. Also there is usually a strong turbulence which brings plankton organisms regularly into deeper water below the euphotic zone, and the residence times are short, which means that plankton populations have little time to develop before they are moving into a zone characterized by a different salinity and temperature. The number of algal cells is usually much lower than the number of particles supplied from other sources, but there is a relatively high influx of plant fragments (fibers, leaves, stems, pieces of wood or peat) supplied from the river, from shore vegetation, adjacent marshes, or mangrove swamps.

Phytoplankton accounts for less than half the total primary production in estuaries (2–45%; Knox 1986), macrophytes and microphytic benthos for the remaining 55–98%. The latter become part of the suspended material mainly as detritus; benthic microphytes can be brought into suspension also when living. The phytoplankton production in estuaries is on the average ca. $100 \, g \, C . y^{-1}$ but shows a considerable variation ranging from $6.8 \, g \, C . m^{-2} . y^{-1}$ to $530 \, g \, C . m^{-2} . y^{-}$. The abundance of phytoplankton in estuaries is related to the level of nitrate (not of phosphorous), to the turbidity of the water and to the ratio of photic depth to mixing depth (photic depth defined as the water depth to which sufficient light penetrates for photosynthesis to be possible). The production follows the gradients in an estuary, which often range from rather turbid water rich in nutrients in the inner estuary to clear water with a low nutrient content in the outer estuary (Cloern 1987). Also there can be strong seasonal variations as well as year to year variations caused by storms and eutrophication. Organic particles are imported from the coastal sea and exported from the estuary: Kennish (1986) discriminates between "American type estuaries" with extensive wetlands and a net export of organic matter, and "European type estuaries" with broad, relatively bare tidal flats that import organic matter. The turbidity in the estuary can be an important factor in this: when the turbidity is less, primary production is higher, and the balance between export and import can shift towards export of particulate organic matter. The question whether an estuary is importing or exporting particulate organic matter is of importance for the size of zooplankton and higher fauna populations that can be sustained. In considering such relationships, however, the nutritional value of the organic particles should be taken into account. Particularly in the coastal sea, where the organic matter may come from a relatively large distance, a large part of the more nutritional material may have been consumed already so that the particulate organic matter may be largely refractory and of low nutritional value. Also when suspended particles carry adsorbed pollutants, the question of import or export can become of great practical importance. Thus in the Seine estuary there is a net import of suspended particles from the coastal sea carrying absorbed radioactive isotopes (^{239}Pu, ^{240}Pu, ^{137}Cs, ^{125}Sb, ^{106}Ru) discharged into the coastal sea from a nuclear plant located at some distance at La Hague. The discharged isotopes are carried with the suspended particles far into the Seine estuary (Jeandel et al. 1981).

2.4 Suspended Matter in the Sea

Estimates of the yearly supply of suspended matter from different sources to the ocean are given in Table 2.6. The principal sources are the production of biogene particles (ca. $125.5 \times 10^{15} \, g . y^{-1}$ or ca. 73% of the total supply) and the supply

Table 2.6. Supply of suspended material to the ocean

Source	Total supply	Supply to the deep sea
	$\times 10^{15}$ g.y^{-1}	$\times 10^{15}$ g.y^{-1}
Terrestrial sources		
Rivers (in suspension)	13.5	< 1.35
(bed load)	1–2	Very small
Atmosphere	0.1–0.5	0.1–0.5
Coastal erosion	0.25	Very small
Seafloor erosion	?	Very small
Ice flows/icebergs	(35–50) (total)	< 0.5–2.0[e]
Marine sources		
Primary production (organic matter)	90	78[a]
CaCo$_3$	8.5	8[b]
SiO$_2$ (opal)	25	12[c]
Submarine volcanism	Small (?)	Small (?)
(hydrothermal SiO$_2$)	(0.02)	(0.02)[d]
Cosmic dust	(10^8–10^9 g.y^{-1})	(10^8–10^9 g.y^{-1})
Total	173–190	100

[a] Primary production on the shelf $10–14 \times 10^{15}$ g org. matter.y^{-1} (Platt and Subba Rao 1975; de Vooys 1979; Spencer 1983).
[b] Production of carbonate suspended matter on the shelf is small; most production is as reefs or relatively coarse and heavy material (shells, shell and coral fragments, algal material, polites).
[c] Opal production on the shelf ca. 13×10^{15} p SiO$_2$.y^{-1} (Spencer 1983).
[d] Ledford-Hoffman et al. (1986).
[e] See p. 28

from rivers (ca. 13.5×10^{15} g.y^{-1} or ca. 8% of the total). The remainder is supplied by a variety of sources: the atmosphere, coastal erosion, seafloor erosion, direct run-off from land, iceflows, and icerafting, submarine volcanisms, from waste discharges in particulate form, and from interplanetary space.

The total supply of particulate matter from land to the sea is ca. 25% of the total sediment supply but dominates on the (inner) shelf, where it accounts for ca. 75% of the particle supply. In the deep ocean < 10% is supplied from land: the data of Table 2.6 indicate that in the deep sea land-supply is ca. 2–4% (4% when the higher figure for supply by ice is used).

The supply of suspended matter from rivers has been estimated to be ca. 13.5 $\times 10^{15}$ g.y^{-1} (Milliman and Meade 1983). As indicated above (Sect. 2.1), this is the best figure available at present, but it should be realized that the supply from many rivers, including several large ones such as the Irrawaddy and the Mekong, is only approximately known. The suspended material supplied by rivers (of which 80% reaches the sea in south and east Asia) remains mainly nearshore or on the inner shelf, because of concentration mechanisms that will be discussed in detail below (Sect. 7.6). Only a small part, less than 10% (Drake 1976), is transported into the deep ocean. Also the bed load remains mostly nearshore and goes into the formation of beaches, channel floors, and banks.

The supply of dust from the atmosphere is in the order of 0.1 to 0.5×10^{15} g.y^{-1} (Judson 1968; Goldberg 1971). The eolian material is mostly blown out of desert and semi-desert areas or comes from volcanoes (Fig. 2.6). Regionally, as in the southern North Sea, dust from industrial and urban areas is of importance. The eolian supply is small compared to the river supply, but most of it is deposited in the deep sea, particularly in large areas off the principal desert regions (Sahara, Gobi, Australia, South Africa, Mexico-California). Windom (1975) estimated that 10–30% of the deep sea deposits consists of eolian material.

Supply from coastal erosion was estimated by Garrels and Mackenzie (1971) to be ca. 0.25×10^{15} g.y^{-1}. This amount is insignificant compared to the supply from rivers, but can be locally important. Also material supplied by erosion of older deposits on the seafloor can contribute to the formation of recent sediments. This supply can be relatively large, as in the North Sea, where it is of

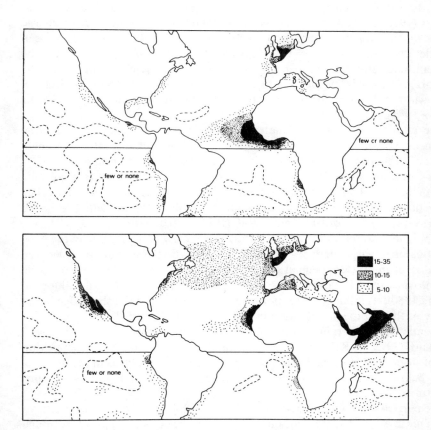

Fig. 2.6. Average frequency (in % of the number of observations) of dust clouds in the atmosphere in summer (*below*) and in winter (*above*). (Windom 1975)

the same order of magnitude as the supply from rivers (Eisma and Irion 1988). Usually, no realistic estimate of seafloor erosion can be made or, if at all, only a minimum value can be given, so that no worldwide estimate exists, but it is likely to be small compared to the total river supply. Since seafloor erosion is the result of wave action and tidal currents, which are both strongest in shallow water, this supply is predominantly located on the inner shelf.

Direct run-off from land through small gullies and depressions comes only from a narrow strip along the coast and is negligible in quantity to the volume of water flowing through rivers into the sea. As with other relatively small sources, the supply may be of local or regional significance.

Iceflows bring sediment directly into the sea where glaciers flow into the water, as occurs both along Antarctica and in the north polar regions, while icebergs transport sediment to lower latitudes. The total supply was estimated by Garrels and Mackenzie (1971) as ca. 2×10^{15} g . y^{-1}. This is of the same order as the amount of material supplied from rivers to the deep sea and the amount of eolian dust that reaches the deep ocean. Lisitzin (1972) and Friedman and Sanders (1978), however, on the basis of estimates of the annual ice discharge and the mean sediment load of the ice, come to a much higher supply, which is in the order of $35-50 \times 10^{15}$ g . y^{-1}. It is not clear, however, what proportion of the sediment load remains in suspension and how much is quickly dropped on the seafloor. At least part of the ice-supplied sediment consists of sand and gravel which will rapidly sink down. It is estimated below (p. 28) that only ca. 1% or $< 0.5 \times 10^{15}$ g . y^{-1} reaches the deep ocean.

The amount of suspended material (episodically) supplied by submarine volcanism, including deep sea vents, is not known, but probably relatively small. Quartz of submarine volcanic origin has been identified in marine sediments as well as clay minerals, feldspars, pyroxenes, and olivines that point to a volcanic origin (Windom 1976), but their distribution is limited to the volcanic areas. The supply of hydrothermal silica from submarine volcanism, including deep-sea vents is given by Ledford-Hoffman et al. (1986) as 0.02×10^{15} g SiO_2 .y^{-1}.

Supply of material from waste discharges is locally very large, as in the New York-Boston areas, where in 1960-1968 waste discharges far exceeded the natural river input (Koppelman et al. 1976). Part of it is transported over the shelf into a canyon, but even where waste is dumped midshelf (as off California; Herring 1980), its influence does not reach very far beyond the shelf edge in the vicinity of the dump site and usually is limited to the inner shelf.

Cosmic particles rain down on the earth to a total of 10^8-10^9 g . y^{-1} (Dohnanyi 1972, in Krishnaswami and Lal 1977). Only where the deposition rates of suspended material from other sources are very small, as in the deep sea basins below the carbonate deposition range, can they be found as a minor admixture.

The particles formed in the ocean itself are predominantly biogenic: particles of organic matter, calcium carbonate, opal (amorphous SiO_2) and, in much smaller quantities, barium sulfate and mineral particles that result from submarine volcanism and hydrothermal activity. Acantharia, which can make up over 30% of the plankton in the Atlantic (Bottazzi et al. 1971), produce skeletons of strontium sulfate that are rapidly dissolved in seawater. The total primary production in the sea, which is restricted to the surface waters (the euphotic zone) was estimated by Koblentz-Mishke et al. (1970) to be 23×10^{-15} g C . y^{-1}, by Platt and Subba Rao (1975), by Spencer (1983) and Berger et al. (1989) as $30-33 \times 10^{15}$ g C . y^{-1}, by Martin et al. (1987) as 42×10^{15} g C . y^{-1}, whereas Bolin et al. (1979), de Vooys (1979), and Wollast (1981) arrived at 45×10^{15} g C . y^{-1}. The higher, more recent values are mainly due to a better estimate of the contribution from picoplankton. Taking the latter value and using a conversion factor of 2, the yearly production of organic matter becomes 90×10^{15} g. The total biomass in the sea is 6×10^{15} g, which gives an average turnover rate (by remineralization) of 24 days. Estimates of the amounts of organic matter that settle out from the euphotic zone (called "export production"; Berger et al. 1989) vary from 0 to 20% and the amounts reaching the ocean floor from 0.01 to 9%, but there is a strong regional and seasonal variation as well as variability on smaller space and time scales. Because of this – and because of horizontal advection – steady state is normally not reached, and a distinction has to be made between the export production and the "new production", which is the production based on inorganic nitrogen supply from below the euphotic zone (Berger et al. 1989). Estimates of export (by settling from the euphotic zone) vary with waterdepth: Berger et al. (1989) give approx. 6×10^{15} g C . y^{-1} at 100 m depth in the world ocean, 3×10^{15} g C . y^{-1} at 200 m, and 1.5×10^{15} g C . y^{-1} at 500 m. Most of the organic matter reaching the seafloor is oxidized at or near to the sediment surface. The total amount that is deposited and buried in the bottom sediment is in the order of 0.2×10^{15} g C . y^{-1}, or less than 1% of the production.

The calcium carbonate and opal particles in the water column are formed by planktonic organisms: calcium carbonate by coccolithophorids, foraminifera, and pteropods, opal by diatoms, radiolaria, and silicoflagellates. Over the shelf, suspended particles of $CaCO_3$ from bentic organisms (mollusks, corals) or from erosion and rock weathering may be present as well. An estimated amount of 8.5×10^{15} g of planktonic $CaCO_3$ is produced per year (Wollast 1981) and the amount accumulated on the seafloor is ca. 1.5×10^{15} g . y^{-1}, which indicates that more than 80% is lost. The surface waters are oversaturated with respect to calcite and aragonite (the two modifications in which $CaCO_3$ is present), but dissolution occurs in deep water, in particular below the lysocline, which is situated at depths varying from 3700 to more than 5000 m. Dissolution may take place in the water column as well as on the bottom.

The biogenic production of silica (opal) in the sea was estimated by Lerman and Lal (1977) to be 10×10^{15} g SiO_2 . y^{-1} and by Wollast (1981), Spencer (1983), and Calvert (1983) to be in the range of $22-32.5 \times 10^{15}$ g SiO_2 . y^{-1}. As seawater is strongly undersaturated with respect to opal, dissolution is rapid also in the surface waters, particularly when Si-concentrations have become very low because of opal production. Only where the silica production is very high, are opal-rich deposits formed, as in the seas around Antarctica, on the Namibian shelf, in the Gulf of California, and in the North Pacific Ocean. Redissolution in the surface waters ranges from 18–58% (av. ca. 32%) in the upper 90–98 m of the Southern Ocean to almost 100% in the upper 60 m of the upwelling area on the NW Africa shelf (Nelson and Goering 1977; Nelson and Gordon 1982). Altogether, probably ca. 40% of the total opal production is redissolved in the euphotic zone.

The total amount of suspended matter in the oceans is $\sim 3 \times 10^{16}$ g (at an average concentration of 0.02 mg. kg^{-1}). Lal (1977) indicates that in the entire ocean ca. 80% is biogenic, although Krishnaswami and Sarin (1976) estimate that less than 5% is crustal material in the Atlantic and, moreover, the supply from land was estimated to be 2–4%. Taking these values as extremes, there is a total amount of $< 1.5–6 \times 10^{15}$ g of silicate minerals. From the data given by Wollast and Mackenzie (1983) the pelagic deposition rate of silicate minerals is approximately 1.34×10^{15} g.y^{-1}, which gives a residence time of the silicate minerals in the sea of 1 to 4 years, which is in good agreement with the residence time of 4 years, estimated by Krishnaswami and Sarin (1976), based on data on particulate Al for the Atlantic. Of the total amount that is deposited, 10–30%, or $0.1–0.4 \times 10^{15}$ g.y^{-1}, is supplied by eolian transport and less than 0.9–1.2 $\times 10^{15}$ g.y^{-1} originates from rivers and iceflows (assuming the supply from coastal and seafloor erosion to be negligible). This means that at most $< 2–4\%$ of the supply by ice reaches the deep ocean. Since also river supply reaches the ocean, it is probably ca. 1%, or $< 0.5 \times 10^{15}$ g.y^{-1} (Table 2.6).

3 Concentration Distribution
and Sampling of Suspended Matter

3.1 Concentration Distribution

3.1.1 In Rivers

Already the first recorded data on the concentration of suspended matter in rivers indicated a wide variation ranging from $g.l^{-1}$ in the Yellow River (Lyell 1830) to $mg.l^{-1}$ in the Elbe and Rhine rivers (Hübbe 1860; van der Toorn 1868). Since then, concentrations have been found to range from less that $0.5\ mg.l^{-1}$ in limestone areas to more than $10\ g.l^{-1}$ in the rivers draining the loess area in northern China. High concentrations are favoured by the presence of steep slopes, a semi-arid or glacial high-mountain climate, and loose sediments and soils that can be eroded easily. Concentrations above $500\ mg.l^{-1}$ are therefore found in the rivers draining the Himalayas and the mountain ranges in and around the Pacific, the Chinese loess area, and some semi-arid areas in Asia (Godavari), Africa (Orange river), N. America (Brazos), and Australia (Murray). The Zaire (Congo) river, although it is, after the Amazon, the second largest river based on its discharge, has a relatively low suspended matter concentration (ca. 20–$40\ mg.l^{-1}$) because the watersheds are low, the river basin is very flat and there are lakes where suspended sediment is trapped. The average concentration of suspended matter in rivers, obtained by dividing the total load by the total river discharge (using the data of Milliman and Meade 1983) is ca. $340\ mg.l^{-1}$. Of the total river supply, ca. 80% is transported by rivers with average concentrations of more than $500\ mg.l^{-1}$, whereas only 7.5% is transported by rivers with concentrations below $150\ mg.l^{-1}$. The highest concentrations ever recorded ($\approx 1000\ g.l^{-1}$) are from the Puerco river in Arizona, an ephemeral stream flowing only a few days a year (Nordin 1985). The data of Table 2.1 give, for the rivers listed there, average concentrations of $4\ g.l^{-1}$ to $143\ mg.l^{-1}$. Many smaller rivers with much lower concentrations are not included in Table 2.1.

Although many data on concentrations and sediment discharge probably are over- or underestimated (Walling and Webb 1981), there appears to be no relation between the suspended matter concentration in a river and the water discharge or the total sediment discharge including the bedload. Some rivers do show a good relationship, but for most rivers the suspended sediment discharge is more determined by the sediment supply than by the capacity of the river flow to transport it. The sediment load is therefore primarily related to climate

(Wilson 1977), rock types, land use, the presence of dams, and catastrophic events like landslides, volcanic eruptions, and earthquakes that bring large amounts of loose material into a river.

In rivers a distinction is made between the wash load, which is transported only in suspension, and the bed material, that can be transported along the bottom as bed load or in suspension as suspended load. There is no relation between the washload material and the bed load material. Thus the material actually transported in suspension is considered to consist of two parts – one that is regularly exchanged with the bottom and another which is not exchanged. The fraction that is not exchanged usually is much finer than the one that is exchanged, but eventually also the wash load is deposited: in reservoirs and estuaries, but also in local pools and on flood plains from where it can be resuspended. For many suspended matter data, it is not clear to what extent only the wash load is meant or whether also suspended bed load material is included. Most of the suspended bed material is sand-sized or larger and will mostly remain near to the bottom, so it will not be collected by some sampling methods, or settles out before the sample reaches the laboratory. Unless stated differently, "suspended matter" here includes all material in suspension, keeping in mind that some suspended matter concentrations may be underestimated because an unknown fraction of suspended bed material is not included.

Rivers usually do not carry suspended material to full capacity. In the arid and semi-arid parts of the central and western United States, concentrations of suspended matter in streams are regularly above $6 \, \mathrm{g.l}^{-1}$ (Meade and Parker 1985) and in the Yellow River can be much higher. External events can easily change the suspended matter concentration. Thus seasonal and more episodic events can result in temporary high concentrations of short duration. A regularly occurring feature in many rivers is a higher suspended matter concentration when the water is rising (and discharge increases) than when the water is falling (and discharge decreases: Fig. 3.1). This is not caused by the higher

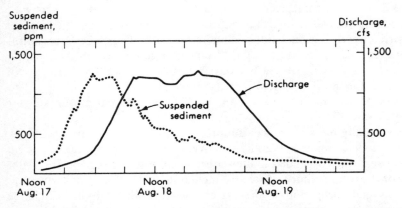

Fig. 3.1. Suspended sediment load and flood discharge for the Enoree River, S.C., Aug. 17–19, 1939. (Morisawa 1968; after Einstein et al. 1940)

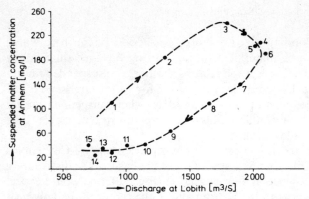

Fig. 3.2. Relation between discharge (at Lobith) and suspended matter concentration (at Arnhem, 1 km downstream) in the Rhine on 15 successive days. (Eisma et al. 1982)

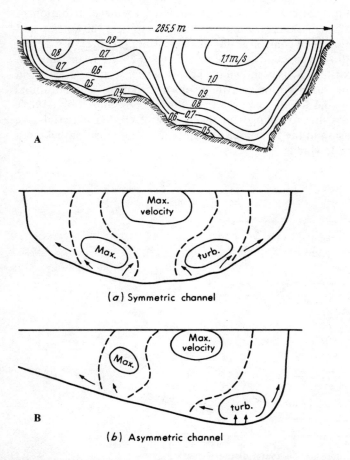

Fig. 3.3. A. Velocity distribution in a river. (Scheidegger 1961). **B** Zones of maximum velocity and turbulence in **a** symmetric stream channels; **b** asymmetric stream channels. (Morisawa 1968; after Leighly 1934)

discharge, but through flooding of river banks or a flood plain, from where newly weathered material, or material deposited during the preceding falling stage of the river, is picked up. In a graph relating discharge to suspended matter concentration this gives a hysteresis loop (Fig. 3.2 for the Rhine River). The reverse hysteresis also occurs as in the Chang Jiang, where suspended matter concentrations increase during falling water because the rice fields are then drained into the river (Meade 1988).

The concentration distribution in rivers has hardly been reported, except for vertical concentration profiles at single stations. On the basis of the distribution of mean velocity and turbulence in a cross-section (Fig. 3.3), one would expect maximum concentrations near to the bottom and in the center of the flow near to the water surface where the currents are highest. The concentration distribution of suspended matter along cross-sections of the Amazon and the Mississippi rivers (Fig. 3.5) can only be partly related to the distribution of current velocity. The higher concentrations generally occur near the bottom and are related to the presence of coarser grains in suspension; however, horizontal gradients also occur (in the concentration of particles $< 62\ \mu$). This distribution can change at short distances, as is shown in Fig. 3.4, where sections of Baldwin Creek are shown, measured within less than 100 m distance. A longitudinal section shown in Fig. 3.6 for the Yellow River (from Long and Xiong 1981), shows abrupt changes due to changes in the supply of eroded material – the concentration increasing in the loess area – and because of deposition behind a dam as well as in the lowlands.

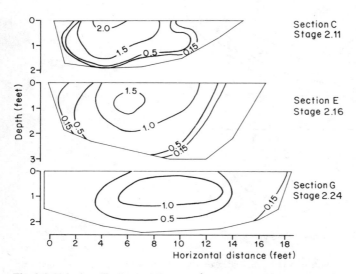

Fig. 3.4. Velocity distribution (in $m.s^{-1}$) in a natural straight river channel, Baldwin Creek Wyoming. (Leopold et al. 1964)

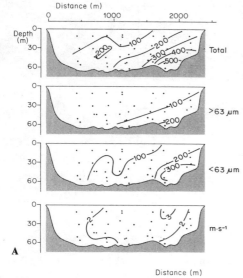

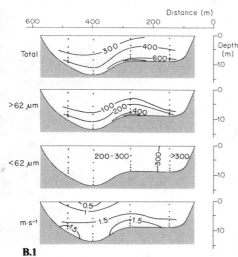

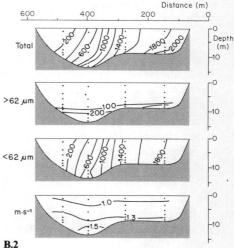

Fig. 3.5. A Distribution of the concentration (in mg.l^{-1}) of total suspended matter, of the fractions > 63 and < 63 μm, and of the current velocity (in m.s^{-1}) in a cross-section of the Amazon River at Obidos, June 15, 1976. (Meade 1979, 1985). **B** Distribution of the concentration (in mg.l^{-1}) of total suspended matter of the fractions > 62 and < 62 μm, and the current velocity (in m.s^{-1}) in a cross-section of the Mississippi River at St. Louis on 24 April 1956 (**B.1** about 20% flow from the Missouri River, which contributes the greater share of suspended matter.) and 9 July 1956 (**B.2** about 50% flow from the Missouri.) (Compiled by R.H. Meade from data of Jordan 1965)

3.1.2 In Lakes

As the supply of suspended matter in lakes comes from the surrounding land, from biological production and chemical precipitation in the lake itself, and to a small extent from the atmosphere, most of the suspended matter will be found nearshore and in the surface waters. In shallow lakes, waves and wind-driven currents keep particles in suspension and may easily resuspend them after they have settled to the bottom during quiet periods.

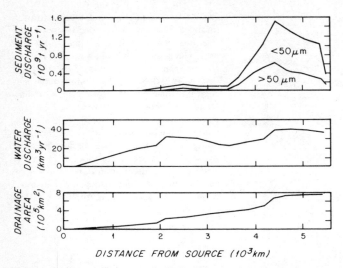

Fig. 3.6. Sediment discharge, water discharge, and drainage area along the Yellow River (Huang He) based on continuous daily measurements at a number of gaging stations during 1965–1974. (Milliman and Meade 1983; after Long and Xiang 1981)

In this way, high suspended matter concentrations can be maintained near to the bottom, particularly when horizontal dispersal is weak. This is shown in Lake Arari, a 2.5- to 4-m-deep lake near to the Amazon river mouth, where concentrations up to 800 mg.l^{-1} occur near the bottom during windy periods, where normally the concentrations are ca. 150 mg.l^{-1}, which is approximately the suspended matter concentration in the river water that enters the lake (Vital and Faria 1990).

In deep lakes, waves and wind-currents will resuspend only material from the shallow areas nearshore: in the deeper parts, settling of particles from the surface dominates. Where the vertical mixing of the surface water is strong and a thermocline is formed, which separates the well-mixed surface water (epilimnion) from the deeper water (hypolimnion), particles may remain in suspension in the surface water for a long time. Figure 3.7 shows the distribution of turbidity in the Lunzer Untersee 1 day and 7 days after a strong flux of suspended matter during high discharge of the principal stream that flows into the lake. Even after 6 days, a strong turbidity remains in the surface water above the thermocline.

When large amounts of sediment flow into the lake in a short time, a turbid bottom flow may develop that goes down the slope from the origin of the influx. It can continue over the bottom as a turbidity current; concentrations may be in the order of 10^2 to 10^3 mg.l^{-1} in such flows. They have been observed, or their existence has been inferred from bottom sediments, in reservoirs and in large lakes like Lake Geneva (see Sect. 7.2). In some lakes (Loch Earn, Lake Geneva),

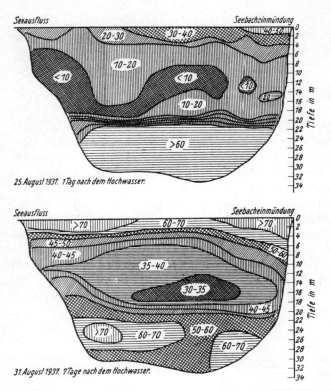

Fig. 3.7. Distribution of turbidity (in % transparency) along a profile in Lower Lake Lunz (Austria). *Upper profile* 1 day after high river inflow; *lower profile* 7 days after high river inflow. Stream inflow on the *right side*; exit of the lake at the *left side*. Depth in meters. (Sauberer and Ruttner 1941)

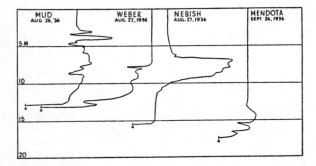

Fig. 3.8. Relative transparency with depth in four Wisconsin lakes. (Whitney 1937)

a subsurface suspended matter concentration or turbidity maximum can be observed which begins near a river mouth and can be followed over large distances (Smith and Syvitski 1982; Duck 1987; Fig. 3.8).

3.1.3 In Estuaries, Tidal Channels, and Coastal Lagoons

From the work of Glangeaud (1938), Gry (1942), Postma and Kalle (1953), Postma (1954, 1961), and others, we know that high suspended matter concentrations, up to 10^2 to 10^4 mg.1^{-1}, can occur in estuaries and tidal channels. These concentrations are much higher than those in the rivers that flow into these estuaries and in the coastal sea. The presence of such high concentrations (turbidity maxima) depends mainly on the residual circulation in the estuary and not on the flux of suspended matter from the river or from the sea. The different types of estuarine circulation are given in Fig. 3.9 (from Postma 1980). In the absence of tides, the river water flows out over the salt water because of its lower density. Mixing between fresh and salt water is limited and mainly occurs because of surface waves and coastal currents and by some entrainment of salt water with the outflow. Most of the suspended matter supplied by the river

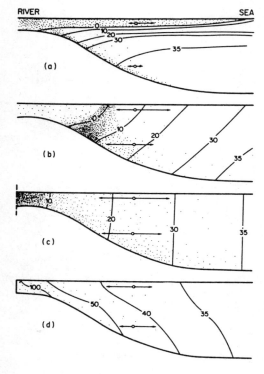

Fig. 3.9a–d. Schematic representation of the salinity distribution, relative residual flood and ebb currents and turbidity in the four main types of estuaries. **a** Saltwedge estuary. **b** Partially mixed estuary. **c** Fully mixed estuary. **d** Estuary with reverse (anti-estuarine) circulation. (Postma 1980)

settles near to the contact with the salt water because of a reduction in the flow velocity, which usually occurs in that area. Because of this influx of river suspended matter, the concentrations in the estuary are usually higher than in the adjacent sea. Sometimes, when large concentrations of sediment are supplied during a short period, a turbid bottom layer may develop that flows out over the bottom of the nearshore sea, may cross the shelf and go down over the continental slope, as regularly happens off the Rhône river mouth (Aloisi et al. 1982).

In tidal estuaries, vertical mixing takes place and a relatively large volume of salt water moves seaward along the surface mixed with the freshwater. In partially mixed estuaries, the vertical mixing is incomplete and strong vertical density gradients occur. In fully mixed estuaries, vertical mixing is virtually complete and the density gradients are mainly horizontal. This occurs where the tides dominate over the river outflow, as occurs in the smaller rivers along the north coast of South America (the Guyanas). In tidal estuaries, the residual current along the surface is outward because of the river water flowing out, while along the bottom the residual current is inward to compensate for the outflow of salt water along the surface. The residual current is the flow that is left over when the flow in the opposite direction has been subtracted.

The suspended matter flowing outward with the surface water settles out where the flow diminishes and is returned towards the inner part of the estuary with the residual bottom current. Where the inward bottom flow meets the outward flow from the river, flow velocities are low, and suspended matter supplied from both directions is accumulated. Very high concentrations can be formed which can result in the formation of fluid mud with concentrations of more than $10 \, g \cdot l^{-1}$. The zone of contact between outward river flow and inward bottom flow moves through the estuary with spring and neap tides and with the periods of high and low river discharge: more inward during springtides and low river discharge and more outward during neaptides and high river discharge. This results in a regular deposition and resuspension of fine-grained material.

In tidal channels and on tidal flats, high concentrations of suspended matter can be formed by inward concentration of suspended matter through the tidal movement alone. In shallow coastal areas, the tidal wave is deformed by interaction with the coastline and the bottom topography, which favours particle concentration (see section 7.4). Often, an asymmetric velocity curve develops where the flood is stronger than the ebb, but of shorter duration. As sediment transport is related to the second or third power of the velocity, more suspended matter is brought inward than is carried back by the ebb, which results in accumulation inward of suspended material. This process is complicated, by the changes in the tidal range, by the action of organisms that promote deposition of suspended matter, and by waves, which resuspend it, but the overall effect is usually an inward increase in concentration, which can be ten fold or more.

Lagoons are formed where the tides are small or absent. Usually, suspended matter concentrations are low, in the order of a few $mg \cdot l^{-1}$ but can be temporarily high during periods of high influx of suspended matter from land, when much bottom sediment is resuspended by waves, or when biological production is high.

3.1.4 In the Sea

From the work of Lisitzin (1959, 1961, 1972), Bogdanov (1968), Gordeev (1963, 1964), Klenova (1959, 1964), Ewing and Thorndike (1965), Jacobs and Ewing (1969), Emery et al. (1973, 1974), Brewer et al. (1976), Krishnaswami et al. (1976), Krishnaswami and Sarin (1976), Krishnaswami and Lal (1977), Lal (1977), and many others, we know that the concentration of suspended matter varies in the ocean from less than 0.005 mg.l^{-1} to ca. 1 mg.l^{-1}. In coastal waters, concentrations of more than 100 mg.l^{-1} can be reached and even higher concentrations occur near river mouths. In coastal areas where large amounts of suspended matter are concentrated and fluid mud is formed, concentrations become more than 5 g.l^{-1}.

On the shelf as well as in the ocean there is usually a more turbid surface layer and often also a more turbid bottom layer (nepheloid layer, e. g., Fig. 3.10). The surface layer contains silicate particles supplied from land by horizontal

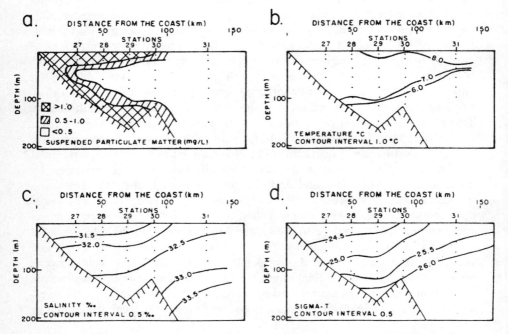

Fig. 3.10a–d. Vertical distribution of suspended matter, temperature, salinity, and density (sigma-t) in the northeast Gulf of Alaska. (Feely et al. 1979)

advection or through the atmosphere, and organogene particles produced in the water itself (organic matter, opal, calcium carbonate). Along the continents, suspended matter concentrations are higher on the eastern side because the large rivers, that drain most of the continental area, flow out on that side (Emery and Milliman 1978). In the ocean, concentrations in the surface water are generally higher at higher latitudes and relatively low around the equator (Krishnaswami and Lal 1977; Lal 1977; Fig. 3.11). In the Pacific, which is much larger than the Atlantic, surface water concentrations are two to three times lower, and the latitudinal differences are less pronounced. This occurs in spite of a relatively large supply of suspended matter from rivers, which, however, is

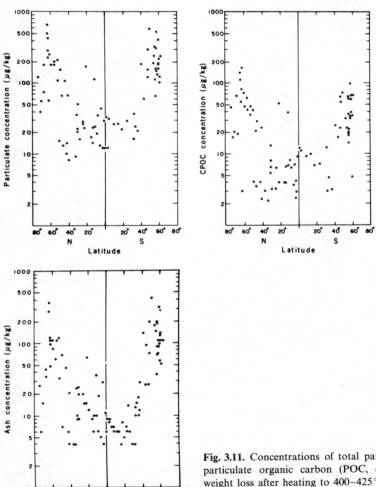

Fig. 3.11. Concentrations of total particulate matter, particulate organic carbon (POC, calculated from weight loss after heating to 400–425 °C), and ash (residue after heating to 400–425 °C), as a function of latitude in Atlantic surface water. (Lal 1977)

mainly deposited on the shelf. With increasing water depth below the surface mixed layer there is a decreasing concentration of suspended particles (Fig. 3.12), but there is an increase again towards the bottom (the bottom nepheloid layer), where particles that settled down from the surface are kept in suspension, or are resuspended from the bottom. In the deep ocean this is related to the presence of bottom currents, on the shallow shelf also to the action of surface waves and organisms. A profile of turbidity with water depth, typical for areas with a strong bottom nepheloid layer, is shown in Fig. 3.13 (from Biscaye and Eittreim 1977).

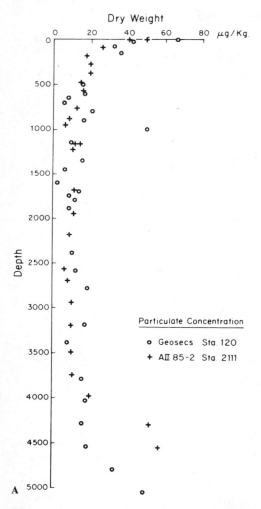

Fig. 3.12. **A** Suspended particles concentration along two vertical profiles in the NW Atlantic Ocean. (Brewer et al. 1976). **B** Particle concentrations along seven vertical profiles in the North Atlantic Ocean at ca. 40°N between 15°W and 45°W, without bottom nepheloid layer. (Gordon 1970)

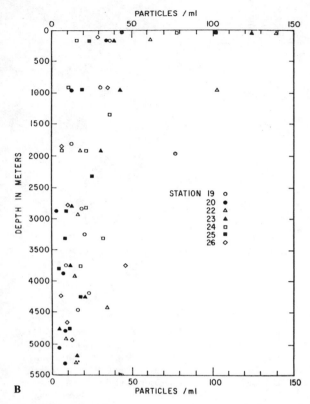

Fig. 3.12. (continued)

In the ocean, the amount of suspended matter below the "clear" water minimum is related to the abyssal current distribution as can be seen for the Atlantic in Fig. 3.14, where the high turbidity at the bottom is related to the relatively strong bottom currents of the western boundary current. Below the large gyres in the North and South Atlantic, concentrations of suspended matter are at a minimum: less than 0.001 mg.l^{-1} in the "clear" water minimum based on light scattering (Biscaye and Eittreim 1977) and less than 0.012 mg.l^{-1} based on filtration (Brewer et al. 1976). Here, the bottom nepheloid layer is small or virtually absent. Bottom nepheloid layers have also been found on many shelves, but can also be absent, as in the North Sea, where their formation is prevented by the tides in combination with initially low suspended matter concentrations. Suspended matter concentrations on the shelf, and particularly nearshore, are strongly influenced by storms, floods, periods of high or low river discharge, and fluctuations in the organic productivity, caused by seasonal fluctuations in wind force and wind direction, precipitation, and available sunlight.

Layers with relatively high suspended matter concentrations are also found at intermediate depths, in the ocean as well as on the shelf. For the Atlantic this

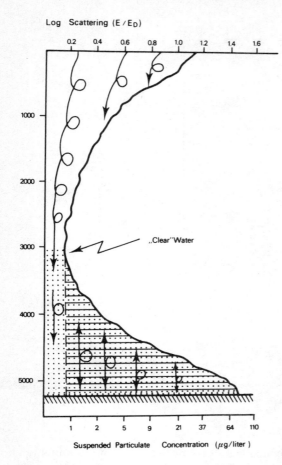

Log Scattering (E / E$_D$)

Fig. 3.13. Schematic depth profile of the concentration of suspended matter in the ocean (in relative units of light scattering). Bottom nepheloid layer can be virtually absent. *Arrows* indicate dominant particle movement: settling in the main water column, resuspension, and settling in the BNL. (Biscaye and Eittreim 1977)

can be observed in a profile measured by Brewer et al. (1976; Fig. 3.15). Concentration maxima occur in the surface water because of organic productivity, and in the bottom water where current velocities are high (Greenland Current, Iceland-Scotland overflow, Antarctic bottom current). At intermediate depths, the maxima are related to the general circulation of water masses: to the Antarctic Intermediate Water that sinks to ca. 800–1000 m at ca. 50° S, to the convergence zones at 30° S and 35° N, and to downwelling Labrador Sea Water at 50° to 60° N. The latter is connected with a plume of high concentrations extending upwards from the bottom at 35° to 40° N. This plume is probably related to the western boundary current flow which reappears in the section at 10° N (cf. Fig. 3.14). The absence of strong density gradients in this zone gives rise to considerable upward diffusion of particles from the bottom. Intermediate maxima in the Indian Ocean and the Bering Sea were also found related to water mass stratification and horizontal advection (Nakajima 1969), but not all maxima at intermediate depth can be explained in this way, and may be related

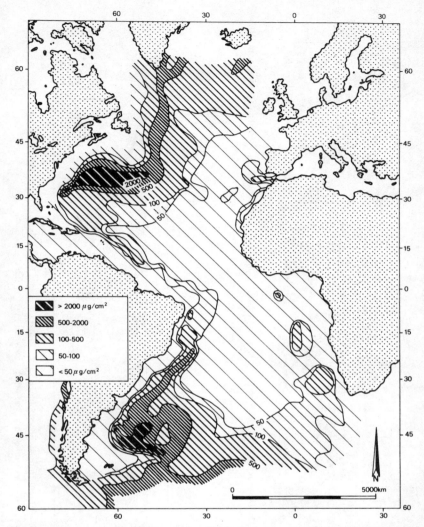

Fig. 3.14. Concentration of suspended matter in the bottom nepheloid layer (BNL) in the Atlantic Ocean. (Biscaye and Eittreim 1977)

to variations in the particle supply from the surface water or to lateral (horizontal) flow along isopycnal surfaces.

On the shelf and continental slope, well-developed mid-water maxima have been found over the inner shelf and in the head waters of canyons (Drake 1971; Drake and Gorsline 1973: California canyons; Pak et al. 1984: Zaire). They are related to the density stratification and explained by the presence of density underflows, which may retard the settling of particles from the surface and may

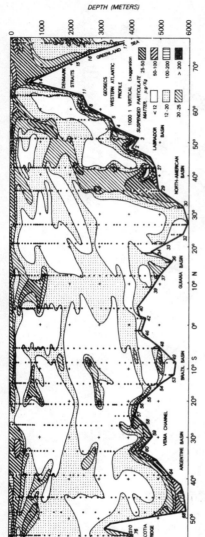

Fig. 3.15. Distribution of the suspended matter concentration along a north-south profile through the Atlantic Ocean. (Brewer et al. 1976)

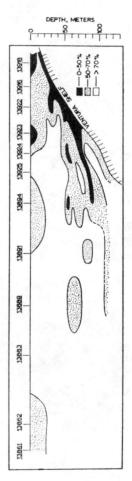

Fig. 3.16. Light transmission profile in Santa Barbara Channel, California. Suspended matter concentration at 0–50% transmission in the order of several mg.l^{-1}. (Drake 1971)

move turbid water seaward by advection from the bottom nepheloid layer on the shelf or slope. Because of the connection between the bottom nepheloid layer and the mid-water turbid zones (Fig. 3.16), Drake and Grosline (1973) believe advection to be the principal process.

3.2 Sampling Suspended Matter and Determination of Suspended Matter Concentration

The techniques used for determination of the suspended matter concentration are essentially the same in rivers, lakes, estuaries, and the sea but usually are adapted to the local conditions such as strong currents, great depth, or very high concentrations. The following techniques are used, most of them regularly, and will be discussed below:

- water sampling followed by filtration,
- in-situ filtration in the water,
- concentration with a continuous flow centrifuge,
- light scattering and X-ray scattering,
- γ-ray and natural radioactivity measurements,
- acoustic methods,
- remote sensing.

The first three methods involve sampling at discrete points, which can be an inadequate approach as suspended matter is nonuniformly distributed in the water. Depth-integrated sampling as employed in rivers gives an averaged depth-integrated concentration, but obscures variations in concentration with depth. Therefore it is often desirable to combine the point-sampling with in-situ profiling of turbidity. The concentrations, also when obtained with a (calibrated) profiling method, are given in units of weight per unit of volume. More detailed descriptions of sampling techniques and references given by Vanoni (1975) for sampling in rivers, by McCave (1979a) for sampling in estuaries, which also holds for sampling in lakes and in the open sea, and by Landolt-Börnstein (1986–1989) and by Sternberg (1989). Particle size measurements based on particle counting (Coutler counter, laser or photographic methods) give concentrations in number of particles per unit volume and will be discussed in Section 6.4. They can be converted into units of particle volume per unit water volume, but reliable conversion into units of particle weight per unit of water volume is usually not possible because the in-situ density of the particles is not sufficiently known. Sediment traps, which can be used floating or fixed to the bottom, are essentially tubes or funnels that are open on one side and with a sediment-collecting device on the other. They do not measure concentrations but fluxes (in unit weight per unit area per unit time), and will be discussed in Section 3.2.5.

3.2.1 Water Sampling and Filtration

Water samples are collected with bottles, or with tubes that are open at both ends, that can be closed at the desired depth with a messenger weight. This triggers a mechanism that closes the bottle or the tube. Usually some time passes between the closing of the sampler and the filtration, and during that time coarser or heavier particles (sand) can sink to the bottom of the sampler. Care should therefore be taken that this is not lost when opening the sampler or is neglected when a subsample is taken for filtration. Therefore a sampler that consists of a horizontal tube instead of a vertical tube is sometimes preferred. A horizontal sampler also has the advantage of sampling a smaller water layer, which can be important when sampling in water with strong vertical gradients.

Arrays of samplers (Niskin bottles) can be mounted on a frame (rosette sampler) lowered on a coaxial cable. The samplers can be closed by giving a signal from above. The rosette sampler is usually combined with a CTD for measuring profiles of salinity [conductivity, temperature, and waterdepth (pressure)] that are registered directly on board, so that the depths at which the samplers are closed can be selected on the basis of the water characteristics. The rosette sampler is difficult to handle in strong currents but is a very suitable instrument for sampling in lakes and in the ocean. Small versions are made for sampling in estuaries and the coastal sea. For geochemical analysis of suspended particles, it can be necessary to use samplers that do not have metal parts or, for organic analysis, samplers made of stainless steel. Rubber is often covered with talcum for preservation and grease is used on moving parts: Go-Flo samplers have been developed that avoid contamination of the sample with talcum or grease. Because sampling at different depths can be very time-consuming in rivers and often only an average value for the suspended matter concentration is needed, depth-integrated samplers have been developed. They accumulate a water-sediment sample while lowered to the river bed and raised to the surface at a uniform rate. The velocity in the intake at all times is almost equal to the local stream velocity. Point-integrating samplers accumulate suspended sediment in amounts that are representative for the mean concentration at that point. Less accurate are pumps with hoses and traps that are designed to collect large amounts of suspended matter for grain size analysis in the laboratory.

For filtration, usually a representative subsample is taken from the sampling bottle and subsequently filtered using either a pressure or a suction filtration system. When suspended matter concentrations are very low, as in clear ocean water, where it is in the order of 0.01 mg.l^{-1} or less, contamination with particles from the surrounding air may become a serious problem. Filtration is then done directly from the sampling bottle by pressure filtration using nitrogen that is cleaned by pressing it through a filter with a pore size that is considerably smaller than in the filter used for the filtration of the water. Particularly with pressure filtration, care should be taken that some large or heavy particles are not left behind on the bottom of the sampler. With suction filtration, care has to be taken that no particles are sucked in from the air.

Mostly three types of filters are used for filtration: cellulose ester membrane filters (e.g., Millipore), perforated polycarbonate filters (Nuclepore) and glassfiber filters (Whatman GFF). By convention, filters with a nominal pore size of 0.4 μm are used to separate the particulate from the dissolved. Cellulose filters have a spongy structure and the pore size is less well defined than in the polycarbonate filters, which have round or oval holes. The effective pore size is determined by the number of larger holes, but occasionally two holes come together so that large particles may pass there. They have a much lower weight than the cellulose filters (ca. 18 mg against ca. 100 mg) and they do not absorb air moisture as much as the cellulose ester filters, so that smaller amounts of suspended matter can be weighed accurately. Also, almost all the particles can be removed from a polycarbonate filter by ultrasonic treatment.

Filtration of 4–20 l of ocean water with very low concentrations over a Nuclepore filter and analyzing microgram quantities of suspended matter give reliable quantitative results: since 1970, when these methods were developed, internally consistent results have been obtained (Krishnaswami and Lal 1977). Glass-fiber filters are made of compressed glass minifibers and retain particles larger than ca. 1.2 μm. They are primarily used for organic matter analysis which cannot be done with the other filters, since they contaminate the analysis with carbon. Other types of filter have been made for special applications, such as silver filters for X-ray diffraction analysis and ceramic filters for collecting large volumes of suspended matter, although for the latter also large diameter Nuclepore filters are used (e.g., of 20 cm ø instead of the normal 4.7 cm ø). All filters become clogged when large amounts of suspended matter are filtered and when this occurs also particles smaller than 0.4 μm are retained. Normally, filtration can be stopped when the filter clogs because sufficient material is then on the filter, but this may not be the case when the filter is clogged by plankton (e.g., by the slimy colonies of *Phaeocystis*, a small flagellate).

Weighing and pre-weighing of the filters to determine the suspended matter concentration can be done on any balance that measures down to 10^{-5} g. For very small amounts a microbalance may be needed: the weight of the filter has to be subtracted from the weight of the filter + sediment. When concentrations are less than 0.01 mg . l^{-1}, the balance must be able to measure less than 0.01 mg in order to obtain a reliable result, even when many liters are filtered. Blanks that follow the same procedures should always be added to the filtration and weighing. Contamination and loss of particles through settling or adherence to the walls of the sampler is almost entirely avoided by in-situ filtration, where only the instrument itself may cause contamination. The water is pumped directly through a filter and systems have been developed that can pump up to thousands of liters at any desired depth in the ocean (Bishop and Edmond 1976). Concentration of large amounts of suspended matter is done with large-diameter filters and pumps. It can also be done with a continuous flow centrifuge by rotating at high speed. The suspension enters through the bottom, by the rotation the suspended material is separated from the water and remains on the walls of the centrifuge, while the clear water is removed at the top. The time needed to deposit a particle in the centrifuge is a function of $[1/(\rho_s - \rho_w)]$

d^2, where ρ_s is the bulk density of the particles, ρ_w is the density of the water, and d is the diameter of the particles. For particles of low density (flocs) and of small size, the time needed is very long so that usually filtration with large-diameter filters is more convenient for sampling large amounts of suspended material.

3.2.2 Light Scattering and Radiation Measurements

Suspended particles in the water scatter light, and the degree of scattering depends on their concentration as well as on their size, shape, and transparency. This allows measuring the concentration of suspended matter in two different ways: by measuring the attenuation of a light beam projected through the water towards a photocell, and by measuring the scattered light at an angle to the beam (forward or backscattering). Beam attenuation is measured with a transmissometer, and a path length is chosen according to the range of concentrations that have to be measured: a short one for turbid water, a long one for clear water. A double-beam transmissometer makes it possible to determine the attenuation coefficient α: when $T_1 = e^{-\alpha L_1}$ is the transmittance over the path length T, their ratio is:

$$\frac{T_2}{T_1} = e^{-\alpha(L_2 - L_1)}, \quad \text{or:} \quad \frac{T_2}{T_1} = e^{-\alpha L_1} \quad \text{(when } L_2 = 2L_1\text{)}.$$

Besides the particle characteristics, also adsorption by the water and the particles influences the attenuation: when concentrations are determined, calibrations should always be carried out with suspensions containing the same particles and the same water as present in the measured suspensions. The advantage of transmissometer measurements is that profiles, or transects combined with point sampling and filtration, give a good picture of the concentration distribution; this cannot be accomplished by collecting and filtering water alone. Beam transmissometers can measure a wide range of concentrations: at 5 cm path length from 0.1 to 500 mg.l^{-1} (Sternberg 1989).

Scattering measurements are carried out with a nephelometer. Often the scattering is measured at a forward angle of 45°, but the ratio between the scattering measured at 45° to the total scattering varies for different water masses. Scattering measurements are sensitive to very low particle concentrations, and therefore very suitable for measurements in the ocean. Backscatter instruments can measure concentrations up to 40 g.l^{-1}. The Secchi disc, first lowered into the Mediterranean by A. Secchi in 1865, is a white disc of 30 cm diameter which is lowered into the water until it becomes (just) invisible. The depth at which this occurs is an indication for the turbidity of the surface water. The visibility is influenced by the light reflected from the water surface, by the decrease in irradiance with depth (which is not only a function of depth and turbidity but also of the characteristics of the suspended particles and the colour of the water), and by the size distribution of the suspended particles. Nevertheless, it gives a good indication. The greatest Secchi disc depth was measured off

Antarctica in a polynya between the mainland and the pack ice (79 m; Gieskes et al. 1987). Smaller, as well as much larger discs have also been used (up to 1 m diameter).

Natural (gamma ray) activity is related to the concentration of the suspended matter and can be measured directly in the water, as was done in the Gironde (Martin et al. 1970). The results also reflect differences in particle size of the suspended matter, and include radiation from contamination, cosmic radiation, and the radiation of the instrument itself. It is a useful technique in areas where the suspended matter concentrations are too high for most optical measurements. It has been applied in the Gironde and the Orinoco river mouth (Martin et al. 1970; Eisma et al. 1978b). Also gamma ray measurements have to be calibrated with the same particulate material as is present in the concentrations that are to be measured. In fluid muds, where concentrations are $5–10 \, g . l^{-1}$ or more, the density (which is a measure for water content and particle concentration) can be measured by γ-ray densitometry, which is based on the attenuation of radiation in a sediment mass. The attenuation in the sediment is measured in situ with a probe, which has californium as a radiation source. When the probe sinks into the soft sediment, the consolidation, resulting in a higher density, is measured. Usually the density increases stepwise with depth in the sediment. The instrument has to be calibrated with mud collected in the same locality.

Transmission-type methods have been developed to measure particle concentrations, based on the attenuation of radioisotopes (^{90}Sr, ^{90}Y, ^{109}Cd, ^{241}Am) by suspended particles. They give good results in concentrations from $100 \, mg . l^{-1}$ to $30 \, g . l^{-1}$ at current velocities of 0.7 to $3.0 \, m . s^{-1}$ (McHenry et al. 1970).

3.2.3 Acoustic Methods

Acoustic methods to measure suspended matter concentrations have been used mainly in estuaries and coastal waters at high suspended sediment concentrations. A 200-kHz echosounder already shows clouds of suspended matter, whereas lower frequently signals can identify fluid muds and stationary suspensions (Parker 1987). With high frequency ($\approx 3 \, mHz$) sound, near-bottom suspended particle concentrations can be measured by backscattering with a resolution of ca.1 cm (Sternberg 1989).

3.2.4 Remote Sensing

Remote sensing from planes or satellites is based on measuring the radiance in the red part of the spectrum ($0.6–0.7 \, \mu m$), which shows suspended matter in the upper ca . 2 m of the water. The $0.7–0.8$-μm range (near-infrared) shows the suspended matter in the upper ca. 10 cm. When planes are used, the precision of the results depends very much on the height of observation, as well as on the type of instrumentation. The Landsat satellite has a horizontal resolution of ca. 70 m. Calibration is difficult and time-consuming; it is complicated by the fact that usually variable mixtures of living plankton and dead suspended matter are present. Where a river discharges large amounts of suspended material (Yellow River, Chang Jiang, Ganges-Brahmaputra), all other sources can be neglected and the turbidity can be interpreted as dispersed river mud. Multi-spectral radiation measurements, corrected for atmospheric interference, allow reasonably accurate estimates of the concentrations of suspended matter, phytoplankton and dissolved "yellow substance" in the surface water. But much computer capacity is needed and additional in-situ measurements are necessary for calibration (Fischer 1983; Fischer et al. 1988).

3.3 Sediment Traps

Sediment traps consist of a tube or a funnel that is open at the upper end and has a collector bottle or tube at the lower end. Traps can be used floating or moored; most of the moored traps now have a rotating series of collector bottles steered by a timeclock so that samples are collected during regular predetermined intervals. Moorings are anchored to a weight on the seafloor with an acoustic release. When an acoustic signal is given from the surface, the release becomes disconnected from the bottom weight and the trap, or series of traps, that is fitted out with floats, comes up to the surface, where it can be picked up. Traps, moored or floating, collect material that settles down from above, but also reverse traps have been used to collect material that moves upwards to the surface (Smith et al. 1989). The material collected per unit time is a measure of the flux of settling material, which can be subdivided into minerals, organic matter, carbonate, opal, etc. Also, particles are collected that are rare or that settle quickly to the bottom and thus are usually not sampled with sampling bottles such as Niskin bottles. Sediment traps give a flux, not a concentration, and the resultant flux can be compared to the deposition rate of material on the bottom. Comparison of the results obtained with traps at different depths shows the changes in flux and composition that take place in the water column.

There is some question as to how representative the data from sediment traps are. Moored traps give only reliable flux data when there is hardly any horizontal flow, as is the case in most of the deep ocean and in stratified lakes. With horizontal flow, increased turbulence is created at the rim of the trap, and suspended matter can either be prevented from settling in the trap, or an increased downward flux can develop towards the quiet water at the bottom of the trap. Where resuspension of bottom sediment takes place by bottom flow, the trap should be placed far above the bottom to be outside the reach of particles that are stirred up, e.g., in the deep ocean above the bottom nepheloid layer (unless, of course, if one wants to study the resuspension). When no additional measures are taken, a special bottom fauna can develop in the trap that consumes the organic matter that comes in, or zooplankton can enter the trap and produce fecal pellets, leading to an overestimation of the flux. To prevent biodegration, poison ($HgCl_2$, azine) or preservatives (formaldehyde, gluteraldehyde) are added, but also when this is done, there is some decay of the labile organic compounds. Also, swimming organisms can come into the trap and die there. To prevent losses by diffusion out of the trap, the collecting bottles are filled with artificial seawater of a slightly higher density that the ambient water. A discussion on the efficiency of traps, comparison of the different types in use, and the interpretation of results is given by Gardner (1980a, b), Reynolds et al. (1980), Gardner et al. (1983), and Fowler and Knauer (1986).

4 Particle Composition

The mineralogical composition of suspended particles reflects their origin: mineral particles like quartz, feldspars, calcite, and clay minerals come from terrestrial erosion, organic compounds, calcium carbonate, and opal from organic production, phytoliths (formed in grasses) from semi-deserts, fly-ash from industry, rare micrometeorites from space. Although there are mineralogical analyses of suspended particles available, they are few and the most comprehensive data are for bottom sediments. Those for fine-grained deposits can be taken to reflect the composition of the material in suspension in the water column, but it should be realized that during or shortly after deposition part of the suspended material may have been removed by oxidation (organic matter) or dissolution (carbonate, opal), and clays may have been altered.

Generally, a distinction can be made between mineral particles that are result of rock-weathering (quartz, alumino-silicate minerals, hydrous oxides of Al, Fe, and Mn, some calcite) and biogenic particles that are formed by organisms in the water or in the bottom (organic matter, carbonate, opal). The mineral particles are discussed in Section 4.1, the biogenic particles in Section 4.2, whereas the elementary and isotopic composition of suspended particles will be discussed in Sections 4.3 and 4.4.

4.1 Mineral Particles

The mineral particles in rivers, lakes, and estuaries, as well as on the shelf and in the ocean, are an assemblage of particles of different types, which reflects the rock types and weathering conditions in the source areas. Mineral particles found in suspended matter and sediments are those that are resistant to weathering (Table 4.1). Quartz, feldspars, and calcite are residues, left over after dissolution and rock break-up. Quartz is more resistant than feldspars, and potassium feldspars are more resistant than sodium feldspars. Clay minerals (Table 4.1) are weathering products of feldspars and other minerals present in rocks and soils. Mica also is a component of many igneous rock types. Because most clay minerals are actually newly formed during rock weathering and in soils, they reflect much more the climatic zones than quartz and feldspars. Kaolinite comes predominantly from tropical and subtropical areas and chlorite from higher latitudes. Illites and micas are the most common clay minerals and

Table 4.1. Composition of the principal minerals in suspended matter

Quartz	SiO_2
Feldspars	
Orthoclase	$KAlSi_3O_8$
Albite	$NaAlSi_3O_8$
Anorthite	$CaAl_2Si_2O_8$
Clay minerals	
Kaolinite	$Al_4 (OH)_8 [Si_4O_{10}]$
Chlorite	$(Al, Mg, Fe)_3 (OH)_2 [(Al, Si)_4O_{10}] Mg_3 (OH)$
Illite	$(K, H_2O) Al_2 (H_2O, OH)_2 [AlSi_3O_{10}]$
Montmorillonite	$\{(Al_{2-x}Mg_x) (OH)_2 [Si_4O_{10}]\}^{-x} Na_x \cdot n\ H_1O$
Calcite/aragonite	$CaCO_3$
Opal	SiO_2 (amorphous)

occur in a wide variety of environments; they dominate where kaolinites and chlorites are present in minor quantities, i.e., at mid-latitudes. Montmorillonites (or smectites) are formed in wet soils at medium or low temperatures by direct weathering of rock, or by transformation of illites or chlorites (through replacement of interlattice K^+ by water molecules). They are easily formed out of volcanic material (ash, basalt) and as such can also be formed in the sea by submarine weathering. In the submarine volcanic areas montmorillonites are the most common clay mineral: the highest concentrations (> 53% of the total clay minerals) are found in the South Pacific, which receives a high supply of submarine volcanic material (Griffin et al. 1968). In the other ocean areas, land-derived clays dominate. The supply and distribution of recently formed clay minerals is often mixed with a supply from erosion of older fine-grained deposits and sedimentary rocks, which have been formed under conditions that differ from those at present in that area. Many drainage areas extend over different weathering zones, which is reflected in the mineral composition of the suspended matter. Thus the Zaire, which drains mainly a tropical basin without large mountains, transports mainly kaolinite together with quartz from the coastal ranges. But the Amazon, which drains the Andes, transports a mixture of clay minerals that is not much different from the mixture transported by the Rio Puerco in New Mexico (USA) (Beverage and Culbertson 1964; Eisma and van der Marel 1971).

Most detrital minerals have a limited size range. Figure 4.1 (from Gibbs 1977a) shows the size distributions of the principal minerals present in material transported by the Amazon. Feldspars are the largest, clay minerals, particularly montmorillonites, the smallest minerals. Silts usually are mainly quartz, feldspars, and micas. The "clay fraction" consists mainly of clays, as the name indicates. As suspended matter is usually flocculated (see Chap. 6), a range of grain sizes is present in the flocs and the flocs themselves are a mineral assemblage. Not included in Fig. 4.1 is the poorly crystallized and amorphous

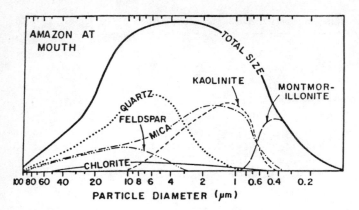

Fig. 4.1. The size distribution of mineral phases transported by the Amazon River. (Gibbs 1977a)

(cryptocrystalline) mineral material (clays or oxides) that is also present in suspension. Its amount is very much underestimated in most mineralogical analyses, and its contribution is not clear. For the Zaire River, van der Gaast and Jansen (1984) estimated that up to 50% of the transported mineral matter consists of poorly crystallized clays.

In the silt and sand fractions, a usually minor admixture of so-called heavy minerals is present, normally less than 1% by weight. These minerals have a density that is higher than 2.8, (quartz, feldspars, and clay minerals have a density of ca. 2.65). They are also erosion residues; usually 10 to 20 mineral types can be identified, with garnet, epidote, amphibole/hornblende, augite, and tourmaline as the most common ones. Associations or assemblages of these minerals (expressed as a percentage distribution by number) are usually a good indication for the origin of the silts or sands they are part of, and thus a good tracer for their dispersal from the source area. During transport, however, size sorting may take place so that the relative percentage of the heavy mineral species in the assemblages may change. Heavy minerals down to a size of ca. 10 μm can be identified with a petrographic microscope. To trace the origin and dispersal of finer-grained material, clay mineral associations can be used. Where suspended material is being dispersed and where mixing of material of different origin takes place, as in many lakes and estuaries, on the shelf, and in the ocean, size sorting can be important but mainly in the sand fraction (see Chap. 5).

On the shelf, quartz and the alumicosilicates usually form the major part of the suspended matter (as long as there is no major plankton bloom) and the bottom sediment, with the exception of those areas where the seafloor is covered with biogenic carbonate sediment (calcareous sands and muds). In oceanic suspensions and sediment, the mineral particles form a minor part, from ca. 20% down to less than 1%, except in sediments along the continental rise, which are supplied directly from the continental shelf by mass movements or turbidity currents, and in the deep sea bottom clays (the "red" deep-sea clays), which contain only few biogenic particles because of the dissolution of carbonate and opal and mineralization of the organic matter. In oceanic pelagic sediments, in general, clays represent 40–80% of the silicate minerals, quartz 4–13% (average 8.5%), and feldspars a similar percentage, the feldspar to quartz ratio being near unity (Wollast and Mackenzie 1983). The clay fraction consists of 30–60% illites, 15–55% montmorillonites (smectites), 10–20% chlorites and 10–20% kaolinites (Windom 1976). Other clay minerals are present in minor quantities, as are glauconite, oxides of Fe and Mn, and pyrite, which can be detrital and supplied from land as well as newly formed in the sea. A regular component of recent origin in all aquatic environments are man-made particles such as fly ash spherules, coal, and aggregates of Fe-oxides and Ti-oxides. They can come from run off, from ships, from ocean dumping, or through the atmosphere (aerosols). Rare types of anthropogenic and natural suspended particles in the sea, which form less than 2.5% to less than 0.02% of the total amount of material in suspension, were described by Jedwab (1980).

Contact with seawater can change somewhat the composition of clay minerals: Ca-montmorillonites, normally present in river material, are transformed into Na-montmorillonites. Other relatively small differences between marine clays and river clays are probably related to uptake of K^+ and Mg^{2+} from seawater, to release of Mn and possibly Fe (Gobeil et al. 1981), and to the formation of illite and chlorite in bottom sediment. In lakes, besides mineral particles supplied from the surrounding land, calcium carbonate particles can be present, formed by inorganic precipitation (see Sect. 2.2.2.).

4.2 Biogenic Particles

Organic matter, biogenic carbonate, and biogenic opal are common in natural waters and in some environments form the bulk of the material in suspension. Photosynthesis of CO_2 and water (primary production) is at the base of all biogenic particle formation and involves the formation of a large variety of compounds ranging from lignin (wood) to proteins and complex enzymes. Rivers, lakes, and to a certain degree also estuaries and the sea, receive particulate organic matter from terrestrial as well as from aquatic origin. Secondary production, because of energy losses, usually is in the order of 10% of the primary production so that by far the largest amount of organic matter in suspension is of primary origin, i.e., from plants. Besides organic matter also biogenic opal and calcium carbonate are formed. In freshwater, opal production (by diatoms) dominates. Biogenic calcium carbonate production is much smaller and includes the formation exoskeletons of higher animals (crustaceae, mollusks). In the sea, biogenic calcium carbonate and opal production can both be high. In freshwater environments the calcium carbonate particles are usually stable, which is also the case in the upper regions of the sea, but in the deep ocean they dissolve below ca. 3500 m waterdepth. Opal particles are easily dissolved in seawater, which is strongly undersaturated with regard to opal. Both biogenic carbonate and biogenic opal particles are embedded in organic matter which can retard dissolution and prevent exchange with elements or compounds dissolved in the surrounding water (see Sects. 4.3 and 4.4).

For the organic matter, a distinction can be made between labile organic material that is easily consumed by microbes and stable (refractory) organic material that is not (Degens and Ittekkot 1985). The labile fraction will rapidly be lost in the water by decomposition; a general scheme of degradation and resynthesis of organic components is given in Fig. 4.2. Rashid (1985) has listed the major differences between terrestrial and marine humic compounds. Marine organic matter is rich in aliphatic compounds relative to terrestrial organic matter (soil humus), has a lower carbon content (45–55% compared to up to 65%) but higher nitrogen content, has nitrate as a nitrogen source (against nitrogen fixed by bacteria on land), more heavy carbon isotopes (see Sects. 4.4 and 7.3), generally more carbonyl groups and high alcoholic hydroxyl contents, and 20–30% functional groups in the humic molecules (compared to up to 60% in terrestrial organic matter). Using carbohydrates and amino acids as indicators, it was found that in rivers 5 to 30% of the organic matter is labile.

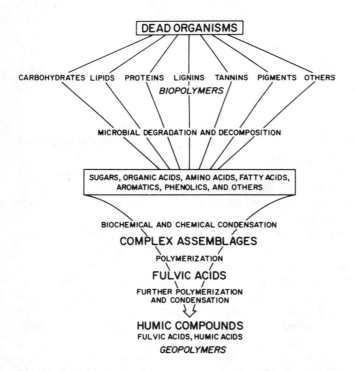

Fig. 4.2. General scheme of degradation and resynthesis of organic components. (Rashid 1985)

Ittekkot (1988) estimated that up to 65% of the river-supplied organic matter that reaches estuaries is refractory and may eventually be buried in the deep sea. From the distribution of organic compounds that have their origin on land, Prahl and Muehlhausen (1989) concluded that in the sediments of the continental margin and in the deep sea more than 20% of the organic matter is land-derived.

Primary production in rivers and particularly in lakes can be high when turbidity is low and nutrients are supplied from the surrounding land. In estuaries, because of the high turbidity and the rapidly changing conditions, primary production is usually lower. Worldwide it is of the same order as the supply of (largely refractory) organic matter from rivers (Williams 1981; Reuther 1981). In the sea, however, primary production can be very high, particularly where suspended matter concentrations are low resulting in favorable light conditions, and where there is abundant supply of nutrients from a nearby river or from deeper water by mixing or upwelling. The highest primary production occurs in upwelling areas along the shelf in the subtropics at the western boundaries of the continents (off West Africa, Namibia, Peru, California, western Australia), where the number of plankton cells may reach billions per liter.

Biogenic particles form the bulk of the suspended material in the surface water of the ocean, while on the shelf biogenic material dominates only in areas of high primary productivity. In the order of 20 to 70% of the total amount of suspended matter may be organic, but it can be more than 90% (Emery et al. 1973; Lal 1977). The production of biogenic particles shows large regional and seasonal variations. Besides in upwelling areas, primary production is strong along fronts between water masses where upward flow occurs, and in shallow seas where there is a rapid exchange between the bottom water and the surface water. It is low in the central parts of the large oceanic gyres. The production of biogenic $CaCO_3$ is high in the tropics and subtropics of the North Atlantic and the South Pacific, whereas opal production (diatoms) is high in the upwelling areas and along the frontal zones in the Southern Ocean, the North Pacific and the North Atlantic. The production of radiolarians is important in the Central Pacific. In the Atlantic, $CaCO_3$ particles constitute < 1 to 30% of the total amount of suspended matter in the surface water. They are formed predominantly between $50°N$ and $50°S$, where their distribution is rather uniform. Silica is particularly present in upwelling areas, where dissolved silica in the surface water is quickly incorporated in opal particles, which sink down. The opal production is limited by the concentration of dissolved silica in the water. The $CaCO_3$ production is mainly limited by the supply of nitrate or phosphate. These nutrients are (1) less quickly used up than silica and therefore more widely dispersed from the upwelling areas, and (2) supplied in large amounts to the surface waters by turbulence and diffusion: the vertical gradients of dissolved N and P are steeper than the gradients of dissolved Si, because N- and P-compounds mineralize more quickly than opal (Fig. 4.3). In deeper waters the concentrations of biogenic particles decrease with increasing water depth,

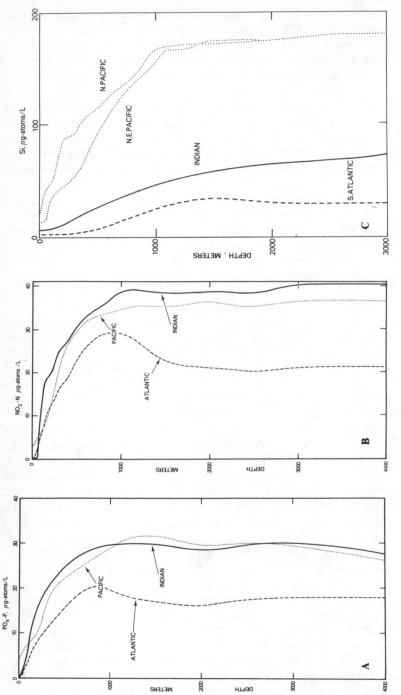

Fig. 4.3A–C. Depth profiles of phosphate (**A**), nitrate (**B**), and Si (**C**) in the Atlantic, Indian, and Pacific Oceans. (Sverdrup et al. 1946)

depending on the degree of remineralization of the organic matter and dissolution of the calcium carbonate and opal (see Sect. 7.7).

Barite ($BaSO_4$) particles that form a regular component of oceanic suspended matter over the entire water column in the Atlantic Ocean, the Pacific, and the Southern Ocean, are probably also of biogenic origin (Chesselet et al. 1976; Dehairs et al. 1980). They are small (≈ 1 μm) and are probably for more than 50% recycled in the surface water so that less than 0.4 μg Ba . cm^2 . y^{-1} (or 1×10^{12} g Ba . y^{-1}) is deposited on the ocean floor. This is very small compared to the amount of biogenic $CaCO_3$ and opal-SiO_2 that are yearly deposited ($\approx 1.5 \times 10^{15}$ g $CaCO_3$. y^{-1} and $\approx 0.4 \times 10^{15}$ g SiO_2 . y^{-1}, respectively; see Sect. 2.4). The distribution of biogenic $CaCO_3$ and opal production in the surface waters is generally reflected in the distribution of bottom sediments (Fig. 4.4).

Specific organic compounds in suspended particles can indicate the origin of the organic material. Lignin and lignin-derived compounds, long-chain hydrocarbons and fatty acids from vascular plants are indicators of terrestrial origin, while the total amount of amino acids and carbohydrate is a measure for the total primary productivity (Lee and Cronin 1984; Saliot et al. 1984). In marine surface waters particles containing proteins are most common and of planktonic origin. High ratios of total amino acids to hexosamine indicate a relatively high contribution of organic matter from primary productivity, as hexosamines are in chitinous material produced by zooplankton (Crustaceae; Degens and Mopper 1976). In small particles in suspension in the surface waters, organic matter is a

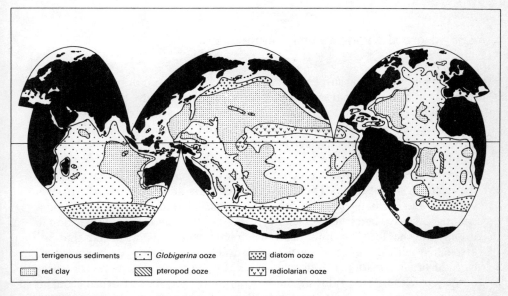

☐ terrigenous sediments	Globigerina ooze	diatom ooze
red clay	pteropod ooze	radiolarian ooze

Fig. 4.4. Distribution of sediment types on the ocean floor

mixture of organic compounds derived from both phytoplankton and zooplankton. In the larger particles that sink down, the organic fraction can be dominated by compounds derived either from zooplankton (Wakeham and Canuel 1988), or from phytoplankton (Ittekkot et al. 1984b). The relative abundance of arabinose and fucose, and of aspartic acid and glycine, indicates the relative contributions from silica and calcium carbonate producers, as arabinose and aspartic acid are relatively abundant when much carbonate (e.g., as coccoliths) is formed (Ittekkot et al. 1984a, b). Ribose concentrations indicate rapid downward transport because it is rather rapidly decomposed. At greater depths, labile compounds are abundant in the small suspended particles; they indicate the presence of undegraded planktonic material that has been sinking rapidly. The larger particles contain compounds that indicate intensive alteration of organic matter (mainly by microbes or zooplankton). This points to the feeding activity of zooplankton and higher organisms, which consume the labile compounds. Wax esters, steryl esters, and sterols are probably animal products (produced by zooplankton and fish; Wakeham et al. 1984). They occur increasingly at greater water depth, whereas the concentrations of labile compounds in the particles decrease with increasing water depth. The zooplankton forms fecal pellets and feces that sink down rather rapidly, and incorporates fine particles and their own biosynthetic products in them. Many fecal pellets and much feces in the water column serve as food for other organisms. On the bottom there is strong degradation of organic matter, which results in total organic matter concentrations of less than 5% in the bottom sediment (in pelagic sediments mostly less than 1%). Resuspended material therefore usually has relatively low concentrations of organic matter.

4.3 Adsorbed Elements and Compounds

The composition of the suspended matter is determined not only by the kind of minerals and organic matter of which it consists, but also by the elements and compounds adsorbed onto the particle surfaces. Thus a distinction can be made between those elements that are immobile and associated with the crystal lattices of the particles (the detrital or residual fraction) and those that are mobile (the nondetrital or non-residual fraction; for the chemistry of such elements and compounds see Salomons and Förstner 1984; Chester 1990, and the references cited there). The mobile elements are adsorbed onto the particle surfaces and exchangeable, or associated with metaloxyde and/or organic coatings. Since chemical (elementary) analysis of material in suspension is usually in the form of bulk analysis, fractions are distinguished that are associated with the mineral particles in suspension, with the carbonates, and with the organic matter. The elements present in the mineral particles were fixed in the particles before or during the weathering process; the adsorbed elements are only superficially associated with the surfaces and can be removed (ex-

Table 4.2. The percentage partitioning of elements in river particulate matter. (Data from Gibbs 1973)

| River transport phase | Element | | | | | |
	Fe	Ni	Co	Cr	Cu	Mn
Amazon River						
Solution	0.7	2.7	1.6	10.4	6.9	17.3
Ion exchange	0.02	2.7	8.0	3.5	4.9	0.7
Metal oxide coating	47.2	44.1	27.3	2.9	8.1	50
Organic matter	6.5	12.7	19.3	7.6	5.8	4.7
Crystalline matrix	45.5	37.7	43.9	75.6	74.3	27.2
Yukon River						
Solution	0.05	2.2	1.7	12.6	3.3	10.1
Ion exchange	0.01	3.1	4.7	2.3	2.3	0.5
Metal oxide coating	40.6	47.8	29.2	7.2	3.8	45.7
Organic matter	11.0	16.0	12.9	13.2	3.3	6.6
Crystalline matrix	48.2	31.0	51.4	64.5	87.3	37.1

changed) by natural aquatic processes. The partitioning of the total content of Fe, Ni, Co, Cr, Cu, and Mn over the different phases in the suspended matter transported by the Amazon and Yukon Rivers is given in Table 4.2 after Gibbs (1973). These data show the importance of the coatings and the organic matter for fixing elements to particulate matter. Both are commonly present in suspended material.

Because the mobile elements are adsorbed onto the particle surfaces, their total content in the particles is related to the total surface area of the particulates, which is expressed as the surface area per unit weight (the specific surface area in $m^2 . g^{-1}$). Related to the specific surface area is the cation exchange capacity, which can be defined as the amount of cations that can be exchanged (expressed in $mEq . 100 \ g^{-1}$). Because the particles normally have an over-all negative change, only the exchange of cations is considered but there can also be some anion exchange capacity, related to the presence of positively charged areas on the particles. The anion exchange is considered small compared to the cation exchange. The specific surface area and the cation exchange capacity of the principal clay minerals, hydroxides, and humic acids are given in Table 4.3 (from Fŏrstner and Wittmann 1979) and for different marine sediments are given in Table 4.4 (after Rashid 1985). For natural suspended particles, the Gironde and Loire estuaries (Garnier et al. 1990) give an increase of specific surface area from about $10 \ m^2 . g^{-1}$ in the freshwater to $30 \ m^2 . g^{-1}$ in the estuary, for particles in the Rhône mouth from about 4 to $13 \ m^2 . g^{-1}$. In the Tamar estuary, the highest values for specific surface area (up to $19.8 \ m^2 . g^{-1}$) occur in the brackish water at 2.2% S and decrease to less than $10 \ m^2 . g^{-1}$ in 28% S (Titley et al. 1987). In the Gironde and the Loire estuaries, the cation exchange capacity increased simultaneously with the specific surface area from about

Table 4.3. Specific surface area and cation exchange capacity of clay minerals and other suspended matter components. (Förstner and Wittmann 1979)

	Specific surface area $(m^2 . g^{-1})$	Cation exchange capacity $(mEq. 100 \ g^{-1})$
Kaolinite	10–50	3–15
Illite	30–80	10–40
Chlorite	—	20–50
Montmorillonite	50–150	80–120
Fresh Fe-hydroxide	300	10–25
Opal (amorphous SiO_2)	—	11–34
Humic acids (soil)	1900	170–590

Table 4.4. Cation exchange capacity of some marine sediments. (After Rashid 1985)

	mEq. 100 g^{-1}	
	Range	Average
Carbonate-free sediment	25–89	41.5
Organic matter in the sediment	10–71	22.8
Humic acids	250–375	301

Note: 23–81% (av. 45.7%) of total cation exchange due to organic matter, 20–77% (av. 54.3%) due to clay minerals. Organic carbon content is 1–13%; clay content is 24–54%.

25 mEq . 100 g^{-1} to 55 mEq . 100 g^{-1} and from about 5 mEq. 100 g^{-1} to 33 mEq. 100 g^{-1} in the Rhône mouth. These increases reflect the different amounts of fine particles in the rivers as well as the relative increase in the concentration of fine particles in suspension in the estuary, either because the larger particles settle out (Rhône mouth) or because the fine particles are concentrated in the estuary (Gironde, Loire; see Sect. 3.1.3). The heat of immersion is also used as a measure. This is the amount of energy (in $J \cdot g \cdot^{-1}$) that goes into the reaction of a particle with the surrounding water, mainly because of hydration of adsorbed ions and binding sites on the particle surface. The heat of immersion therefore shows a positive relation with the cation exchange capacity (Garnier et al. 1990). The influence of the organic matter on the cation exchange capacity of some coastal marine sediments is shown in Table 4.5 (from Picard and Felbeck 1976). Because of the relation of specific surface area, cation exchange capacity and heat of immersion to particle size, the fine-grained fractions have much higher contents of adsorbed elements than the coarse fractions. In addition, the clay minerals, and the organic particles, which are concentrated in the fine size fractions, are much more surface-active and therefore more cohesive than the particles of quartz and feldspar in the coarser fractions. The high adsorptivity of clay minerals, organic matter, and coatings is

Table 4.5. Effect of humic compounds on the cation exchange capacity of Narragansett Bay sediments. (Picard and Felbeck 1976)

Cation exchange capacity mEq. 100 g^{-1}	% org. C.
5.2–22.1	1–5 Natural sediment
2.4–15.6	0.7–3.5 After 0.5 N NaOH extraction
(16–54%)	(28–30%) (% decrease)

caused by the presence of open bonds along their surfaces and edges. There are areas with negative as well as with positive open bonds, particularly on clay minerals, but measurements of electrophoresis of suspended matter particles in natural waters have shown that the overall particle charge is negative (Hunter and Liss 1982). In freshwater the particles are more negative than in salt water, which points to a certain compensation of the negative charge by the larger concentration of positive ions in the salt water. Probably all natural particles in water are covered with an organic coating (Neihof and Loeb 1974; Loeb and Neihof 1977) and usually mineral particles are closely associated with organic matter in flocs (see Chap. 6). There are indications that the reactivity of suspended particles is determined by these organic coatings (Loder and Liss 1985; Martin et al. 1986; Garnier et al. 1990).

The mineral particles themselves do not change very much when transported from a river through an estuary into the sea. Some clay minerals may change somewhat through exchange of Ca with Na and adsorption of K and Mg (see Sect. 4.1), but the main process is the desorption or exchange of elements that are adsorbed onto the particle surfaces. This occurs mainly in estuaries: here the composition of the surrounding water changes drastically and the equilibria between the amounts that are dissolved and that are adsorbed shift. This is usually shown in a diagram relating the element concentration in solution in the water or its content in the particles to the salinity of the water. Deviations from the changes in concentration caused by physical mixing indicates desorption or adsorption (Fig. 4.5). The interpretation of such curves is not always as straightforward as it may seem, because physical mixing takes time and the concentrations in the supply from the river as well as the concentrations in the coastal sea may have varied during that time. Also, supply from other sources (e.g., dissolution of river-supplied diatom frustules) or loss because of processes within the estuary (e.g., primary and secondary production) have to be considered. Table 4.6 (after Chester 1990, data from Martin and Whitfield 1983) gives the concentrations of elements in river suspended matter and in deep sea clays. For the principal elements present in aluminosilicates and quartz (Si, Al, and Ti) the differences are small. Na, K, and Mg are higher in the deep sea clays because of uptake by clay minerals in contact with seawater, Fe and Mn are higher because of the formation of Fe and Mn hydroxides in the sea and Ba is higher because of barite formation (see Sect. 4.2). Ca, however, is lower in the

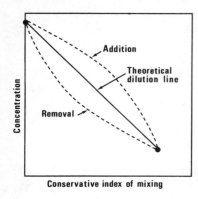

Fig. 4.5. Theoretical relation between the composition of estuarine water and salinity (chlorinity) as a conservative index of mixing. End members are considered to remain constant over the flushing time of the estuary. (Salomons and Förstner 1984)

Table 4.6. Average compositions of crustal rocks, soils, particulate river material, and deep sea clays. (Data from Martin and Whitfield 1983; after Martin and Meybeck 1979)

	Continents		Rivers	Ocean
	Rock ($\mu g \cdot g^{-1}$)	Soils ($\mu g \cdot g^{-1}$)	Particulate ($\mu g \cdot g^{-1}$)	Deep sea clays ($\mu g \cdot g^{-1}$)
Ag	0.07	0.05	0.07	0.1
Al	69300	71000	94000	95000
As	7.9	6	5	13
Au	0.01	0.001	0.05	0.003
B	65	10	70	220
Ba	445	500	600	1500
Br	4	10	5	100
Ca	45000	15000	21500	10000
Cd	0.2	0.35	(1)	0.23
Ce	86	50	95	100
Co	13	8	20	55
Cr	71	70	100	100
Cs	3.6	4	6	5
Cu	32	30	100	200
Er	3.7	2	(3)	2.7
Eu	1.2	1	1.5	1.5
Fe	35900	40000	48000	60000
Ga	16	20	25	20
Gd	6.5	4	(5)	7.8
Hf	5	—	6	4.5
Ho	1.6	0.6	(1)	1
K	24400	14000	20000	28000
La	41	40	45	45
Li	42	25	25	45
Lu	0.45	0.4	0.5	0.5
Mg	16400	5000	11800	18000
Mn	720	1000	1050	6000
Mo	1.7	1.2	3	8
Na	14200	5000	7100	20000

Table 4.6. (continued)

	Continents		Rivers	Ocean
	Rock $(\mu g \cdot g^{-1})$	Soils $(\mu g \cdot g^{-1})$	Particulate $(\mu g \cdot g^{-1})$	Deep sea clays $(\mu g \cdot g^{-1})$
Nd	37	35	35	40
Ni	49	50	90	200
P	610	800	1150	1400
Pb	16	35	100	200
Pr	9.6	—	(8)	9
Rb	112	150	100	110
Sb	0.9	1	2.5	0.8
Sc	10.3	7	18	20
Si	275000	330000	285000	283000
Sm	7.1	4.5	7	7.0
Sr	278	250	150	250
Ta	0.8	2	1.25	1.0
Tb	1.05	0.7	1.0	1.0
Th	9.3	9	14	10
Ti	3800	5000	5600	5700
Tm	0.5	0.6	(0.4)	0.4
U	3	2	3	2.0
V	97	90	170	150
Y	33	40	30	32
Yb	3.5	—	3.5	3
Zn	127	90	250	120

deep sea clays because of exchange of Ca with Na in the seawater in estuaries and coastal waters and because of dissolution of $CaCO_3$ at great water depths. For the other elements, similar processes of adsorption or desorption are responsible for the differences found, as well as biological uptake or release and particle size sorting with the fine particles being deposited on the deep sea floor.

4.4 Stable and Radioactive Isotopes

Suspended particles contain stable and radioactive isotopes, which can be of importance as indicators for their involvement in transport, deposition, and resuspension processes, as well as for their age. Those present in particle lattices include the oxygen and carbon isotopes in carbonate and organic matter. The adsorbed isotopes include the Th, Pa, and Pb isotopes, ^{10}Be and the fall-out isotopes from bomb tests and industrial discharges. For a more detailed discussion than is presented here, see Bowen (1988), Broecker and Peng (1982), Cochran (1984), and Mook and Tan (1991).

$^{18}O/^{16}O$: in carbonate and organic matter the ratio reflects the ratio in the water but with a small difference because uptake of ^{16}O is slightly higher than that of ^{18}O. The ratio is usually expressed in relation to the ratio of some standard material:

$$\delta^{18}O = 1000 \left[\frac{^{18}O/^{16}O \text{ sample } - \ ^{18}O/^{16}O \text{ reference}}{^{18}O/^{16}O \text{ reference}} \right]. \tag{4.1}$$

The $\delta^{18}O$ in water is related to the temperature of the water (evaporation) and to the amount of ice that is present: ice contains a relatively large amount of ^{18}O so that ice formation or ice melting results in fluctuations in the $\delta^{18}O$ of the water.

$^{12}C/^{13}C$: a $\delta^{13}C$ is defined in the same way as the δO^{18}. Air, rainwater, and river water have a relatively low concentration of ^{13}C, while seawater, where much water is evaporated, has a relatively high concentration. Organic matter of terrestrial origin therefore has very low (very negative) $\delta^{13}C$ values and marine organic matter higher (less negative) $\delta^{13}C$ values (for a recent discussion, see Mook and Tan 1991). The difference in $\delta^{13}C$ makes it possible to distinguish terrestrial organic matter from marine organic matter. Analyses of organic matter in suspension and in bottom deposits in river mouths and on the adjacent shelf have shown that, in general, on the basis of the $\delta^{13}C$ values, a large amount of terrestrial organic matter is mineralized in the estuary and the coastal sea, while the remainder is quickly dispersed in an excess of marine organic matter. The isotopic ratios, however, are highly variable so that they are probably less adequate tracers than the composition of the organic matter (Ittekkot 1988).

^{14}C is a radioactive isotope that is formed in the atmosphere out of ^{14}N by cosmic radiation. It has a half-life ($T^{1/2}$) of 5730 years and a mean lifetime (T_M) of 8700 years. ^{14}C is only a 10^{-12} part of the total amount of carbon that is present on earth. As it takes some time before it reaches the water surface, the ^{14}C in the water already has a certain age. In the water it is mixed with older ^{14}C so that the ^{14}C activity in the water is related to the mixing rate and the residence time of the ^{14}C in the water. In the past, the ^{14}C production has not been constant because of the variations in the intensity of the cosmic radiation and because of ^{14}C production during bomb explosions. These variations have been calibrated by dating materials of known age (historic materials, tree-ring sequences). The age of carbon-containing material is given by

$$T = \frac{T_{1/2}}{\ln 2} \cdot \ln \frac{A}{A_0}, \tag{4.2}$$

which is based on the exponential radioactive decay:

$$A = A_0 \, e^{-\lambda t}, \tag{4.3}$$

where A_0 is the initial activity, A the measured activity, λ the decay constant,

T the time, and $T_{1/2}$ the half-life of the isotope for which holds:

$$\lambda = \ln \frac{2}{T_{1/2}}. \tag{4.4}$$

$T_{1/2}$ and A can be measured. A_0 has to be inferred from the locality where the carbon-containing material was formed: $A_0 = A_i + n$, where A_i is the moment of ^{14}C formation and n the number of years that has passed before the ^{14}C was incorporated in the carbon-containing material. For land plants n can be taken as zero; for ocean surface water in the Atlantic n is ca. 475 y. Also a correction has to be made for fractionation because ^{14}C is less easily taken up than ^{13}C or ^{12}C. For land plants this correction, with respect to ^{12}C, is equivalent to ca. 16 years of decay.

^{210}Po and ^{210}Po are radioactive isotopes that are formed during the decay of ^{238}U (Cochran 1984) (Fig. 4.6). They are formed out of ^{222}Rn with short steps between the ^{222}Rn and ^{210}Pb. ^{210}Pb has a half-life of 22.3 y, its daughter ^{210}Po of 138 days. ^{222}Rn is a gas which comes out of ^{238}U-containing rocks. It is present in the atmosphere and in seawater. ^{238}U is present dissolved in seawater, as well as in suspended particles. The ^{210}Pb that is formed in the atmosphere comes down with the rain and is adsorbed onto the suspended particles together with the ^{210}Pb formed in the seawater and supplied from rivers. This is called the excess ^{210}Pb present in the suspended matter, in contrast to the supported ^{210}Pb, which is formed out of the uranium present in the particles. Because the distribution of U and ^{222}Ra is not uniform over the earth's surface, the excess ^{210}Pb in the suspended particles varies. This effect can be enhanced when the ^{210}Pb activity is measured by measuring the activity of its daughter ^{210}Po. ^{210}Pb and ^{210}Po are not everywhere in equilibrium because ^{210}Pb is rather easily adsorbed by organic matter so that the removal or addition of organic material can influence the amount of ^{210}Po present in suspension. ^{210}Pb is mainly used for dating fine-grained sediments over periods of 10 to 100 years. For A_0, the value is taken of the ^{210}Pb activity in the surface sediment, but this value is influenced by the rate of sediment supply and by the particle size of the deposited sediment. For both effect corrections can be made. The deposition (or accumulation) rate is given by the slope of the line relating the log ^{210}Pb activity with depth in the sediment. Reworking of the sediment by bioturbation or otherwise should be considered carefully, when interpreting ^{210}Pb profiles.

^{230}Th, with a half-life of 7.5×10^4 y, is also a product of ^{238}U decay. Uranium has a residence time in the sea of ca. 0.5×10^6 y. The Th that is produced is much more reactive than the uranium and is adsorbed onto particles within decades. As with ^{210}Pb, a distinction is made between supported ^{230}Th and excess ^{230}Th that is adsorbed. ^{230}Th has been used for sediment dating with the proviso that all the ^{230}Th is removed quickly from solution and deposited on the seafloor with the settling particles. This, however, is rarely the case: differences in particle size of the suspended matter influence the degree of scavenging and biogenic particle formation (carbonate formation) may interfere.

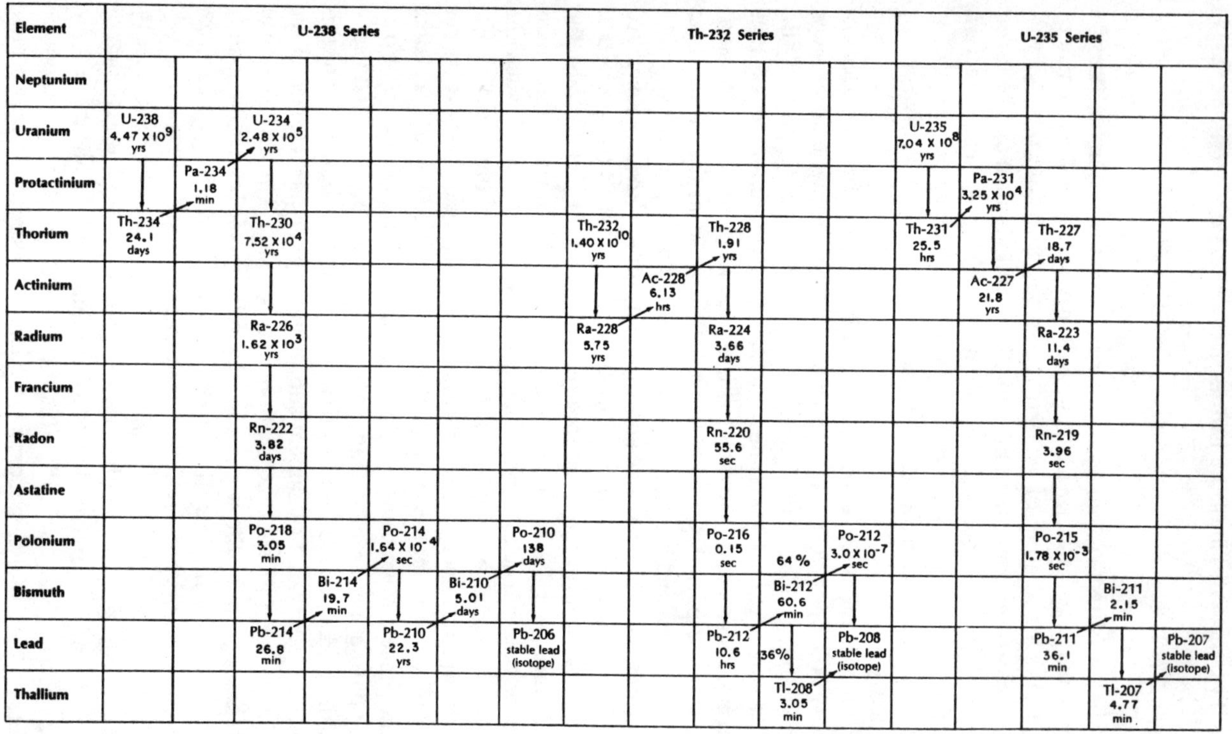

Fig. 4.6. The decay chains of the uranium and thorium series isotopes and the half-lives of each isotope. *Vertical arrows*: α-decay; *diagonal arrows*: β-decay. (Broecker and Peng 1982)

Bacon and Anderson (1982) found that in the deep sea only ca. 17% of the ^{230}Th is associated with particles and the remainder is in a dissolved state. Therefore often the ratio ^{230}Th/^{232}Th is used: the ^{232}Th is chemically equivalent to ^{230}Th, but has a half-life of 0.4×10^{10} y.

^{234}Th and ^{228}Th with half-lives of 24 days and 1.91 y respectively can be used as indicators for rapid processes like particle settling, deposition, and resuspension, and reworking of bottom sediment (bioturbation). ^{234}Th is often used in relation to ^{238}U, of which it is a daughter. ^{238}U has a half-life of 4.47×10^9 y.

^{231}Pa has a half-life of 34×10^3 y, or about the half-life of ^{230}Th. Like ^{230}Th it is also a decay product of ^{238}U and is very insoluble in water. It is removed from the water column within ca. 100 years after its formation. The same factors that affect the behavior of ^{230}Th influence also the deposition of ^{231}Pa, but ^{231}Pa is less abundant and disappears more quickly. With ^{231}Pa ocean bottom sediment can only be dated to a depth of ca. 3 m in the sediment instead of ca. 7 m with ^{230}Th.

^{10}Be, like ^{14}C, is produced in the atmosphere by cosmic radiation and comes down with rain and as aerosols. It has a half-life of 1.5×10^6 y, but like ^{230}Th and ^{231}Pa, its flux to the ocean floor is influenced by the sediment deposition rate, the suspended particle size, and biological activity.

Fallout isotopes and discharged isotopes from nuclear industry reach the atmosphere and the aquatic environment in a large variety. The most important are ^{137}Cs, which is mainly distributed through the atmosphere, and ^{134}Cs and the plutonium isotopes, which are predominantly dispersed through the water. The ^{137}Cs activity, which began in the early 1950s with the bomb tests, shows a maximum in 1963 because of the high number of bomb tests just before the atmospheric tests were stopped. A small peak occurs in 1980 because of the French and Chinese tests and a strong peak in 1986 in Europe and adjacent regions because of the Chernobyl reactor accident. This allows dating of relatively undisturbed bottom sediments.

^{134}Cs has a half-life of 2.06 y and is an important component of discharges of the nuclear industry. The plutonium isotopes, of which ^{239}Pu with a half-life of 24.1×10^3 is the most important, are artificial and appeared in the aquatic environment in 1954. They are also adsorbed onto particles and provide a means for sediment dating.

4.5 The Scavenging Process

All isotopes discussed above, except the principal oxygen and carbon isotopes, are trace elements; they are adsorbed onto particles and therefore give indications about the suspended particle behavior. These and other isotopes of the natural radioactive decay series (the ^{238}U series, ^{232}Th series, and ^{235}U series; Fig. 4.6) have been studied to understand the processes that regulate the partitioning of elements between the particulate and dissolved phases by sorption processes (scavenging). Comparing the production of an isotope (calculated from the measured concentrations of the parent isotope) with the actual concentration in the water and in the particles gives an estimate of the degree and rate of scavenging (Broecker and Peng 1982).

Sorption of elements or compounds can be a physical process by Van der Waals forces or ion-dipole and dipole-dipole interactions (Salomons and Förstner 1984). This can occur on the outer surfaces of the particles as well as the inner surfaces of the pores of a porous particle. Sorption occurs as a chemical process by ion exchange through which positive or negative charges on the mineral lattice of the particles are compensated by ions of opposite charge, which are exchangeable with ions in solution. The various mechanisms of adsorption of trace elements and compounds on particle surfaces are summarized by Salomons and Förstner (1984) and Chester (1990). They include electrostatic adsorption, adsorption after hydrolyzation of the ions (which lowers the ionic charge of the ions and decreases their interaction with the solvent and makes it possible to come closer to the particle surface), ion exchange, surface complex formation, and formation of metal-organic complexes. The adsorption of metals such as Zn and Cd (but not e.g. Cu and Pb) is strongly regulated by the pH of the water and shows large pH-dependent changes in the range of pH 6 to 9, which covers most natural waters. The interaction of metal species with suspended particles is schematically given in Fig. 4.7 (after Salomons and Förstner 1984).

There is a relationship between the residence time of an element, the residence time of the suspended particles, and the partitioning of the element over the particles and the water, expressed as the ratio between the concentration in the particulate phase and the total concentration of the element in water and suspended matter. For many trace metals, Bacon and Anderson (1982) have shown that in seawater the removal is controlled by a population of particles with a residence time of ca. 5 to 10 years. This indicates that the scavenging process is dominated by small particles of ca 10 μm or less, as suggested earlier by Lal (1977). Most of the scavenging is through a reversible exchange. Resuspension, or increased turbidity, gives enhanced scavenging, which is probably also the case in the turbidity maximum present in many tidally mixed estuaries. The small particles themselves can be scavenged by the larger ones (see Chap. 6), resulting in a relatively quick removal towards the bottom of particles as well as of the adsorbed elements.

Trace element removal by complexation with functional groups on the particle surfaces implies that the rate of adsorption varies with the particle concentration. Biological removal implies a relation with particle production by primary productivity. Both relationships have been found for the removal of dissolved Th (Nyffeler et al. 1984; Coale and Bruland 1985, 1987). With radioactive elements, the decay rate of the isotopes, besides the reaction rate, determines the scavenging rate. Honeyman et al. (1988) suggested that a combination of complex formation on the particle surfaces and colloidal (< 0.4 μm) particle-to-particle aggregation, whereby the latter serve as a new adsorption substrate,

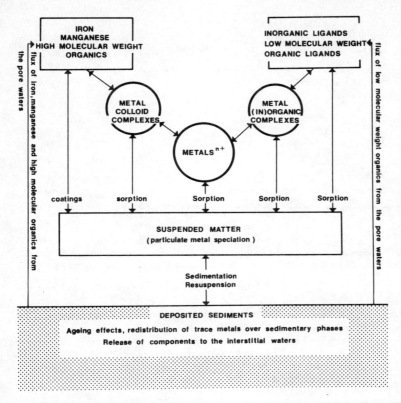

Fig. 4.7. Summary of major processes and mechanisms in the interactions between dissolved and solid metal species in surface waters. (Salomons and Förstner 1984)

controls the adsorption process. In deep sea bottom clays and nodules, K, Rb, Cs, Be, Sc, and Ga are mainly associated with aluminosilicate particles (Li 1981); associated with Fe (hydr)oxides are elements that form anions or oxyanions (P, S, As, B, Sn, I, Br, F, U, Pb, Hg) and hydroxide complexes (Ti, Zr, Hf, Cr), while those associated with Mn(hydr)oxides are those that form mono- and divalent kations (Mg, Ca, Ba, Co, Ni, Cu, Zn, Cd) and oxyanions (Mo, W, Sn). This applies to bottom sediments where the organic phase is insignificant. In the water in suspension, a large number of elements is associated with the organic matter, as was found, e.g., in the North Sea, where Cu, Zn, Cr, Pb, Ni, and Cd are associated with the particulate organic carbon (Nolting and Eisma 1988).

The partitioning of elements between the dissolved phase and the particulate phase is expressed by the K_D value, which is the ratio between the concentration in solution and the total concentration in particulate + dissolved state. K_D values can be determined by collecting water samples and analyzing these for particulate and dissolved elements. But also when the amount present in the crystal lattices, which does not take part in the aquatic processes, is taken into

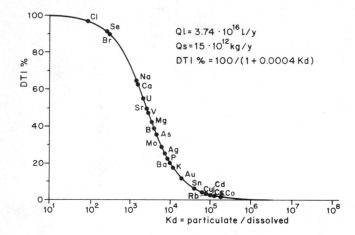

Fig. 4.8. Relation between the dissolved transport index (DTI) and K_D for a number of elements (after Martin pers. comm.). Most of the elements not indicated have K_D values higher than 10^4

Table 4.7. Percentage of transport in particulate form, relative to total transport in rivers. (Data from Martin and Meybeck 1979)

99–99.9%	Ga, Tm, Lu, Gd, Ti, Er, Nd, Ho, La, Sm, Tb, Yb, Fe, Eu, Ce, Al.
90–99	P, Ni, Si, Rb, U, Co, Mn, Cr, Th, Pb, V, Cs.
50–90	Li, N, Sb, As, Mg, B, Mo, F, Cu, Zn, Ba, K.
10–50	Br, I, Cl, Ca, Na, Sr.

account, the K_D values, do not necessarily indicate equilibrium values, as equilibrium may not have been reached. This is particularly so for rivers and estuaries, where the residence times of the particles in the water or of both water and particles may be less than the time needed to reach equilibrium.

Particulate matter is an important means of transport for many elements. Martin and Meybeck (1979) calculated a dissolved transport index for a number of elements, which is the percentage of the amount transported in solution in relation to the total transport. Reversing this index gives the percentage of the amount transported in particulate form (Table 4.7; Fig. 4.8). There is considerable variation, however: rare earths are the most constant, but the percentages for Ca, Cs, Cu, Mo, Ni, Pb, Si, and Zu are highly variable (Martin and Meybeck 1979; Förstner 1986). The most variable are those elements that are transported mainly in solution or whose concentrations are strongly influenced by human activities (Zn, Pb, Cu).

Also in the sea, the concentrations may be influenced by the amount of pollution. Table 4.8 gives the percentages that are transported in particulate form in the coastal waters of the North Sea (Eisma 1990). At mid- and high

Table 4.8. Percentage of trace metals in particulate form in suspension (% of particulate + dissolved). (Eisma 1990)

	% Particulate
German Bight	
Cd	20–30
Pb	66–84
Cu	10–20
Co	6–15
Ni	6–14
Elbe Estuary	
Cd	80
Pb	97.5
Cu	59
Co	67
Ni	16
Northern North Sea	
Cd	16
Pb	< 10
Cu	7
Skagerrak	
Cd	13
Pb	50
Cu	12
Co	6
Ni	3
Zn	11

latitudes seasonal variations are caused by seasonally varying sources that supply the dissolved and particulate material.

4.6 Determination of Suspended Particle Composition

Prior to any compositional analysis, the suspended matter is filtered off, usually over a 0.4 μm pore-size filter, but also special filters are sometimes used such as glass fiber filters for some organic analyses, and silver filters for X-ray diffraction. Filtration can be carried out insitu with an underwater pump system or after collecting a water sample with a sampling bottle. In the latter case, filtration can be carried out directly from the sampling bottle, which is done to avoid contamination (necessary when collecting samples for analysis of trace constituents), or after taking a subsample to the laboratory. The filtration is essentially the same as for total suspended matter (see Sect. 3.2). There are two main groups of analysis that can be carried out: bulk analysis of the total suspended matter sample and analysis of individual particles.

4.6.1 Analysis of Suspended Matter Composition

4.6.1.1 Bulk Analysis

For *wet chemical analysis*, the suspended matter sample is digested in strong acid (HF, HCl, H_2SO_4, and mixtures of these). Usually high temperatures are applied to ensure complete distruction of the minerals particles in the sample. The concentration of elements in the solution that is obtained can be determined in various ways. The method was most commonly applied at present is atomic absorption spectrometry (AAS; Ramirez-Muñoz 1968; Angino and Billings 1967; see for this and other instrumental analysis also Skoog 1985). For this the solution is brought into a flame where the elements dissolved in the solution are strongly heated and absorb light of a wavelength that is characteristic for each species of atom. When light of that wavelength is put through the flame, the amount of light that is absorbed can be measured and is an indication of the concentration of that species/of atom in the solution. In this way, for each element a concentration can be measured. A high precision can be reached and concentrations in the ppb range can be measured accurately. For each element a special lamp has to be used that emits light of the required wavelength. *ICP* (*inductively coupled plasma analysis*) is very suitable for determining the elementary composition of rocks and mineral particles. The sample is completely dissolved in strong acids and then is brought into a flame of very high temperature (6000–10000 K; Walsh et al. 1981). At these temperatures ionic species rather than atoms emit light of a characteristic wavelength for each element, while the intensity is related to the quantity of that element present in the flame. Also, a mass spectrometer can be coupled to this system so that the gases coming out of the flame can be analyzed for different isotopes.

To determine the composition of a suspended matter sample, often a stepwise production is followed which gives the composition of the carbonate fraction, the adsorbed fraction, and the mineral fraction separately. Complete separation and analysis of the contents of the different fractions is difficult and usually achieved by using a series of acids of different composition and strength, followed by separate analysis of the solutions.

A less accurate method than wet analysis by AAS or ICP, but one that allows determination of the concentration of a large number of elements simultaneously is *emission spectrography*. For this analysis the filtered suspended matter sample is powdered and burned between carbon electrodes. While burning, the atoms emit light, and each species of atom emits light of a characteristic wavelength. This can be measured for each wavelength separately and simultaneously, which gives a direct measurement of the concentration of the different atoms in the sample.

Neutron activation analysis is also carried out on a dry powdered sample. This is brought in a neutron beam which activates the atoms in the samples into radioactive species. This activity can be measured and the types and concentrations of radioactive species that are measured are an indication for the concentrations of the original elements. This method depends very much on the possibility of forming radioactive elements out of the original elements in the sample. It is particularly suitable for the determination of rare earth element concentrations.

X-ray fluorescence spectrometry (Jenkins 1976) is based on the fluorescent radiation that is produced when elements are radiated with X-rays, or with an electron beam, as happens in an electron microscope. The particles in suspension are filtered off and homogenized by fusing them to a glass (mainly with sodium or lithium tetraborate). The fluorescence spectrum has to be corrected to give the real concentrations because the intensity of radiation differs for equal amounts of different elements, and because of mutual interference between the radiation of different elements.

For *α-spectrometry* (Siegbahn 1966; Cochran 1984) the radionuclide to be measured is concentrated by chemical separation and collected on a plate which is put into an α-counter for measurement. For *β- and γ-radiation measurements* (mainly used for fallout isotopes) usually no pretreatment is necessary but a considerable volume of sample may be needed to obtain accurate results on radionuclides that occur in low concentrations. *β*- and *γ*-counters have to be shielded from environmental radiation (including cosmic radiation), but it is possible to measure the environmental radiation separately and substract all counts that coincide with the environmental counts so that only the counts coming from the sample are recorded. Recently, small quantities of radionuclides (e.g., of ^{14}C) can be measured using an *accelerator* in combination with a mass spectrometer, provided the isotope to be measured can be brought into gaseous form (Elmore and Phillips 1987).

Bulk mineralogical analysis is usually done by *X-ray diffraction* (XRD; Klug and Alexander 1974), but this is only possible for crystalline material. The content of amorphous or microcrystalline material can sometimes be estimated from background radiation (e.g., opal) and sometimes by measuring the crystalline form that is produced on heating the sample. Samples are usually ground, mixed, and deposited on a glass slide or ceramic tile, resulting in a random-oriented sample. Small amounts of suspended matter can be filtered directly over a silver filter and analyzed, but here the orientation of the platy minerals can lead to over- or underestimation of the amounts that are present. Also in samples assumed to be randomly oriented, orientation of platy minerals may give problems. Identification of minerals is based on the diffraction patterns and on the changes produced at higher or lower humidity, temperature, or on adding glycol: these changes are specific for certain (clay) minerals. "Amorphous" materials like allophanes can be detected by measuring at very low angles. Refinements in the technique make it possible to obtain results that are reproducible within a few percent.

Bulk *organic analysis* varies from relatively simple ways to measure the total organic matter in a sample to complex systems based on gas chromatography in combination with mass spectrometry (Skoog 1985).

A commonly used method for determining the total organic matter content is by heating the sample to ca. 500 °C over ca. 8 h. The organic matter content (or content of combustibles) is considered to be equal to the loss of weight after heating. It is assumed that carbonates are not dissociated into CaO and CO_2 during the heating, and that no water is lost from the minerals, which would also result in a loss of weight. The method is rapid but not very accurate. More precise analysis is done by determining the amount of organic carbon in the sample by converting the organic matter into CO_2. This, however, depends very much on the composition of the organic matter and also here dissociation of carbonate (or avoiding this) may influence the results considerably. Analysis at low temperatures may not oxidize part of the more resistant organic matter.

Analysis by *gas chromatography* and *mass spectrometry* is based on separation of the different organic compounds by the temperature at which they evaporate. Further separation is done on the basis of the differences in molecular weight. Complete analysis of a sample gives a detailed picture of the composition of the organic matter but is time-consuming and usually involves difficulties in interpretation. Therefore, organic analysis is often directed to determination of specific compounds or groups of compounds (carbohydrates, amino acids, humic acids, lipids, lignine, steroles, etc.). Another relatively rapid method is Curie point-flash pyrolysis coupled to mass spectrometry (Saliot et al. 1984). In this analysis the sample is heated quickly to a high temperature (350 to 700 °C) whereby organic compounds are broken up into fragments. The fragments are identified and through statistical analysis the original compounds (or groups of compounds) are reconstructed and their relative amounts estimated.

4.6.1.2 Single Particle Analysis

The mineralogy of individual particles can be determined with a *petrographic (polarizing) microscope* (Milner 1962), with *X-ray diffraction* and, in an *electron microscope*, by electron diffraction, which is based on the diffraction of an electron beam by a crystal lattice. A representative mineralogical composition of the entire sample can be obtained by identifying randomly 100 to 300 particles (for the statistics see Van der Plas and Tobi 1965). Amorphous or microcrystalline material can sometimes be identified by the shape of the particle (e.g., biogenic opal), but can be identified better by the elementary composition. Much used for this is the *microprobe* that can be attached to an electron microscope or a scanning electron microscope (SEM; Chandler 1977). It measures the fluorescence that the elements emit in an electron beam. For SEM analysis the particles have to be filtered over a 0.4 μm pore-size Nuclepore filter in small amounts so that the particles do not overlap. A piece of the filter is glued to a sample holder and coated, preferably with carbon, to obtain a surface that scatters back enough electrons to obtain a picture. As the probe-beam has a diameter of ca. 0.5 μm or less, individual particles can be measured. Automated systems have been developed that can measure large numbers of single particles automatically. Also the distribution of an element in the entire sample can be measured by scanning the sample for that element. When single particles are measured, there are particle size and orientation effects. Wet analysis of single particles is possible after selection of such particles (usually a number of particles of the same type), but this is very time-consuming.

Organic analysis on single particles of suspended material has not been done, to the author's knowledge, but there is an in-situ method to identify organic suspended particles by *flow cytometry* (Chisholm et al. 1988). This is based on measuring the fluorescence of organic matter in UV-light and allows identification of different groups of algae and other suspended organisms down to submicron size. Also, particles that do not show fluorescence can be identified and measured. By adding fluorescent compounds that are adsorbed onto the particles and by combining this with staining techniques, the organic matter in the nonliving particles can be identified to some extent.

Note added in proof: A recent book edited by Hurd and Spencer (1991) deals with analysis and characteristics of marine particles, part of which also applies to particles in other aquatic environments.

5 Transport of Suspended Matter

Transport of suspended matter takes place through the water column, as opposed to bottom transport. Sand may partly move in an intermediate mode by jumping over the bottom, which is called saltation. Suspended matter is usually fine-grained/low density material while the bed load is usually coarser-grained/high density material. There is, however, no sharp distinction between transport in suspension and bottom transport: also sand grains that are normally transported over the bottom can be transported in suspension in large quantities where currents and turbulence are sufficiently high, as is the case in rivers, in the surf along beaches, and on sandy shelves during storms. Transport in suspension is primarily related to the turbulence in the water but the conditions under which the turbulence is generated strongly influence the transport pattern. In this chapter, unidirectional or quasi-unidirectional transport as occurs in rivers and streams and suspension by surface waves will be discussed in Sections 5.2 and 5.4, preceded by a short general introduction on turbulence (in Sect. 5.1). Transport takes place between two moments: the initiation of particle motion by scour (erosion, resuspension), and the settling of particles where the turbulence is too weak to keep them in suspension. These will be discussed in Sections 5.3, 5.5 and 5.6 followed by brief discussions of suspended matter transport in stratified waters (Sect. 5.7) and of autosuspension of suspended matter in downward flow (Sect. 5.8).

5.1 Turbulence

Water has a certain viscosity: the general dynamics of flow are based on the relations between the velocity and the viscosity of the water, and gravity. These relations have been expressed in a set of equations, known as the Navier-Stokes equations, which date from 1845. They apply only to incompressible fluids; water is considered incompressible although it can be compressed somewhat, but this only has a small effect (on the water temperature) in deep sea basins and trenches below ca. 4500 m depth, and need not be considered here: at less pressure the compressibility is so small that it can be taken as zero. A complication in applying the Navier-Stokes equations to natural flow is that at higher flow velocities the flow becomes turbulent, while at lower flow velocities, in smooth surroundings without any obstacles, the flow remains laminar. Turbulent flow is characterized by eddies that are carried downstream in the direction of the general flow, whereas in laminar flow the fluid moves in layers without local fluctuations. A criterion for the transition from laminar to turbulent flow is formulated in terms of Reynolds number:

$$\text{Re} = \frac{\rho u d}{\eta}, \tag{5.01}$$

which relates the velocity u to the density ρ of the fluid, the viscosity η, and a characteristic diameter d of the flowsystem (in large rivers the mean depth; for flow around particles the particle diameter). The critical value for Re varies for different flow systems. Laminar flow can become turbulent without an increase in large-scale flow velocity because of obstacles in the flow (e.g., bottom irregularities) which locally change u and d and result in a series of eddies. Stable flow can persist at values for Re well above the value for transition when there are no disturbances present. In nature there are always disturbances, but in pipes, where the transition occurs at Re = ca. 2200, flow can remain stable up to values of 40 000 for Re when the pipe is very smooth.

Turbulence is also initiated by shear in the flow, without any obstacles, which is called "free turbulence" in contrast to "wall-turbulence" and can be caused by inertial forces, centrifugal forces, the Coriolis force, thermal instability (buoyancy), and surface tension (Lugt 1983). Turbulence is an irregular rotational motion; eddies are formed and their energy is dissipated by transfer to smaller and smaller-scaled eddies until a size is reached where the energy dissipation by molecular viscosity is almost immediate and the energy is transformed into heat. The amount of rotation, i.e., the vorticity of eddies, is proportional to the reciprocal of the time scale: most of the energy in the flow is associated with the large scale motion and most of the vorticity with the small scale motion. In the large eddies the viscous forces are small in relation to inertial forces: the largest eddies with a length scale near to the scale of the flow system (d in Reynolds number) are almost independent of viscosity and almost completely determined by inertia, which gives them an almost permanent character (for as long as the turbulent system is not changed). Von Helmholtz (1858) has shown that the transfer of kinetic energy from the basic flow through shear forces or outside forces (e.g., heating, or a disturbance) and through larger eddies to increasingly smaller eddies is possible only along a pathline in three-dimensional flow, not in two-dimensional motion: turbulence is essentially three-dimensional.

Between stable (laminar) and unstable (turbulent) flow there can be a state of intermediate flow consisting of stable layers of ordered vortices. When Re increases, more vortices appear until the flow is fully turbulent. Near to a solid boundary (the bottom or a particle surface) the flow is laminar in a thin layer (the laminar sublayer) also in turbulent flow. The boundary is considered rough when the projections from the boundary (protruding grains, sharp protruding edges) disrupt the laminar film. With increasing Reynolds number, the laminar film becomes thinner: whether the bottom becomes "rough" depends on the relative values of the linear dimension of the projections into the flow, the density and viscosity of the water, and the shear stress (Rouse 1938).

Where the bottom is "rough", turbulent eddies develop at the bottom that dissipate upwards into the main flow. Thus, near the bottom, the turbulence is usually more intense, and of smaller scale, than higher in the flow but this near-bottom turbulent region may reach upwards over the entire water depth when the bottom is very rough and the water shallow. Where the turbulent flow is over a (movable) sand bottom, bottom ripples are formed that strongly influence the near-bottom flow characteristics by inducing the formation of local turbulent eddies. Also on a flat rough bottom small local eddies can be formed. The local eddies can become separated from the bottom and develop into large irregular whirls or vortices that burst upwards and can reach the surface (so-

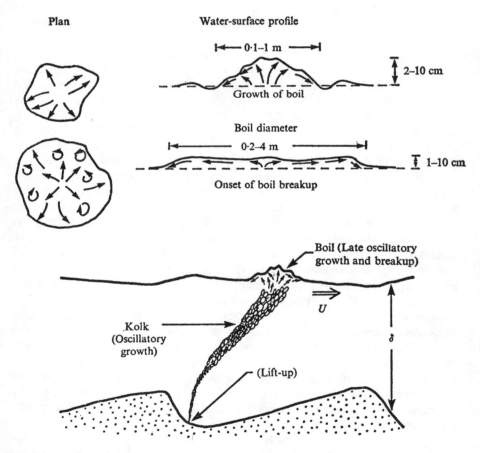

Fig. 5.1. The formation of boils and kolks in relation to the process of bursting. U Mean flow; δ water depth. The bed can also be flat. (Jackson 1976)

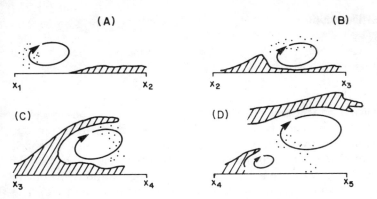

Fig. 5.2A–D. Schematic sequence of suspension caused by a burst. *Shaded area* is a zone of low velocity. (Dyer 1986; after Sumer and Oğuz 1978)

called surface boils; Figs. 5.1 and 5.2). In this way, suspended matter can be entrained quickly from the bottom into the main flow (Jackson 1976). Bagnold (1966) reasoned that to bring particles is suspension from the bottom, the upward turbulent movement must be stronger than the downward movement to compensate the higher density of the particles as compared to the fluid. This can only be maintained (because of conservation of energy) when the upward movement is stronger but of shorter duration than the downward movement. Observations in nature have demonstrated that particles are brought upward from the bottom into the main flow predominantly during moments of strong turbulence or "bursts".

Turbulence is characterized by oscillations in time of velocity and pressure, which has led to the development of a statistical approach, whereby the movement is reduced to average quantities. But turbulence is not entirely statistical, since coherent, long-lasting vortices exist in it (Lugt 1983), which led to the recognition of a large-scale more coherent flow pattern and a small-scale more chaotic pattern. The deterministic approach to turbulence takes this coherence into account. The theory and mathematics involved in both approaches, which cover turbulence not only in flowing water but also in air or in any fluid or gas, are quite complicated and outside the scope of this book. Here only those aspects are discussed that seem relevant – in a rather applied sense – to suspended particle transport. If, nevertheless, the reader feels unsatisfied and has the impression that some basic notion concerning turbulence has not been expressed here, it should be realized that in spite of its universal occurrence in flowing matter, and in spite of hundreds of years of scientific approach and probably hundreds of thousands of years of observation (Lugt 1983), turbulence is not completely understood and no satisfying theory exists that can derive turbulent motion from first principles. For further study, the reader is referred to general works on turbulence and hydrodynamics (Lamb 1932; Tennekes and Lumley 1972; Lugt 1983; Lesieur 1987).

5.1.1 Statistical Description

Applying Cartesian coordinates, let the velocity components in the x, y and z directions of the flow be given by

$$u = \bar{u} + u'; \quad v = \bar{v} + v'; \quad w = \bar{w} + w',$$

where $\bar{u}$, $\bar{v}$, and $\bar{w}$ indicate the mean flow along the x, y and z directions and u', v', and w' are the turbulent fluctuations, which by averaging many instantaneous values, become zero. Steady or stationary flow is characterized by values for $\bar{u}$, $\bar{v}$, and $\bar{w}$ that do not change with time at any fixed point through which the fluid moves. The flow is uniform when $\bar{u}$, $\bar{v}$ and $\bar{w}$ do not change as the x-coordinate in the flow is varied (but change when the y- and z-coordinates are varied). The difficulty with accurately describing the dynamics of turbulent flow is that real turbulent flow is usually nonhomogeneous and nonsteady. Although in the analysis of turbulence many quantities can be expressed as averages of stationary variables, the averaging tends to mask the characteristics of the variables that are averaged (Tennekes and Lumley 1972).

As a measure for the intensity of turbulence the root mean square values of the turbulent fluctuations are used:

$$\sqrt{\overline{(u')^2}}; \quad \sqrt{\overline{(v')^2}}; \quad \sqrt{\overline{(w')^2}}$$

When the mean flow is in the x-direction, the turbulent fluctuations are stronger in that direction than in the other directions. As seen above, the turbulence is related to the environment of the flow.

Measurements in flumes, summarized by Vanoni (1975), indicate that $\sqrt{\overline{(u')^2}}$, normalized to the friction velocity u_*, varies with the ratio y/d, where y is the distance to the bottom and d the waterdepth. The velocity gradient u_* is influenced by the bottom shear and therefore called the shear velocity. It is defined as τ_0/ρ where τ_0 is the bottom shear stress and ρ the fluid density. The intensity of turbulence in the x-direction is strongest at some distance from the bottom, at a smaller distance when the bottom is smooth than when the bottom is rough. The values for $\sqrt{\overline{(u')^2}}/u_*$ do not differ much for smooth and rough bottoms, but at a smooth bottom, u_* has a smaller value than at a rough bottom, so that above a rough bottom $\sqrt{\overline{(u')^2}}$ has a higher value. The measured turbulence fluctuations in the y- and z-directions are considerably smaller than in the x-direction, when the mean flow is unidirectional.

5.1.2 Size of Turbulent Eddies

The mean size of the eddies in a turbulent flow is given by

$$l = \int_0^\infty R(\xi)\, d(\xi), \tag{5.02}$$

where $R(\xi) = \overline{u'(X)u'(X + \xi)}/\overline{(u')^2}$ and $u'(X)$ and $u'(X + \xi)$ are simultaneous values of the turbulence fluctuations in the x-direction at positions X and (X + ξ). The bar denotes mean values of the products of many (simultaneous) measurements (Vanoni 1975). Around the mean size, a spectrum of eddies is present ranging from large eddies – the largest one being as large as the width of the flow – to very small eddies.

The size of the smallest eddy is called the Kolmogorov length $\gamma = (v^3/\varepsilon)^{1/4}$, where v is the kinematic viscosity and ε the energy dissipation rate. ε is proportional to u^3/l, where l is the size of the largest eddy and u is the flow velocity. As was seen above, at scales below the Kolmogorov length scale, the kinetic energy disappears by viscous dissipation.

The large eddies are anisotropic in unidirectional flow because of the flow strain, the smaller eddies become more isotropic and at small scales a "local isotropy" develops. The small eddies are at equilibrium with the local conditions and respond quickly to changes in the nearby flow. The large eddies need more time to adjust and these are said to have a "memory" of previous conditions.

5.1.3 Diffusion by Turbulence

In turbulent flow there is transport of material because of the turbulent velocity fluctuations as well as through mixing. The result is diffusion of material whereby the transfer is in the direction of decreasing concentration. When a volume V_1 of water moves from y_1 to y_2, a similar volume V_2 moves in the opposite direction and there is no net flow ($V_1 = V_2$). Over a length L during a time T at a mean speed v', the transferred volume of unit width is $V_1 = V_2 = v'$ LT. The rates of turbulent dispersal and mixing are much larger (and the time scales shorter) than the rates of molecular diffusion. Much transport of suspended matter can be described by approximation as a diffusion process. The representative diffusivity (eddy diffusivity or exchange coefficient) should be chosen in such a way that the time scale of the hypothetical diffusion process is equal to that of the actual mixing process (Tennekes and Lumley 1972). The intensity of the turbulent movement is given by a coefficient f that relates the turbulent shearing stress (τ) to the velocity gradient:

$$\tau = f\frac{d\bar{u}}{dy}.$$
(5.03)

It should be realized that while viscosity is a property of the fluid itself, independent of the motion, f varies widely from point to point in the flow.

5.2 Transport of Suspended Matter

The transport, deposition, and erosion of suspended matter until very recently was not considered in terms of turbulent motion but mainly in terms of average flow characteristics and, after 1937, in terms of diffusive mixing. These aspects will be treated here only in a limited way; for a more general discussion the reader is referred to the more detailed treatments by Vanoni (1975), Richards (1982), and Dyer (1986). In two-dimensional flow (in the x-direction and the y-direction) the suspended sediment transport per unit width (Q_{ss}) is given by

$$Q_{ss} = \int_{y_0}^{d} C u \, dy, \tag{5.04}$$

where C is the suspended sediment concentration, u is the mean flow velocity at a distance y above the bottom, d is the flow depth, and y_0 a small value of y taken as the upper limit of bed transport (including saltation) and the lower limit of suspension transport. In steady flow the average concentration at any level is constant: the upward dispersal (because of diffusion) is balanced by settling of the sediment because of its weight, for which holds

$$Cw + \varepsilon_s \frac{dC}{dy} = 0, \tag{5.05}$$

where Cw is the settling rate per unit area (w is the settling velocity of the individual particles) and ε_s is the diffusion coefficient for the suspended material. $\varepsilon_s = \beta_1 \sqrt{(v')^2} \cdot l_1$, where $\sqrt{(v')^2}$ is a measure for the intensity of the turbulence in the y-direction as seen above. β_1 is a correlation coefficient defined as

$$\beta_1 = \frac{\overline{c'v'}}{\sqrt{\overline{(c')^2}} \cdot \sqrt{\overline{(v')^2}}}, \tag{5.06}$$

where c' denotes the concentration fluctuations. l_1 is a length scale, defined as the average distance that a unit of suspended sediment moves in the x-direction before it mixes. The diffusion coefficient ε_s generally is a function of y. In an unsteady, nonuniform suspended sediment distribution in a two-dimensional steady and uniform flow, the flow of sediment into an element of volume during a small unit of time Δt, minus the flow that goes out, is equal to the change in concentration in that volume.

Relation (5.05) can be integrated to give

$$\frac{C}{Ca} = \exp\left[-\frac{w}{\varepsilon_s} y - a \right], \tag{5.07}$$

where C_a is the concentration at level y = a. This relationship was found to agree with experimental results (Rouse 1938). Also it was shown that the mixing of fluid and sediment are not the same (for fluids $\beta = 1$).

If the velocity distribution is known, ε_s can be expressed as a function of y. Close to the bottom there exists the so-called constant stress layer in which the velocity distribution over the vertical is given by the von Karman-Prandtl equation:

$$\frac{u}{u_*} = \frac{1}{k} \ln \frac{y}{y_0}, \tag{5.08}$$

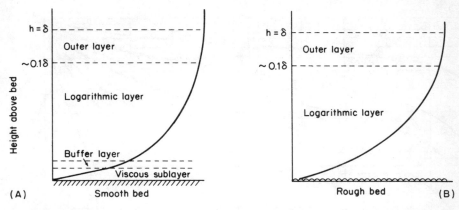

Fig. 5.3A,B. Schematic representation of the velocity profiles for **A** smooth turbulent, and **B** rough turbulent flow. The thickness of the layers is not to scale. (Dyer 1986)

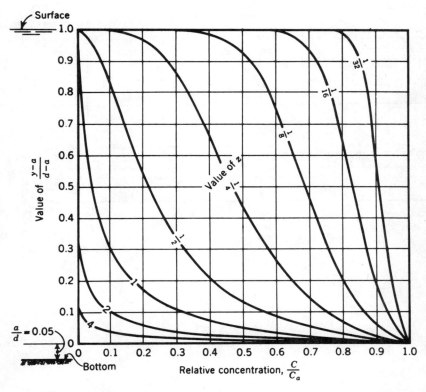

Fig. 5.4. Calculated distribution of the relative concentration C/C_a of suspended matter for different values of z. (Vanoni 1975)

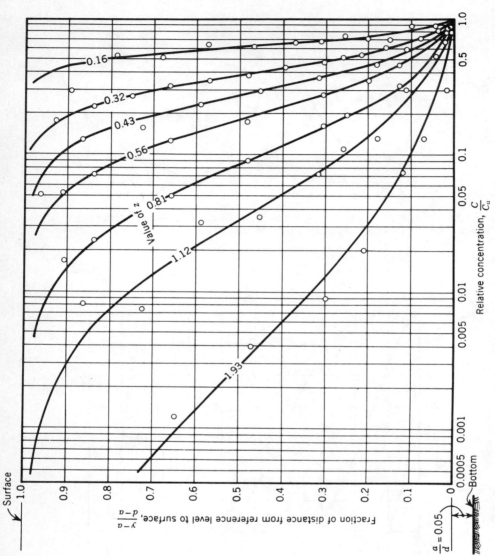

Fig. 5.5. Measured vertical distribution of the relative concentration C/C_a with depth compared with the calculated distribution for a wide range of stream sizes and of values of z (Vanoni 1975)

where y_0 is the bed roughness length (which depends on the height of the bottom irregularities), k is the von Kárman constant, which has a value of 0.4 in water, $u_* = \sqrt{\tau/\rho}$, and k is the mixing length, i.e., the length over which turbulent eddies penetrate upwards. Based on this, the velocity distribution over the vertical can be given as

$$\frac{u}{u_*} = 2.5 \ln \frac{yu_*}{v} + 5.5 \qquad \left[\frac{u_*D}{v} < 5\right] \tag{5.09}$$

for a smooth (Fig. 5.3) and

$$\frac{u}{u_*} = 2.5 \ln \frac{y}{D} + 8.5 \qquad \left[\frac{u_*D}{v} > 70\right] \tag{5.10}$$

for a rough bottom, where D is the grain diameter, assuming the grains to be standing on the boundary. By combining Eq (5.07) with the sediment diffusion coefficient ε_s, and assuming that the ratio between ε_s and the diffusion coefficient for the fluid ε_m is constant (or unity), Rouse (1937) arrived at the relation

$$\frac{C}{C_a} = \left(\frac{d - y}{y} \cdot \frac{a}{d - a}\right)^z, \tag{5.11}$$

where C_a is the concentration of sediment with a settling velocity w_s at level $y = a$, and z is $\dfrac{w_s}{\beta.k.u_*}$ where β is a numerical constant relating ε_s and ε_m

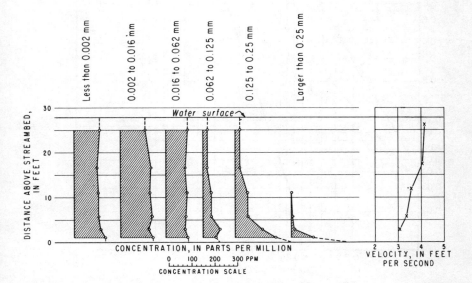

Fig. 5.6. Concentration distribution of different size fractions of suspended matter with depth. (Colby 1963)

($\beta = \varepsilon_s/\varepsilon_m$). This relation is only valid for well-sorted sediment and $\varepsilon_s > \varepsilon_m$. Ghosh (1986) found in the Mahanadi river (fine sediment $< 200\ \mu m$) that β varied between 1.4 and 5.8 at suspended sediment concentrations up to $250\ \text{mg}.1^{-1}$. For particles $< 62\ \mu m$ (Fig. 5.6) and for particles whose density is almost equal to the density of the water, ε_s can be taken as equal to ε_m.

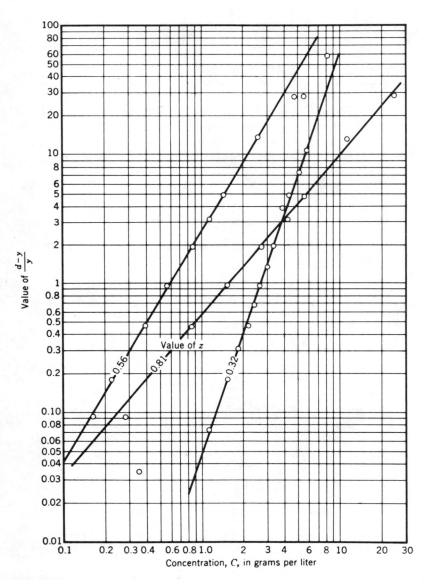

Fig. 5.7. Measured distributions of suspended matter concentration with depth for different values of z. (Vanoni 1975)

The calculated suspended load distribution for several values of z and for $a/d = 0.05$ is given in Fig. 5.4 (after Vanoni 1975). Measured vertical distributions of C/C_a for flow depths ranging from ca 0.10 m (in flumes) to 3 m (Missouri River) and different values of z (suspended particle size ranging from 44 to 295 μm) are given in Fig. 5.5. For a given shear stress, z_z is proportional to w_s, so that particles with a low value for w_s (small or low density particles) will have a rather uniform distribution with height above the bottom whereas particles with a high value for w_s are more concentrated near to the bottom. Figure 5.6 (from Colby 1963) gives typical distributions over the vertical for different particle sizes, measured in the Mississippi River. For particle sizes below 62 micron the distributions are almost uniform. Based on the observations of Coleman (1969, 1970), that the values for the sediment diffusion coefficient increase from zero at the bottom to a maximum at about mid-depth and remain constant up to the surface, two-layer transport models have been developed (e.g. van Rijn 1984; for references see Bechteler 1986).

Refinements of the Rouse equation (5.09), producing (sometimes) a better fit with observed distributions, are discussed by Vanoni (1975), Bechteler and Vetter (1983) and Vetter (1986). The best fit was obtained by also considering the particle settling velocity in relation to the rate of upward and downward particle transport, as was done by Einstein and Chien (1952). C/C_a is expressed as a function of $z_1 = w_s/\beta \cdot k \cdot u_*$, where k is a constant and $\beta = \varepsilon_s/\varepsilon_m$. For low values of w_s the influence of w_s becomes very small.

A different way to show the relation between C and the measured distribution is to plot $(d - y)/y$ against the concentration for different values of z. An example is given in Fig. 5.7 (after Vanoni 1975). As the suspended sediment load influences the flow, the von Kármán constant k, which is 0.4 in clear water, was found to change when the suspended sediment concentration increases. Velocity profiles with and without a suspended load of particles of ca. 100 μm are given in

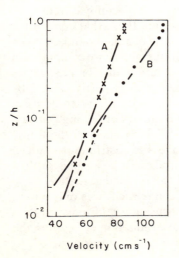

Fig. 5.8A, B. Velocity profiles obtained at the same flow depth and surface slope, i.e., the same pressure gradient, but with different suspended matter concentrations. *A* Clear water, with $K = 0.4$; *B* concentration 15.8 g.l^{-1} of 100 μm diameter particles, with $K = 0.21$. (Dyer 1986; after Vanoni and Nomicos 1959)

Fig. 5.8. The particle concentration in Fig. 5.8 (curve B) is in $g.l^{-1}$, a concentration that is hardly ever reached in natural rivers, (only in a few exceptional streams such as the Yellow River), and in some tidal estuaries and coastal waters. Vetter (1986) has shown that the decrease of k with increasing concentration becomes important only at concentrations higher than ca. $2 g.l^{-1}$. At lower concentrations, other influences such as bed forms are more important, resulting in values of k well above 0.4 (up to 2.0). The lower value of k, which indicates reduced turbulence, is caused by the loss of energy that goes into keeping the supended particles in suspension. The effect is largest near to the bed, where larger and heavier particles are present and concentrations are usually highest. Coleman (1986) found no change in k in flow ranging from clear water to suspensions loaded to full capacity but indicated that the deviation of the suspended matter concentration profile from the logarithmic form (as expressed by the wake-strength coefficient) varied with the sediment concentration.

5.2.1 Estimating Suspended Matter Transport

The total suspended matter transport can be estimated by combining Eg. (5.04) with the Rouse equation for the suspended matter distribution [Eq. (5.07)] and the velocity distribution given by Eq. (5.08). Bechteler (1986) and Schrimpf (1986) compared the different ways of obtaining the vertical concentration distribution and found that for fine sediment ($< 62 \mu m$) all models agree equally well with the measurements and that for material of this size different values for ε_s result only in small deviations. The mean transport velocity of particles of this size can be regarded as equal to the mean flow velocity (Schoelhamer 1986). With larger (sand-sized) particles, the settling velocity predominates, particle shape becomes increasingly important and significant differences appear between the models and the measurements in nature (the two-layer models not being an improvement); the mean transport velocity is significantly less than the mean flow velocity, and the fundamental processes are those that control the exchange between bed load and suspended load in the benthic boundary layer. Samaga et al. (1985) compared observed concentrations with those computed on the basis of the Rouse equation, and found that only 56% of the observed values were within 40% of the predicted value. This deviation is partly caused by uncertainties about the particle settling velocity values that have to be used. Kineke et al. (1989) found that when using in-situ estimates of settling velocity, the calculated suspended sediment concentration profiles (based on the Rouse equation, slightly modified to account for particle size distribution) agreed within 25% with the observed concentration profiles. When using settling velocities estimated from the size distributions of the suspended matter, using Stokes' law or the approach of Dietrich (1982), the agreement was only within 40–50%. The calculation of total transport in suspension (Q_{ss}) from such data results in rather complicated formulae that allow determination of Q_{ss} from the suspended matter concentration at mid-depth, the exponent z (proportional to

w_s) and $k\bar{u}/u_*$, where u is the mean flow velocity (Brooks 1965). Einstein (1950) modified this by taking into account the bed roughness and calculated the discharge for separate particle size fractions. As the Rouse equation [(5.11)] is based on the relation between C and C_a, the value of C_a has to be calculated. This was done by Lane and Kalinske (1939), assuming that the vertical turbulence fluctuations near the bed are normally distributed, that the particles are picked up at a rate that is proportional to their relative amounts by weight at the bottom sediment surface, and proportional to the fraction of time that a velocity occurs that is capable of picking up the particles, and lastly that only velocities larger than the settling velocity of the particles can pick them up. Einstein (1950) defined $Y_0 = 2D_s$, as the upper limit of the bottom sediment (the "bed") and as the lower limit of the material in suspension (D_s is the mean particle size of the size fractions that are considered). By assuming a constant relation between bed load discharge per unit area, $2D_s$, the mean particle velocity of the size fraction that is considered, and the concentration at y = a for that particle size fraction, the suspended matter discharge can be calculated for a total sediment load consisting of a range of particle sizes.

It should be realized that the existing theory does not completely describe the suspended sediment transport processes. Measurements in nature as well as in flumes show relations not predicted by theory and conditions near to the bed, whereas the actual settling velocities in the water are insufficiently known. Sediment discharge can differ by a factor of 10 to 20, depending on whether the bottom is rippled or flat. A large concentration of suspended matter can increase the capacity to transport coarser fractions. This was found at a "hyperconcentration" of 40–60% of suspended load (by weight; Beverage and Culbertson 1964): a suspended matter concentration of $10\,\mathrm{g.l^{-1}}$ can reduce the settling velocity of sand grains with ca. 60%, and the presence of fine suspended material (clay) can reduce the friction along the bottom. Temperature has an effect on the suspended matter transport because it influences the viscosity of the water: the viscosity increases at lower temperatures so that also the particle settling velocity will be lower. Vanoni (1975) summarized data showing this relation. In Fig. 5.9 (from Lane et al. 1949), this effect is shown for suspended particles smaller than 295 μm, whose settling velocity is perceptibly influenced by the water temperature. Another complication is that diffusion models, based on depth-averaged calculations with constant turbulent exchange coefficients throughout the flow, lead to erroneous results when the vertical flow is not uniform. This is the case when secondary flow occurs, or interaction between outflow (discharge) with river flow, or in near-bank river flow. Therefore k-ε models were developed: the eddy diffusivity Γ_t is taken as proportional to the eddy viscosity

$$\Gamma_t = \frac{\eta}{\beta^{-1}}, \tag{5.12}$$

where $\beta^{-1} = \varepsilon_m/\varepsilon_s$, which in the more simplified relations, such as, e.g., [Eq. (5.11)] is taken to be 1. The eddy viscosity is related to a value k indicating the

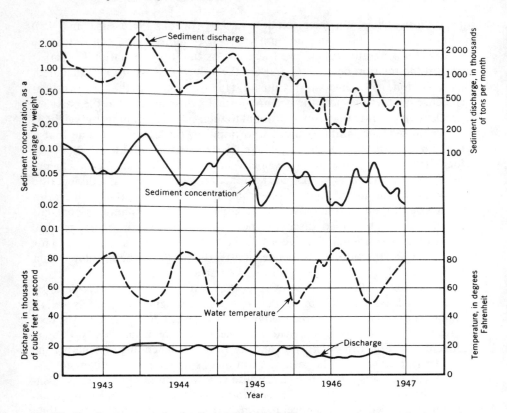

Fig. 5.9. Simultaneous records of sediment discharge, sediment concentration, water temperature and water discharge in the Colorado River. (After Lane et al. 1949)

turbulent kinetic energy, and to ε_m:

$$\eta = C_\mu \frac{k^2}{\varepsilon_m},\qquad(5.13)$$

where C_μ is a constant. Equations for k and ε_m are developed balancing the rate of change in k and ε_m with convective and diffusive transport, eddy formation by mean velocity gradients and buoyancy forces, and eddy destruction by energy dissipation. The equation for ε is highly empirical and based on a series of constants whose value depends on conditions in the flow, whereas the equation of k is more theoretical. An extended version takes the directional influences on the turbulence into account, such as gravity or the dampening of the turbulence against a surface bordering the flow. A full discussion is given by Rodi (1980, 1986). The vertical turbulent mixing counteracts the settling of the particles which results in an equilibrium particle concentration profile for a particular flow situation. For a given flow distribution, eddy diffusivity and particle settling velocity, the particle concentration C can be calculated from

$$u\frac{\partial C}{\partial k} + v\frac{\partial C}{\partial y} = \frac{\partial}{\partial y}\left(\frac{\eta}{\beta^{-1}} \cdot \frac{\partial C}{\partial y} + w_s C\right),$$ (5.14)

where w_s is the particle settling velocity. Experiments with sandsized particles demonstrated that the development of a realistic equilibrium concentration profile could be simulated.

5.2.2 Other Approaches To Suspended Matter Transport

The preceding models to describe the suspended sediment concentration distribution with depth, that forms the basis for estimating the suspended sediment transport, were derived by considering suspended sediment dispersal as a diffusion process, as first done by Rouse (1937). Average concentrations and average flow conditions were used. An extension of this approach was developed by Hunt (1954), who took the particle size of the suspended matter into account expressed as the volume occupied by the particles. Other types of transport models have been based on an evaluation of the energy balance in the flow (Bagnold 1962), by considering the forces acting on the particles and on the fluid (Velikanov 1944), and on a stochastic approach, whereby the pattern of motion of the particles is considered and a suspended sediment concentration distribution is obtained by simultaneous simulation of the movements of many single particles (Chiu 1967; Bechteler and Färber 1985).

5.2.2.1 The energy Model

Bagnold (1962, 1966) has based a theory of suspended matter transport (as part of a theory of total sediment transport) on the energy that is needed to support the suspended matter in suspension and on the power supplied by the vertical component of the stream flow:

$$W_s = q_{ss}\frac{w_s}{\bar{u}_s} = e_s \cdot E_p(1 - e_b),$$ (5.15)

where W_s is the "work rate", q_{ss} is the suspended sediment load, w_s is the settling velocity of the transported particles, $\bar{u}_s$ is the average transport velocity of the suspended particles, E_p is the available energy, e_s and e_b are the fractions of E_p used for respectively the suspended matter transport and the transport of bottom sediment. From Eq. (5.15), q_{ss} can be calculated, when e_s, e_b, and $\bar{u}_s/w_s$ are obtained from measurements. In a more practical form, it can be given as

$$q_{ss} = \int_{y_0}^{d} \bar{C} \cdot u\, dy,$$ (5.16)

where y_0 is the value for y where the region of bed load and suspended load are separated $\bar{C}$ is the mean particle concentration, and d is the water depth. Assuming that C does not vary very much with y, which is approximately true for fine-grained low density particles, and that the particles travel with the same velocity as the water, this can be simplified to

$$q_{ss} = \bar{C} \cdot \bar{u}_m, \tag{5.17}$$

where u_m is the mean flow velocity. Measuring the mean particle concentration and the mean flow velocity and obtaining the transport rate by multiplication is, in fact, often done. Chang et al. (1967) have refined Eq. (5.17) by taking into account the variation of C with y and developed a relation for q_{ss}, which only needs determination of C at the level $y = y_0$. To do this, it was needed to assume that the suspended load rate is proportional to the bed load rate:

$$r_{yo} \cdot q_{ss} \sim \frac{1}{y_o} q_{sb} \cdot (\bar{u}_m; u_*), \tag{5.18}$$

where r_{yo} is a proportionality factor at the level $y = y_0$. Chang et al. (1967) found $r_{yo} \approx 0.8$.

5.2.2.2 The Gravitational Theory

Velikanov (1955, 1958) arrived at a concentration distribution by starting from the relation $\overline{u' \cdot c'} = -\bar{u} \cdot \bar{c}$, which indicates that the average amount of particulate matter displaced upwards by the turbulence is equal to the amount of particulate matter that is settling down. The amount of energy lost in bringing the sediment up must be equal to the energy gained when it is brought down by gravity. Not only particles are brought upwards but also fluid and there is internal friction to overcome. This leads to complicated relations, but when $\bar{c}$ (average particle concentration) can be regarded as small, the relation becomes $e_x = \gamma_s w_s / \bar{u}\bar{\tau}$ where e_x is the potential energy, γ_s is the specific weight of the sediment particles, w_s the settling rate, $\bar{u}$ the average flow velocity and $\bar{\tau}$ the average shear stress. From this, a relation for C/C_0 is obtained which is very similar to the one obtained by Rouse (1937) based on diffusion (5.09) and which gives a similar series of curves, depending on the value of a parameter $(w/u_*) \cdot (k\gamma_s / \gamma J)$, where γ is the specific weight of the fluid and J the surface slope, i.e., the driving pressure gradient [Fig. 5.10 after Yalin 1977: $\eta_0 = 0.05$ is comparable to a/d = 0.05 in Fig. 5.4 and η is the height above the bottom ($\eta = 1$ is the surface)]. The curves agree well with the measurements (Fig. 5.10). This approach has the same drawback as the diffusion approach in that only a relative suspended matter concentration distribution can be obtained, not an absolute one (Vetter 1984).

5.2.2.3 Stochastic Models

Stochastic models describe the dispersion of suspended particles in a flow by considering the simultaneous probability of the movement of a large number of

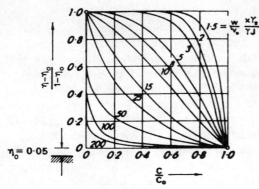

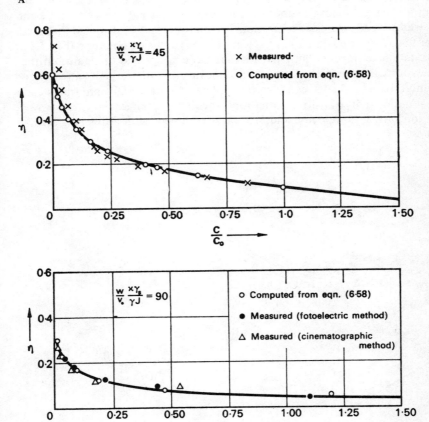

Fig. 5.10. A Calculated distribution of C/C_0 in relation to water depth using Velikanov's equation. **B** Measured and computed distributions of C/C_0 using Velikanov's equation (**A, B** Yalin 1977)

particles in a certain direction and with a certain velocity, based on measured turbulence parameters (Bechteler 1980, 1986). As the turbulent velocity fluctuations that move the particles are random in nature, a probabilistic approach appeared more effective to describe the process of suspended sediment transport than diffusion. The movement of single particles and groups (mixtures) of particles was discussed by Chiu (1967) and Soo (1967), who also summarized earlier work. By using a Monte Carlo simulation method the random movement of particles released from a point source could be traced (Fig. 5.11). Yalin (1973) considered a mobile bottom sediment over the entire length of the flow as a continuous source and sink, from where the particles are picked up by the flow. He reached good agreement with measured values in stationary and uniform two-dimensional flow. Bechteler and Färber (1985) used a stochastic model as described by Chiu (1967), Bayazit (1972), and Li and Shen (1975), and built up a concentration profile from a concentration gradient and a dispersion coefficient. In this way, a steady state concentration distribution can be calculated that gives a better fit to measured data than the Rouse (1937) diffusion equation (Fig. 5.12). Steady state here implies that particles that settle are picked up again but that no new particles are picked up, which means that there is no erosion or sedimentation. The theory can also be applied to non-equilibrium (nonsteady state) situations or transition situations involving changes in erosion rate, deposition rate, flow velocity, and erosion of sediment with a different settling velocity.

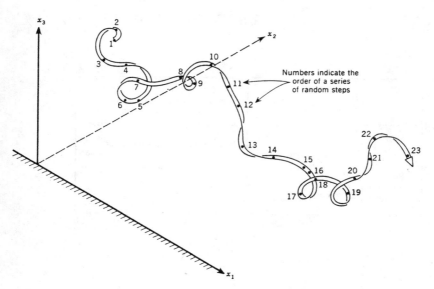

Fig. 5.11. Pattern of motion of a suspended particle in turbulent flow. (Chiu 1967)

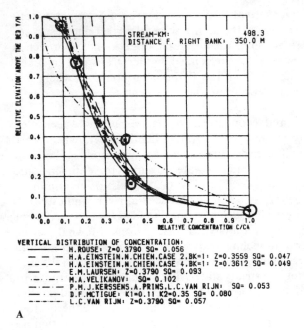

VERTICAL DISTRIBUTION OF CONCENTRATION:
  ———— H.ROUSE: Z=0.3790 SQ= 0.056
  — — — H.A.EINSTEIN,N.CHIEN.CASE 2.BK=1: Z=0.3559 SQ= 0.047
  - - - - H.A.EINSTEIN,N.CHIEN.CASE 4.BK=1: Z=0.3612 SQ= 0.049
  — — E.M.LAURSEN: Z=0.3790 SQ= 0.093
  —·—·— M.A.VELIKANOV: SQ= 0.102
  —·—·— P.M.J.KERSSENS,A.PRINS,L.C.VAN RIJN: SQ= 0.053
  — ·— D.F.MCTIGUE: K1=0.11 K2=0.35 SQ= 0.080
  ------- L.C.VAN RIJN: Z=0.3790 SQ= 0.057

A

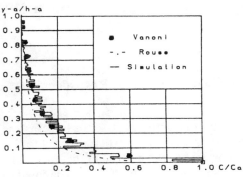

B

Fig. 5.12. A Suspended sediment concentration profiles obtained from different theories. Measured points *circled*. (Bechteler 1986). B Concentration distribution (in steady flow) with depth showing Vanoni's (1975) data, the distribution according to Rouse's (1937) equation and using a stochastic simulation model. (Bechteler and Färber 1985)

5.2.2.4 Three-Dimensional Models

Applying the theory for two-dimensional transport to real rivers and streams gives problems other than those inherent in calculating suspended matter transport in turbulent flow. Rivers are seldom straight channels but have curves, bends, meanders, sandbanks in combination with a sinuous path of the main flow, and irregular topographic disturbances. This means that there is flow and dispersal in a direction normal to the main flow, which also influences the suspended matter transport. The river bed morphology, which creates large problems in estimating bed load transport rates, also influences the suspended matter transport, although to a smaller degree. In addition to this, suspended matter transport is usually intermittent, deposition alternating with resuspension, whereby deposition is related to the falling stage of a river and resuspension to the rising stage. Therefore some form of three-dimensional approach is necessary. This is even more true for suspended matter transport in lakes, estuaries, and the sea where flow is even less two-dimensional and where bed morphology, and alternation of deposition and resuspension play an equally large role.

Until recently, technical (and economical) difficulties limited suspended matter transport modeling to vertical or horizontal (depth-integrated) models. Examples of the latter kind for rivers are given, e.g., by Rodi (1986) and Tingsanchali and Rodi (1986). An-intermediate towards three-dimensional modeling is a superposition of several horizontal models or a combination of different types of modeling, as was made for several rivers and estuaries by combining a large-scale physical model, numerical models for flow, sediment transport and wave propagation, and analytical calculations (McAnally et al. 1986), which is called hybrid modeling. Advances in computer speed and memory size now make three-dimensional models possible that also take into account that flow usually is not in a steady state. All suspended matter transport models, however, are based on a model of the vertical suspended matter distribution. The fact that the in-situ fall velocity of the particles, the bottom shear stress, and the values for k are usually insufficiently known, means that no calculation of the suspended matter transport can be made with a high reliability.

5.3 Initiation of Particle Motion (Scour, Erosion, Resuspension)

The force the flow exerts on a particle on the bottom (the drag force) is the product of the shear stress τ_0 and the surface area of the particle exposed to this stress. This force is counteracted by the weight of the particle in the water. The critical shear stress τ_c occurs when the upward component of the drag is equal to the particle weight. On a downward sloping bed τ_c is smaller, on an upward sloping bed higher. $\tau_c \sim u_c^2$, where u_c is the flow velocity near to the bed, when $\tau_0 = \tau_c$, and $u_c^2 \sim d_s$, (d_s is the particle diameter) This gives $u_c^6 \sim d_s^3$, which was already expressed by Brahms in 1753 (Leliavsky 1955). This relation is only valid when the flow around the particle is turbulent, i.e., when d_s is large compared to the thickness of the laminar sublayer (Rubey 1938).

As τ_c is related to the critical shear velocity $u_{*c} = \overline{\tau_c/\rho_w}$, this gives:

$$\frac{\tau_c}{(\rho_s - \rho_w)d_s} = f\left(\frac{u_* d_s}{v}\right) \quad \text{or} \quad \tau_{*c} = f(Re_{*c}), \tag{5.19}$$

which was first formulated by Shields (1936; Fig. 5.13), and where v is the kinematic viscosity ($= \mu/\rho$ where μ is the viscosity of the fluid), τ_{*c} is the

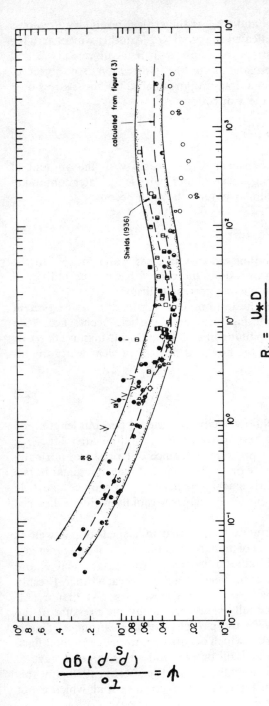

Fig. 5.13 The relation between Shield's (1936) entrainment function and Reynolds number. Data are for unidirectional flow over noncohesive beds. (Larsen et al. 1981)

dimensionless critical shear stress and Re_{*c} is the critical boundary Reynolds number. When $\tau_0 \neq \tau_c$, they are indicated as τ_* and Re_*. Another way of arriving at Re_* is by starting with the stress in a flow with a mean velocity $\bar{u}$ and fluctuations u' in the x-direction and v' in the y-direction. An object of dimension l in this flow experiences a drag proportional to the shear stress over an area proportional to l^2, or in the x-direction:

$$Re_* \sim l^2 . \rho . \overline{(u')^2} = \frac{\pi}{8} C_D . l^2 . \rho . \overline{(u')^2}. \tag{5.20}$$

The coefficient C_D is the drag coefficient. C_D increases with the amount of particles in suspension. When a drag is exerted on a particle by water containing uniformly dispersed particles of similar nature, the drag becomes

$$C_{D_w} = \frac{w_0}{w_c} . C_{D_0}, \tag{5.21}$$

where w_0 is the single particle settling velocity in clear water and w_c in the suspension. At high particle concentrations the drag force on the individual particles may become even higher, e.g., by particle collisions.

For particles to be lifted from the bottom, a lifting force must be present (Vanoni 1975; Sleath 1984). The lifting force on a particle occurs near to a boundary, or when the particle is spinning in a uniform flow (Magnus effect), or when it is in a shear flow, or when the flow is distorted. In clear water the lift coefficient is:

$$C_L = \frac{L}{\frac{1}{2}\rho u_0^2 d}, \tag{5.22}$$

where U_0 is the amplitude of the fluid velocity, L is the lift per unit length, and d the particle diameter. There can be positive lift (C_{LA}) and negative lift (C_{LB}). Measurements have shown that a pressure difference (lift pressure) normally exists near the particles when they are on the bottom and that infrequent bursts of larger positive lift forces apparently entrain the particles. As soon as a particle is separated from the bottom, the lift force decreases rapidly with the distance from the bottom.

The fine particles that are normally transported in suspension, as well as large ones of low density, that consist of (mainly) organic material or aggregated (flocculated) organic matter and mineral particles, are surface active and cohesive. Cohesion is the tendency of particles to adhere to each other. Freshly deposited mud sticks together because of this cohesiveness. At first it still contains much water which is gradually squeezed out by the pressure of the sediment while the cohesion between the particles becomes stronger. This is enhanced when the sediment falls dry, which occurs (temporarily) on tidal flats, where the squeezed out water can flow off or evaporate. For the first stage of consolidation the term "gelation" is sometimes used (e.g., Mehta 1988), but this term presupposes that the suspended matter behaves as a colloid, which is not so, as will be discussed in Chapter 6.

Because of this consolidation, fine sediment is increasingly difficult to resuspend after deposition, so that the critical shear stress can become as high as for sand or even gravel. Resuspension or erosion of fine sediment can go particle by particle (or aggregate by aggregate), but consolidated stiff clays usually break apart into pieces that are moved over the bottom, are broken further or, while rolling over the bottom are rounded into mud pebbles. During this process, part of the material goes into suspension in the form of separate particles. When high concentrations of suspended material are involved, in the order of $1 \ g \ . \ l^{-1}$ or more, a loosely consolidated mud layer can be formed (fluid mud or hyperpycnal layer; Wright et al. 1988) which acts as a fluid and can be re-entrained as a high density suspension. This occurs when the settling rate is high enough to prevent rapid dewatering, but the settling of the individual particles (flocs) is reduced by the high concentration (hindered settling; see Sect. 5.5; Mehta 1989). The fluid mud, when loosing water, can develop into a stationary mud (which can be horizontally mobile) and then grade into a cohesive consolidated mud layer. This general profile, which usually is present, is shown in Fig. 5.14 (from Mehta 1989). The fluid mud tends to have a smooth upper surface which reduces the shear stress and even in high energy environments it can persist for some time as

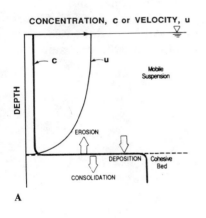

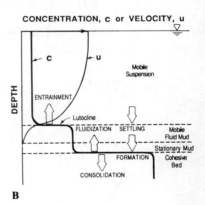

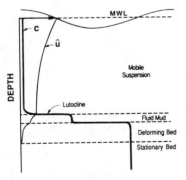

Fig. 5.14. A Classical definition of sediment bed and suspension-related processes. **B** Schematic profile of fluid mud formation. **C** Schematic profile of cohesive bed response to waves. *c*, concentration; *u*, velocity. (Mehta 1989)

a separate layer before the mud is resuspended. Nichols (1985) explained this by the strong density gradient at the fluid mud surface, which tends to stabilize the fluid mud. In recently deposited mud, usually a soft mud, is lying on top of a semiconsolidated layer which is on top of a consolidated layer. Mehta (1988) indicated that in the upper zone τ_c increases only a little with depth in the sediment, in the second (soft) layer the increase in τ_c with depth is much stronger, and in the layer below this one (a dense bed) τ_c remains constant with depth in the sediment. The principal characteristic for fine sediment consolidation is the density of the sediment. Measurements by Kirby and Parker (1977) in the Severn and at the Rhine mouth indicated a stepwise increase in mud density from less than 1.1 near the fluid mud surface to more than 1.4 at ca. 1 m depth in the sediment (Fig. 5.15). The steps can reflect an original layering of the mud resulting from original differences in water content, and/or intermittent deposition with periods of deposition alternating with periods of nondeposition. Another explanation may be that the steps reflect a stepwise change in aggregate structure, caused by the pressure of the overlying mud. Consolidation results in lowering of the sediment surface. During deposition, steps are formed (Fig. 5.16), while at the same time the bed comes up, only to go down again when the mud consolidates (Fig. 5.17). The semi-consolidated mud (a stationary suspension) is first resuspended as a layer, as Parker (1987) has shown, and then is further dispersed upwards (Fig. 5.18). The degree of consolidation (and the velocity of the process) depends mainly on the initial concentration and the sand content of

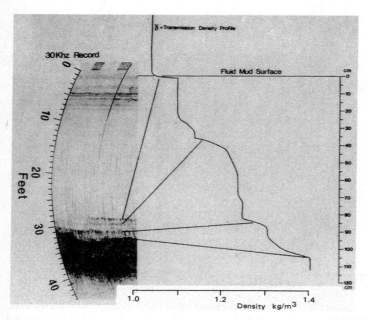

Fig. 5.15. Density steps in a fluid mud layer in the Severn estuary measured in situ with a γ-ray densitometer, together with a 30 kHz echosounder record, showing density interfaces. (Kirby 1988)

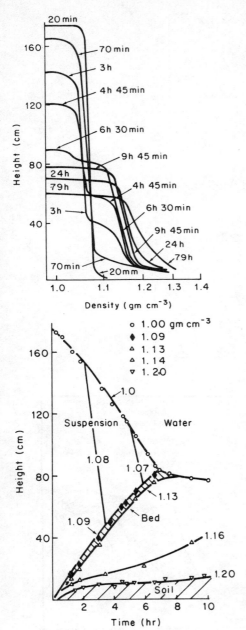

Fig. 5.16. Density profiles in a settling suspension with an initial density of 1.09 g.cm^{-3}. (Dyer 1986; from Been and Sills 1981)

Fig. 5.17. Settling of a suspension interface and growth of the bed with time. Initial density 1.09 g.cm^{-3}. (Dyer 1986; from Been and Sills 1981)

the sediment [Fig. 5.19; after data of Migniot (1968) and Terwindt and Breusers (1972); from Terwindt 1977]. The behavior of semi-consolidated mud and the degree of its resuspension is one of the major problems involved in suspended matter transport. It is important in estuaries or coastal seas where high

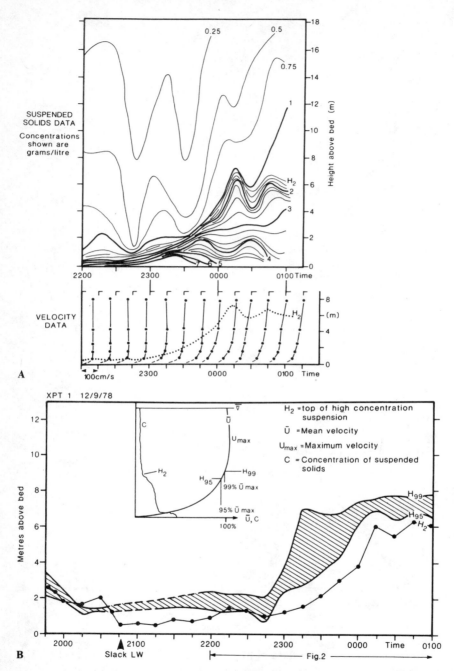

Fig. 5.18. A Time series of velocity and suspended matter concentration during redispersion of an ephemeral stationary suspension. See **B** for explanation of H_2. **B** Elevation of suspension surface related to thickening of the boundary layer. (**A, B** Parker 1987)

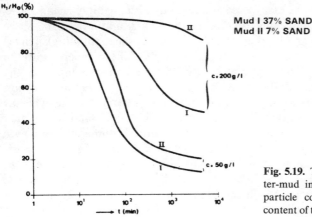

Fig. 5.19. The lowering of the water-mud interface (H_1) with time, particle concentration, and sand content of the mud. (Terwindt 1977)

concentrations of suspended matter develop (see Chapter 7). Cohesion is also a factor in erosion (scour) of soils containing a significant amount of fine-grained material (which can be as little as 5%), and of consolidated fine-grained deposits of geological age, as long as they have not hardened into stone.

Freshly deposited mud, even when it is still very fluid, does not behave like a normal (Newtonian) fluid under stress. In a Newtonian fluid (or a suspension of noncohesive particles), when the shear rate (the rate of deformation of the material) increases, the shear stress of the fluid (the force applied to cause the flow) increases likewise. This relation follows a straight line with the shear stress being zero when the shear rate is zero. The slope of the line is equal to the molecular viscosity of the fluid. (Fig. 5.20). Clay suspensions behave differently: at increasing shear rate the shear stress first increases rapidly and then levels off to a constant rate of increase, the slope of which is equal to the plastic viscosity. The initial rapid increase is attributed to the breakdown of flocs to a basic stable unit. The residual stress T_B at zero shear rate is called the Bingham yield stress which is related to the residual net attraction between the particles. A diagram

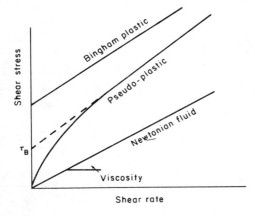

Fig. 5.20. The relation of shear stress with shear rate for suspensions and deposits with different rheological properties. The *slope of the curves* equals the effective viscosity. (Dyer 1986)

as shown in Fig. 5.21 shows first increasing resistance against increasing shear, (which is called shear thickening or dilatant behavior) and where the slope flattens at increasing shear rate there is less resistance (shear thinning). This is probably related to a change in the internal structure of the sediment, with the particles probably more densely packed during shear thickening and the structure becoming looser during thinning. Shear thickening reduces erosion because more shear needs to be applied to make the material mobile, whereas thinning increases the mobility of the mud. Generation of fluid mud is probably a first step in the erosion of mud, as was indicated during flume tests of waves acting on natural muds (Maa and Mehta 1987). Resuspension of a mud deposit as shown in Fig. 5.18 shows how the mud comes up as a separate high concentration layer instead of being eroded particle by particle. These observations were made on rather fresh, ephemeral mud deposits. If muds become more consolidated, they will become too stiff for any fluidization and will break apart into pieces, with some material being eroded particle by particle from the surfaces exposed to the flow. This breaking apart and the surface erosion can be easily observed in tidal areas where older mud deposits are eroded by waves or moving channels and small cliffs are formed in the mud layers.

The resistance against scour can be related to the shear strength and the plasticity index of the sediment. The shear strength is the degree of resistance against horizontal pressure and is related to the percentage by weight of fine material. The plasticity index is the difference between the liquid limit and the plastic limit of the sediment (Sundborg 1956). The liquid limit is the water content (in weight percent) of the sediment at which a shear strength is present;

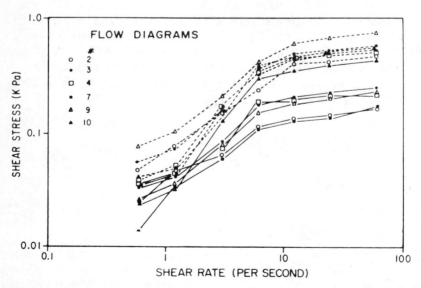

Fig. 5.21. Flow diagram for Cape Lookout Bight mud from 1 h settling (*solid lines*) and 2 h settling (*dashed lines*). (Faas and Wells 1990)

the plastic limit is the water content at which the sediment starts to crumble. Good relations have been found between the critical shear stress τ_c, the plasticity index, and the percentage (by weight) of clay (Smerdon and Beasley 1961). Where scour is not particle by particle, as occurs in more consolidated material, scour is more a function of time, of the structure of the deposit (and of its thickness) than of the shear stress. An admixture of sand (which is not cohesive or only slightly) also can strongly influence the erodibility. When a consolidated clay layer, lying on top of sand, is broken, it will crumble into pieces once a hole is formed and the pieces are entrained separately. Partheniades (1965) inferred that τ_c did not depend on the degree of consolidation but on the bond strength of the flocs or aggregates, and on a combination of the percentage of fine-grained material, the water content, the salt content, and the exchange capacity (Vanoni 1975) Martin (1962) suggested that there is a range of values for τ_c depending on the condition of the sediment and time. The force needed to break the bonds between the aggregated particles, the yield stress (τ_y), generally is correlated with τ_c (Krone 1963; Migniot 1968). The yield stress is defined as the shear stress at the initiation of movement. Krone (1963) found that for concentrations (by weight $= C_w$) between 10 and 100 g $.1^{-1}$ the mud behaves like a

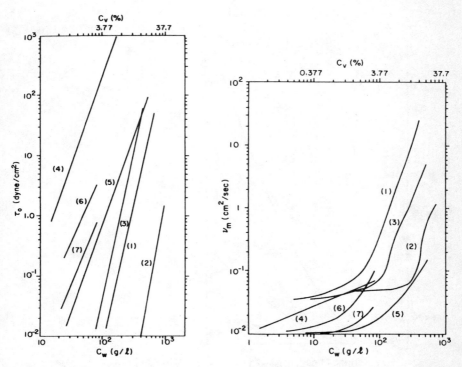

Fig. 5.22. Relation between the yield stress τ_0 and the concentration by weight (C_w) and the concentration by volume (C_v) (*left part*), and the relation between the viscosity v_m and C_w and C_v (*right part*), for different clay/water mixtures. (Mei and Liu 1987)

Bingham plastic for which τ_0 (the critical shear stress or yield stress) and v_m (the Bingham viscosity) depend on $C_w = C_v\rho_s$ (where C_v is the concentration by volume and ρ_s the sediment particle density; Fig. 5.22; from Mei and Liu 1987). Most measurements have been done in laboratories where the mud is no longer in a natural state, and where particularly the natural bioturbation and biological aggregation are absent. In-situ measurements by Young and Southard (1978) indicated that the values for u_{*c}, found in situ, were about half the values found in the laboratory. The influence of organisms can result in an increase as well as in a decrease in erodibility.

Organisms that produce fecal pellets and pseudofeces, diatoms living on the surface of the mud, and microbes producing sticky material like polysaccharides, aggregate the mud particles and protect the mud against erosion. Paterson (1989) found a significant increase in cohesive strength because of the formation of an extracellular matrix of pōlysaccharides by benthic diatoms during a 7.5-h period of low tide. Only very little residual stability remained, however, after inundation during the following flood tide. Observations in coastal (intertidal) areas in the Dutch-German Wadden Sea indicate that this protection is only

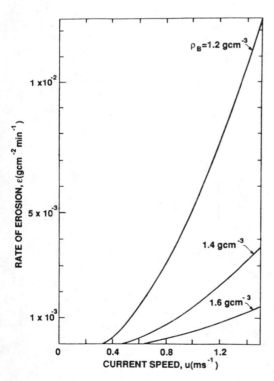

Fig. 5.23. Rate of erosion variation with current speed at three bed bulk densities for San Francisco Bay mud. (Mehta 1988; after Villaret and Paulic 1986)

efficient at low shear stresses; at high shear stresses such as occur at high current velocities and large waves (storms), the protection is not effective and the mud goes into suspension anyway. Organisms that burrow through the bottom (there are also worms and echinoderms that move more or less horizontally), and organisms that produce tracks or mounds on the surface of the mud, may either loosen the sediment and increase its water content, or may fix it by excreting glue-like material. The results of Young and Southard (1978), therefore, depend

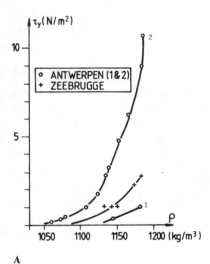

Fig. 5.24. **A** Effect of density on the yield stress τ_y for Zeebrugge and Antwerpen harbor mud. **B** *Left* Relation between the yield stress for Zeebrugge harbor mud with increasing amounts of pure mud ($< 2 \, \mu m$) added. *Right* Relation between log τ_y (log yield stress) and the log % $< 1 \, \mu m$ in the particulate fraction. (**A, B** Verreet et al. 1986)

A

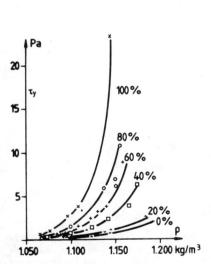

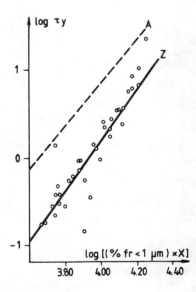

B

on the kind of organisms that were present in the area during the experiment (or just before it), and cannot be generalized.

Some measurements of mud properties, necessary for predicting mud behavior, can be made in situ. The density can be measured by γ-ray densitometry (see Sect. 3.1) so that τ_c can be estimated. When uniform sediment is present, the relation between current velocity and erosion rate can be worked out for muds of different density, as was done for San Francisco Bay (Fig. 5.23, from Mehta 1988 after Villaret and Paulic 1986). With a digital viscosimeter the maximum shear stress can be measured just after the initiation of motion (Fig. 5.24 from Verreet et al. 1986). This gives the yield stress τ_y, which is related to the concentration of the particles X:

$$\tau_y = AX^n, \tag{5.23}$$

where τ_y is in $N \cdot m^{-2}$, X is in $mg \cdot l^{-1}$, and A and n have to be determined empirically for different types of mud (differing mainly in granulometry, clay mineralogy, and organic matter content and composition, which can be determined in the laboratory). Relations between τ_y and density, particle concentration and percentage of particles smaller than 1 μm are given in Fig. 5.24 for harbor muds from Antwerpen and Zeebrugge (after Verreet et al. 1986), together with the relation with consolidation time. The increase of τ_y with consolidation is not only caused by an increase in density but also, probably, by rearrangement of the particles.

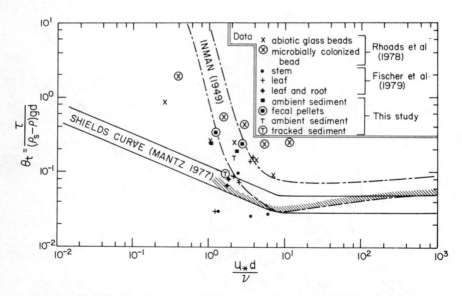

Fig. 5.25. Shields curve with nondimensional shear stress related to Reynolds number. *Hatched area* covers the curve suggested by Shields (1936), with extensions by Mantz (1977), and calculated on the basis of Inman's (1949) curve. Data for biogenic material and biologically altered sediments are added. (Nowell et al. 1981)

Also other aspects of the bottom sediment than the presence of living organisms can influence the erodibility, such as the admixture of plant remains like stems, leaves, and roots. Some effects of this on the entrainment are shown in Fig. 5.25 (after Nowell et al. 1981).

5.4 Transport by Surface Waves

Waves are very efficient in stirring up bottom sediment and can form suspensions at much lower equivalent velocities than unidirectional (steady) currents. Net sediment transport, however, is very limited and only takes place by Stokes' drift, i.e., the small net flow and movement of the water in the direction of wave propagation, and is proportional to $\pi^2 . H^2 . L^{-1}$, for surface waves at the water surface. The velocity in the forward direction is ca. 10 $(H . L^{-1})^2 . C$ (H = wave height, L = wave length, and C = velocity of wave propagation). This occurs with tidal waves as well as with surface waves. In tidal areas the tidal current velocities and the effects of tidal asymmetry in shallow areas (see Sect. 7.4) are usually much larger than the Stokes' drift, but in areas where tidal currents do not reach high velocities, the continuous effect of Stokes' drift can become an important factor for suspended sediment transport. When surface waves stir up sediment from the bottom, there is considerable variation in particle concentration within each wave oscillation, related to wave phase, height above the bottom, and the maximum oscillatory velocity. The initial motion of particles on the bottom which are moved by the oscillatory motion of surface waves is related to the amplitude U of the horizontal velocity component u_0 at the upper limit (y_0) of the oscillatory bottom boundary layer. The maximum horizontal force per unit area of the bottom, $\hat{\tau}_0$, is a function of $U\infty$ (the velocity far away from the bottom), which is a function of $(\rho_s - \rho)\rho^{-1} . g . d^3 . v^{-2}$ but this function is not constant. Mainly because of this, the predicted values for $U\infty$ for initial noncohesive particle motion vary widely (Sleath 1984, who mentions 18 different formulas). Measurements made by Sternberg and Larsen (1975) indicated a good agreement with a somewhat modified version of the relation given by Komar and Miller (1973, 1974) for fine sediment:

$$\frac{u\infty}{\left(\frac{\rho_s - \rho}{\rho}\right)^{2/3} g^{2/3} D^{1/3} T^{1/3}} = 0.24 \quad \text{for} \left(\frac{\rho_s - \rho}{\rho} . \frac{g}{\gamma^2}\right)^{1/3} . D \le 12.5, \quad (5.24)$$

where D is the median grain size of the particles, g is the acceleration because of gravity, and T is the period of oscillation. For the initial transport of non-cohesive particles it is of importance whether the bottom is rippled or not. Ripples generally form at velocities that are a little higher than the critical velocity for particle movement (u_c). Manohar (1955) found that ripples are formed at a velocity of 1.24 u_c, but if the bottom is already rippled when wave motion starts, particle movement may start at velocities as low as 0.2 u_c because over the crests of the ripples the velocities are much higher than the average velocity. Rippled beds also induce a lifting of the particles from the bed into the water above the bed, where even weak currents can carry the particles away downstream. This upward movement is caused by the vortex that develops at the lee-side of a ripple, which changes in strength and direction with the oscillation of the waves. The maximum upward movement occurs when the flow reverses and the vortex suddenly increases in strength, which occurs every half-wave cycle (Fig. 5.26). At this maximum upward movement the concentration at a height y above the top of the ripple is given by

$$C = C_0 \exp\left(-\frac{y-\delta}{\sigma}\right)^2, \tag{5.25}$$

where C_0 is the concentration at the height $y - \delta$ above the crest of the ripple), δ is the thickness of the viscous sublayer and σ the coefficient of surface tension. σ varies with the wave phase. The height to which particles are being moved upwards is in the order of several cm. Analytical models of this process have not yet become sufficiently accurate and are still highly empirical. In-situ measurements of suspended particle concentrations (particle size = 400 μm) at 15 mm above the bottom above the crest of a ripple are shown in Fig. 5.27. The mean concentration is estimated from

$$\text{(a)} \quad \bar{C} = \bar{C}_a y^{-m} \quad \text{or} \quad \text{(b)} \quad \bar{C} = \bar{C}_0 e^{-ay}, \tag{5.26}$$

where y is the height above the bottom and the other parameters are coefficients that do not depend on y and are determined from experiments. In Eq. (5.26a) the diffusion coefficient is assumed to be directly proportional to y, while in Eq. (5.26b), it is assumed to be constant. Using (a) or (b) does not give significantly different results (Swart 1976).

Wang and Liang (1975) give

$$\frac{\bar{C}}{\bar{C}_a} = \left(\frac{y}{a}\right)^{-R} \tag{5.27}$$

based on the assumption that the diffusion coefficients for both turbulence and sediment particles are proportional to the vertical velocity amplitude. Here, a is the amplitude and R = $W_s/\sigma k u_m$, where k = $2\pi/\lambda$, λ is the wave length, u_m is the near-bed maximum orbital velocity, and σ is the angular frequency = $2\pi/T$, where T is the wave period. Field measurements showed a reasonable agreement. Nielsen (1984) found from field experiments that

$$\bar{C} = \bar{C}_0 \exp(-y \cdot l_s^{-1}), \tag{5.28}$$

where C_0 is the mean particle concentration in the bed and l_s approximates the bottom ripple height (3–5 cm). Concentration decreased exponentially by a factor of 100 away from the bed over a distance of 15–25 cm. The different

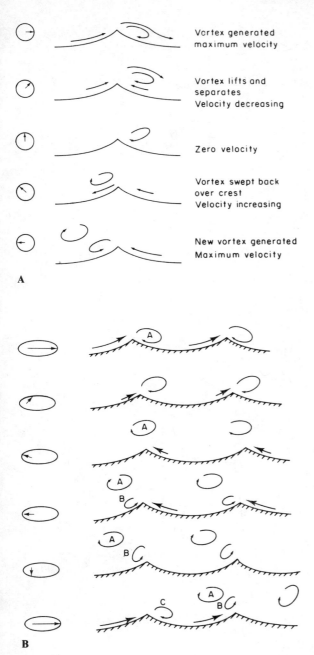

Vortex generated
maximum velocity

Vortex lifts and
separates
Velocity decreasing

Zero velocity

Vortex swept back
over crest
Velocity increasing

New vortex generated
Maximum velocity

A

B

Fig. 5.26. A Generation, separation, and advection of vortices during a wave oscillation. The rotating vector on the *left* indicates the wave phase. **B** Vortex generation, separation, and advection under combined waves and currents. The vector on the *left* indicates the wave phase. (**A, B** Dyer 1986)

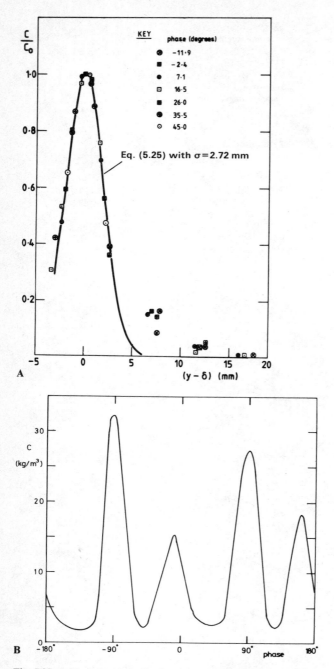

Fig. 5.27. A Variation in sediment concentration C/C_0 with height $(y - \delta)$ just downstream of a ripple. **B** Example of the variation in time of the suspended sediment concentration at a fixed point above a ripple crest. (**A, B** Sleath 1984)

equations give very different concentration distribution profiles related to the concentration at a reference level a = 1 cm (Dyer 1986). Particularly the fine size fractions are resuspended (winnowed out) so that a very precise definition is needed for the concentration at reference level.

Strong waves can bring particles in suspension to a much higher height above the bottom, as is the case with breakers, which disperse particles from the bottom over the entire waterdepth. This is caused by the strong increase in turbulence when the wave breaks. Plunging breakers bring more particles in suspension than spilling breakers (Fairchild 1977) but there are no reliable models for the process of particle suspension by breakers.

The discussion up to now has been limited to progressive (surface) waves. In standing waves, maximum concentrations of suspended particles occur at the antinodes and minimum concentrations at the nodes: a circulation develops (Fig. 5.28) whereby the particles are moved towards the antinodes, where they are dispersed upwards, while at the nodes the particle movement is downward and away from the node.

When the waves are asymmetric, the peak velocity is higher in the direction of wave propagation than in the reverse direction. This has a strong effect on particle transport and gives a net transport in the direction of wave propagation and the transport rate is proportional to a power of the horizontal velocity component. Solitary waves have a strong forward flow under the crest but no return flow, which can be regarded as an extreme case of asymmetry. They can develop in shallow water. No reliable relation has been established between solitary wave characteristics such as the ratio of wave height to water depth, and the motion and dispersal of particles (Wells and Kemp 1986).

For very fine-grained, cohesive particles the critical shear stress for initial motion is the same in wave motion as for steady non-oscillatory flow, but

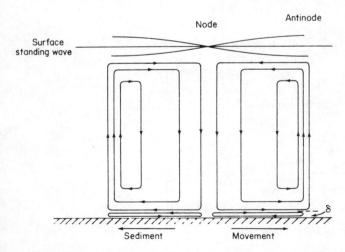

Fig. 5.28. Mean water circulation beneath a surface standing wave. (Dyer 1986; after Liu and Davies 1977)

characteristic for oscillatory flow are the fluctuations in pressure (cyclic loading) which can weaken the cohesive sediment structure, break the aggregate structure, and result in fluidization. The rate of erosion (particle pick-up) can be given as

$$\frac{E}{E_w} = \frac{\tau_b - \tau_{sR}}{\tau_{sR}}, \tag{5.29}$$

where E is the erosion rate (in units of mass per units of time), E_w is a rate constant, τ_b is the shear stress, and τ_{sR} is the critical shear. During the process of erosion, the shear strength can change considerably with time and with depth of erosion, as the sediment becomes increasingly cohesive with depth in the sediment. The waves cause the mud bottom to oscillate; the oscillations of the bottom become quickly out of phase with the oscillations in the water at increasing depth in the bottom so that the shear stress increases with ca. 30% as compared to a non-oscillating, rigid bed (Maa, in Mehta 1988). Even under progressive nonbreaking waves, erosion therefore tends to be ca. one order of magnitude higher than under steady flow, under breaking waves even higher.

Because waves (re)suspend bottom material with a limited horizontal displacement, this results in high suspended matter concentrations near to the bottom (Fig. 5.29). At a high erosion rate and a low upward diffusion rate, and in the absence of horizontal currents, a fluid mud with concentrations of 10^4 to

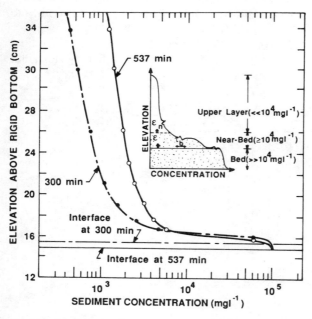

Fig. 5.29. Suspended sediment concentration profiles during erosion of mud by waves in a flume tank. Elevations are measured above rigid flume bottom. (Mehta 1988; after Mehta and Maa 1986)

10^5 mg . l^{-1} easily develops. This also occurs when high concentrations are formed through dumping of dredge spoils, or when suspended matter is concentrated by estuarine flow (turbidity maximum). An existing fluid mud will be moved by waves, but most of the mud moves over or just above the consolidated bed. Wave energy is quickly dissipated in fluid muds. In shallow water, waves can be reduced in height by 50% because of this energy dissipation (as can be observed, e.g., on the Guyana and Mississippi delta coasts; Tubman and Suhayda 1976; Wells 1978).

Particle transport in suspension by waves can quantitatively probably best be approximated by averaging the concentration distribution over a wave period. Much suspended matter transport, however, occurs under the combined influence of currents and waves, particularly in estuaries and coastal seas, and takes place over a sandy bottom that is flat or rippled, the ripples ranging in wave length from several cm to hundreds of meters. This means that the velocity profile is determined by the roughness and morphology of the sandy bottom, by the oscillatory movements and short distance horizontal pressure gradients associated with waves, and by the long distance horizontal gradients driving the nonoscillatory flow. The waves usually come from a different direction than the currents, and there is a strong interaction between waves and currents, particularly in the surface water. It is likely that the capacity to transport suspended matter increases by this interaction (and observations in nearshore waters support this), but the complexity of this interaction and its effect on the distribution and intensity of the turbulence and on particle transport has not yet allowed a reliable quantification. Theoretical work indicates that the drag on a particle increases when oscillatory flow is superposed on steady flow.

5.5 Particle Settling

When the turbulence is not able to keep the particles in suspension, they settle to the bottom. The underwater weight W_u of a particle is

$$W_u = V_g(\rho_s - \rho), \tag{5.30}$$

where V is the volume of the particle, ρ the density of the water, ρ_s the density of the particle and g the gravity accelaration. The settling is opposed by the resistance or drag of the water, which for a sphere with radius a is:

$$R_D = 6\pi a\eta\, w_s, \tag{5.31}$$

where R_D is the drag, η in the viscosity of the water (for η also the symbol μ is used), and W_s the relative velocity between the water and the sphere (the Stokes' velocity). Equilibrium is reached when $W_u = R_D$ or

$$\frac{4}{3}\pi a^3 g(\rho_s - \rho) = 6\pi a\eta\, w_s$$

or

$$W_s \equiv \frac{2 \pi a^2 (\rho_s - \rho) g}{9 \eta}, \tag{5.32}$$

which is Stokes' law. Natural particles are not perfect spheres, so a shape factor has to be introduced, but more important is, that the drag on a particle not only depends on the viscosity of the water but also on the influence the falling particle exerts on the fluid. A drag coefficient has to be introduced and since the type of flow around the settling particle is related to Reynolds number $Re = w_s d/v$, relation (5.33) can also be written as (with $d = 2a$ the diameter of the sphere):

$$R_D = 6 \pi a \eta w_s = \frac{24v}{w_s d} \cdot A \frac{\rho w_s^2}{2}, \tag{5.33}$$

where A is the surface area of the particle normal to the direction of motion. The drag on a particle was given by Newton in 1687 as

$$F_w = C_D A \rho \frac{w_s^2}{2}, \tag{5.34}$$

where C_D was considered a constant (Leliavsky 1955). As the drag was found to

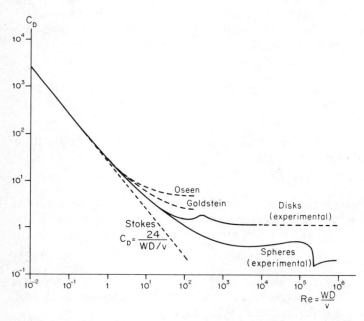

Fig. 5.30. Relation between the drag coefficient C_D and the Reynolds number Re according to the theories of Stokes, Oseen, and Goldstein, and from experiments with sphere-shaped and disc-shaped particles. (Graf 1971)

depend on Reynolds number, C_D is not a constant and, combining (5.32) and (5.33):

$$C_D = \frac{24}{Re},\qquad(5.35)$$

where C_D is the drag coefficient. Equation (5.36) is valid as long as Stokes' law holds, which is for $Re < 0.1$. With increasing values of W_s the relation of C_D with Re deviates increasingly from the Stokes' relation mainly because of the formation of a boundary layer around the particle which separates from the particle surface towards the rear. C_D then follows approximately the relation found by Oseen (1927):

$$C_D = \frac{24}{Re}\left(1 + \frac{3}{16}Re\right).\qquad(5.36)$$

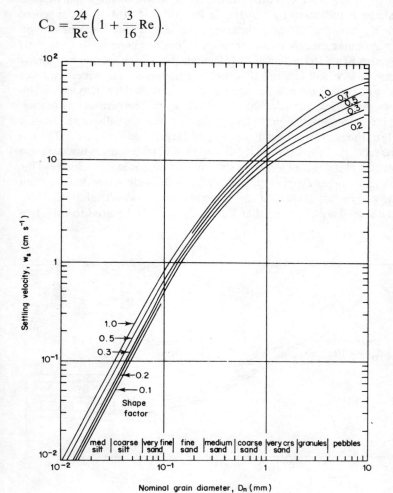

Fig. 5.31. Settling velocities of grains in water at 20 °C as a function of shape factor. (Dyer 1986; after Komar and Reimers 1978)

A further refinement, by increasing the number of terms within the brackets, was given by Goldstein (1929). Figure. 5.30 shows that the experimental curves, which also demonstrate a pronounced effect of particle shape, are between the Stokes' relation and the theoretical curves of Oseen and Goldstein. At large Reynolds numbers C_D becomes gradually independent of Re when Re increases, and the drag becomes proportional to the square of the fall velocity. A wake is formed behind the particle which is (still) laminar. When the wake becomes turbulent, the particle starts to leave a trail of eddies behind it and C_D drops suddenly to a lower value. This occurs for spheres at a Re number of ca. 3×10^5. From there, C_D remains constant again at higher values of Re.

In Fig. 5.31 the influence of shape on the fall velocity of single particles is shown. The shape is indicated by a shape factor $SF = D_1/(D_2 + D_3^{1/2}$ where D_1, D_2, and D_3 are the shortest, intermediate, and longest diameters on mutually perpendicular axes. A shape factor of 1 does not necessarily indicate a sphere: a cube also has a shape factor of 1. Besides particle shape the roughness of the particle surface influences the drag: a rough surface, which induces turbulence, gives a higher drag. Since the viscosity of the water increases when its temperature decreases, particles fall faster when the temperature is higher. This effect is larger on smaller particles. A heavier particle falls faster than a lighter particle of the same size because $(\rho_s - \rho)$ is larger [see Eq. (5.31)]. The fall velocity can be related to the fall velocity of standard particles: an equivalent size can be given which is the size of a quartz sphere with the same fall velocity. This can also be done for other types of particles whose characteristics deviate from those of quartz spheres. It should be remembered, however, that C_D also depends on the size of a particle so that Eq. (5.32) is not to be used to calculate

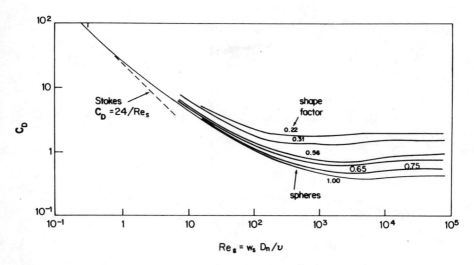

Fig. 5.32. Drag coefficients (C_D) as a function of Reynolds number (Re_s), based on the nominal diameter D_n for grains with different shape factors. (Dyer 1986; after Komar and Reimers 1978)

the equivalent size, but Eq. (5.33); a value for C_D has to be estimated from Re using Fig. 5.32.

In natural environments, suspended particles seldom occur alone, and there is usually some turbulent flow in the water. Clouds of particles settle faster than single particles because a large number of particles in the cloud settle in the wake of others, but the fall velocity of a single particle in a uniform suspension is lower than in clear water because the falling particle is hindered by the other particles

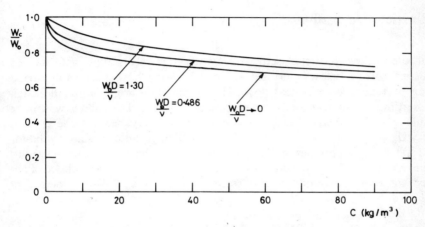

Fig. 5.33. Effect of particle concentration on fall velocity for different Reynolds numbers. (Sleath 1984; after McNown and Lin 1952). W_c fall velocity at concentration c; Wo fall velocity in clear fluid; D grain diameter; τ kinematic viscosity)

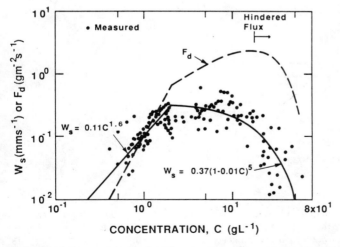

Fig. 5.34. Sedimentation velocity W_s and sediment flux variations F_d with particle concentration for Tampa Bay mud. (Mehta 1989; after Ross 1988)

in suspension. This is enhanced when the particles flocculate during settling (see also Chapter 6). This is already notable at concentrations of less than $1 \text{ g} . 1^{-1}$ (Fig. 5.33) but becomes very strong at high concentrations. At concentrations of ca. $2 \text{ g} . 1^{-1}$ this begins to have a marked effect (hindered settling) and the particle settling velocity reaches a maximum at ca. $5 \text{ g} . 1^{-1}$ (Fig. 5.34). The particle flux reaches a maximum at ca. $20 \text{ g} . 1^{-1}$. At concentrations above $100 \text{ g} . 1^{-1}$ the settling is negligible.

5.6 Particle Deposition

In turbulent flow particles settle out when the mean flow velocity – or turbulence – drops below a critical value. Diagrams have been constructed relating particle size to critical flow velocity and flow regions are indicated where settling, steady transport, and erosion occurs. The classical one was made by Hjulström (1935) and based on mean flow velocity because the friction velocity U_* was seldom available (Fig. 5.35). Such diagrams are usually valid for quartz spheres, and particles of different shape and/or density are normalized to an equivalent size. Observations in nature have shown that (flocculated) fine-grained suspended material usually starts to settle out at flow velocities of

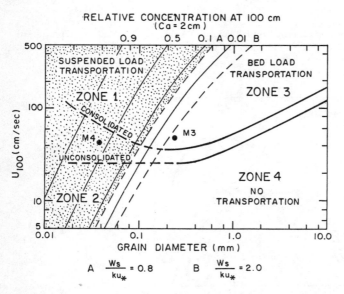

Fig. 5.35. Relation between grain diameter, current velocity at 1 m above the bottom, and concentration at 1 m above the bottom relative to the concentration at 2 cm. *Dashed line boundaries A and B* based on the ratio between particle settling velocity (W_s) and the product of von Karman's constant K and the friction velocity U_* (Sternberg et al. 1985; after Sundborg 1967 and Smith and Hopkins 1972)

10–$15\ cm.s^{-1}$, while suspended sandgrains of ca. $500\ \mu m$ already settle out ca. $1\ m.s^{-1}$.

Partheniades and Kennedy (1966) found in experiments with kaolinite particles that after a few hours of deposition a concentration was reached that remained constant during the remaining time of the experiments (33-h duration; Fig. 5.36). This concentration C_{eq} was related to the initial concentration C_0 and the ratio C_{eq}/C_0 was found to be a function of the average shear stress. The actual values of C_{eq}/C_0 ranged from 0.5 to 0.7. Data of Einstein and Krone (1962) and Partheniades (1965) suggested that this constant concentration was only reached at velocities somewhat higher than $18\ cm.s^{-1}$. Below that velocity (or below a certain critical shear stress) deposition of all particles in suspension occurred, and was increasingly more rapid at lower velocities (or at lower bottom shear; Fig. 5.37). An explanation can be found in the results of Wood and Jenkins (1973), who demonstrated that particles can remain in suspension indefinitely within turbulent eddies. Another explanation was given by Partheniades (1971), Mehta (1973) (and Mehta and Partheniades (1975) based on the observation that suspended particles are flocculated (see Chap. 6), and that suspended particles for which the critical shear stress is smaller than the actual shear stress will remain in suspension, whereas those for which the critical shear stress is higher, will deposit. This implies that only those flocs that are strong enough to withstand the maximum shear near to the bottom will deposit, but that the weaker ones will break apart and be re-entrained into the main flow

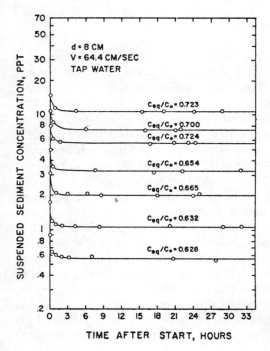

Fig. 5.36. C_{eq}/C_0 as a function of time (kaolinite in tap water). (Etter et al. 1968)

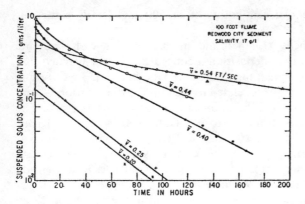

Fig. 5.37. Deposition of sediment from flowing water. (Einstein and Krone 1962)

(Mehta and Partheniades 1975). The latter ones are the particles that continue to remain in suspension. For nonflocculated particles (sand, possibly also silt grains) the critical shear stress, which is a measure for turbulence intensity, will apply as a criterion for remaining in suspension.

5.7 Stratification

Besides the complications indicated in Section 5.2.2.4 (which arise when applying suspended matter transport calculation models to the actual conditions), complications of a different kind appear when lakes, estuaries, and the sea are considered.

Estuaries and coastal seas, with a few exceptions (such as the Baltic and the Mediterranean, except the northern Adriatic), are characterized by tidal flow, which is unsteady with often a marked asymmetry of the tidal velocity curve. During the accelerating and the decelerating phases the characteristics of the turbulence are markedly different (Vittori 1989). In estuaries, the coastal sea, and in many lakes, because of the shallow water depth, the interaction of the bed and the flow is strong, and where the bottom is muddy, the interface between bed and flow is poorly defined. A variety of bottom sediments can be found as well as a variable bottom topography (ripples, dunes, flat beds, ridges, banks, etc.) and a variety of values for bottom roughness to which also bottom fauna and flora contribute (Table 5.1). Waves are usually of importance in lakes as well as in many estuarine areas and the coastal sea; current patterns are determined by the tides, wind forcing, and surface waves. At latitudes higher than 15° N or S, the earth's rotation (the Coriolis force) is a factor influencing the direction of the flow. Estuaries and coastal seas also often show strong mixing of fresh and saline

Table 5.1. Bed roughness lengths y_0 and drag coefficients C_{100} for typical seabed types. The drag coefficient C_{100} based on a velocity measured 100 cm above the bed. (Dyer 1986)

Bottom type	y_0 (cm)	C_{100}
Mud	0.02	0.0022
Mud/sand	0.07	0.0030
Silt/sand	0.002	0.0014
Sand (unrippled)	0.04	0.0026
Sand (rippled)	0.6	0.0061
Sand/shell	0.03	0.0024
Sand/gravel	0.03	0.0024
Mud/sand/gravel	0.03	0.0024
Gravel	0.03	0.0047

water and high suspended matter concentrations. The differences in salinity, temperature, or suspended matter content usually cause strong density gradients, and result in a horizontal, vertical, or oblique stratification with the water masses of different density separated by a halocline, thermocline, or lutocline. All these types occur in estuaries and in the sea; in lakes often a thermocline and sometimes a lutocline can be found. Rivers are usually too well mixed to show a thermocline, but a more or less vertical or oblique lutocline may be present where a river with a heavy load of sediment enters a clear-water river (or the reverse). Interfaces between different water masses can be very stable.

In this section the effects of density gradients on suspended sediment will be discussed. It is a complex issue: the stratification influences the flow and thus the sediment transport, while the suspended sediment concentration distribution itself causes a density gradient. Autosuspension of suspended matter, which can develop on subaquaceous slopes under the influence of gravity, is discussed in Section 5.7. The other factors influencing suspended matter concentration and transport will be discussed in relation to the processes operating in the different environments in Chapter 7.

5.7.1 Density Gradients

In the preceding sections, the flow was considered to be homogeneous. This is hardly ever true in nature, but in rivers the density differences are so small that they can be neglected except when considerable amounts of particles of different settling velocity are present resulting in a concentration gradient, or when water from a tributary is incompletely mixed. In all aquatic environments, density gradients are present at least during part of the year. The effect of density gradients is to damp the turbulence; when density gradients are strong over a short distance, the water is called stratified. The relation between the stabilizing effect of the density gradient and the destabilizing effect of the turbulence is given by Richardson's number

$$\text{Ri} = \frac{g/\rho \, \dfrac{d\rho}{dy}}{\left(\dfrac{d\bar{u}}{dy}\right)^2},$$

(5.37)

where g is the gravity acceleration, ρ the density, and $\bar{u}$ is the turbulent mean horizontal velocity. $\left(\dfrac{d\bar{u}}{dy}\right)^2$ denotes the upward transfer of momentum, $g/\rho \, \dfrac{d\rho}{dy}$ indicates the downward flux because of gravity. For $\text{Ri} > 0$ there is a stable stratification, for $\text{Ri} = 0$ the stability is neutral and there is no stratification, and for $\text{Ri} < 0$ the stratification is unstable. At $\text{Ri} \approx 0.25$ turbulence is damped out and the flow becomes laminar. Usually flow is nonsteady and nonuniform so that various critical values for Ri are found: in the sea a neutral situation exists for $0 > \text{Ri} > 0.03$ (Dyer 1986). On the basis of Ri (the gradient Richardson's number), some other forms of Richardson's number have been formulated that include other parameters:

- The flux Richardson's number (Ri_f) includes a factor for the degree of mixing: the Prandtl number $= v_t k_t^{-1}$, where v_t is the eddy viscosity and k_t the eddy diffusivity (Abraham 1988).
- The layer Richardson's number (Ri_L) includes the thickness of the intermediate layer (Abraham 1988). For $\text{Ri}_L = 0.32$ no mixing occurs: there are no stable internal waves (see Sect. 5.6.2) and there is no turbulent mixing so that the flow within this layer is laminar.
- The equilibrium Richardson's number (Ri_e) refers to the conditions in the stable interface layer and has a value of ≈ 0.1 (Abraham 1988). When $\text{Ri}_L < \text{Ri}_e$, shear dominates and the density gradient is reduced; when $\text{Ri}_L > \text{Ri}_e$, the density gradient dominates and turbulence is suppressed.
- The threshold Richardson's number (Ri_h) has a value of ≈ 20: when $\text{Ri}_L < 20$, the bottom turbulence is effective in mixing the water column; when Ri_L is higher, the internal stratification is also effective (Dyer and New 1986).
- The rotational Richardson's number (R) includes a factor for the coriolis force (McClimans 1988).
- The bulk Richardson's number (Ri_0) is based on the bulk properties of the flow and includes the depth-averaged suspended matter concentration as well as the slope (Stacey and Bowen 1988).

For those suspended particles that can be assumed to have the same momentum transfer as water (see Sect 5.2) the stratification interferes with particle diffusion: for values of $\text{Ri}_L > 0.32$, no particles can move across the intermediate layer, whereas complete mixing of particles can occur only when $\text{Ri}_L \approx \text{Ri}_e \approx 0.1$. For $0.32 > \text{Ri}_L > \text{Ri}_e$, some particulate matter can be transferred but not as effectively as when mixing is complete. Because of stable density gradients, very turbid waters can continue to exist for a long time next to very clear water.

The sediment in suspension increases the density of the water in which it is suspended and therefore produces a vertical density gradient above the bottom. This gradient has a stabilizing effect on the flow by reducing or inhibiting the vertical exchange of turbulence. Therefore, as soon as particulate matter goes into suspension, the vertical velocity distribution changes and deviates from the logarithmic von Karman-Prandtl equation [Eq. (5.08)]. (Dyer 1986). The values for u are always greater and the values for u_* always smaller when sediment is in suspension than when the flow is free of suspended sediment. When the suspended sediment remains close to the bed (as is the case with sandy sediment and/or at low current velocities) the velocity profile is only modified close to the bed and the profile becomes concave upwards. When the suspended particles are dispersed over a large part of the water column, the velocity profile becomes convex upwards. This is shown in Fig. 5.38 (from Dyer 1986) as being grain size-dependent. For $z/L = 0.03$, where L is the length scale associated with the stratification, the profile will show a minimal stratification effect. A and B in Fig. 5.38 are constants that relate the particle settling velocity to u_*, the von Karman constant k, the relative particle density in the water, and to the particle concentration at $y = y_0$ (Dyer 1986). Figure 5.38 indicates that there are four regimes,which apply equally to the vertical velocity distribution as to the vertical concentration distribution:

  I: the stratification decreases with height above the bottom but is not significant and the velocity distribution differs little from the distribution in clear water;
 II: the stratification is significant near the bottom: the logarithmic velocity profile is concave upwards;
III: the flow is hardly stratified near the bottom but becomes increasingly stratified with height above the bottom: the logarithmic velocity distribution is convex upwards;
IV: the flow is stratified throughout but becomes increasingly stratified with height above the bottom: the velocity distribution is strongly convex upwards.

The influence of other density gradients caused by particulate matter in suspension, e.g., along interfaces of water masses, has not been investigated, but they may have a strong influence on the suspended matter transport.

Where freshwater with a high concentration of suspended matter flows out over saline water with a low suspended matter concentration – or where in lakes interflows are formed – vertical suspended sediment concentration gradients are associated with a vertical salt gradient or temperature gradient (or both). The gradient of the faster-diffusing component will diminish and disappear more quickly than the gradient of the slower-diffusing component. In this way, an originally stable layering can become unstable and vertical convection currents can develop associated with "fingers" of surface water moving downward from

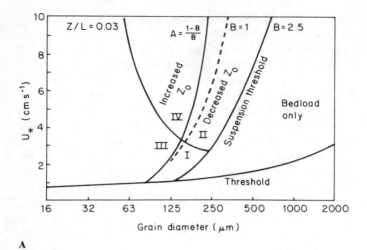

A

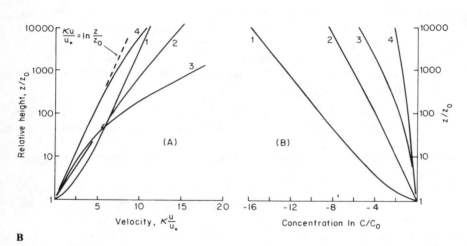

B

Fig. 5.38. A Regimes of velocity and concentration profiles in a suspension. **B** Velocity and concentration profiles for the four regimes indicated in **A** (numbers *1* to *4* correspond to the areas indicated by *Roman numbers* in **A**). (**A, B** Dyer 1986)

the upper layer. A criterion for "fingering" of surface suspensions is given by Green (1986): it occurs when the particle concentration (in mg . l^{-1}) is higher than 20 $D^2 \rho_s$ where D is the particle diameter (in μm) and ρ_s the particle density. Thus at concentrations of ca . 100 mg.l^{-1}, the particle size has to be $< 1.5\ \mu$m to prevent fingering, which is not commonly the case, while for particles of 20 μm the concentration has to be above 21 g.l^{-1} which occurs, e.g., in the Yellow river. At its mouth, however, instead of fingering dense underflows are formed that flow over the bottom while the concentrations in the surface water are ca. 50 mg . l^{-1}. It is as yet uncertain to what extent double-diffusive fingering is an important factor in the downward transport of suspended matter.

5.7.2 Internal Waves

Because of velocity shear, internal waves may develop in a horizontal or oblique density interface or interlayer in the same way as surface waves are formed by wind stress along the water surface. Internal waves may break also in a similar way as surface waves, whereby denser fluid from the lower layer is mixed with the lighter fluid in the upper layer. This is one-way upward mixing and is called "entrainment". It occurs wherever the turbulence in the lower layer is considerably less than in the upper layer, as is the case in some river estuaries (e.g., the Zaire). The criterion for the development of internal waves is given by the interfacial Froude number

$$F_1 = \frac{u}{\sqrt{\frac{\Delta\rho}{\rho} gd}}, \tag{5.38}$$

where $\Delta\rho$ is the density difference between the two layers, ρ is the density of the lower layer, u is the velocity of the upper layer relative to the lower layer, and d is the depth of the upper layer. When F_i approaches unity, internal waves form; they break at $F_i = 1$.

5.8 Autosuspension

On steep underwater slopes and in underwater valleys and canyons, sediment can transport itself as an avalanche when the density gradient becomes equal to or exceeds the internal friction. Suspended matter behaves in a similar way when

$$w_s \cos\beta/u \sin\beta \le e, \tag{5.39}$$

where w_s is the settling velocity of the particles in suspension, β is the angle of the slope, u is the transport velocity of the suspension, and e is a constant. The effect is larger when the particle concentration is higher until at very high concentrations particle interactions dominate, turbulence is damped out, the flow becomes viscous and the particles settle out. Such flows – turbidity currents – have been observed or postulated in lakes and reservoirs, in submarine canyons, and along submarine slopes. They can be triggered by high discharge from a river, earthquakes, storm waves, edge waves along the shelf, slumps and heavy discharges of mine tailings. At such moments large amounts of sediment that have piled up, are loosened. When condition (5.39) is reached, a turbidity current can be formed and can maintain itself over large distances. A slope of only a few degrees can already be sufficient for a self-generated turbidity current (Eidsvik and Brørs 1989). Considerable speeds in the order of 10 to 30 m.s^{-1} can be attained by such currents, particularly when the concentration increases on the way by entrainment of particulate material from the bottom. The current can continue over the ocean floor at a slope of only 0.5° for a long distance, and yield sufficient bottom stress to transport sand and gravels in deep sea fan channels (Komar 1970).

For initiating a turbidity current, the particles have to be suspended in such a way that a vertical density gradient is produced which gives a downslope gravitational component. For the current to become self-sustaining, an upward turbulent diffusive flux has to be generated that is larger than the downward settling flux, but with time the upward flux will decrease because of the density stratification, and a kind of equilibrium will be reached. Then the current will continue downslope, its thickness will remain constant, and an equal amount of particulate matter that is picked up from the bottom will be deposited. At this stage the current is constant (uniform).

The currents are driven by the excess density of the suspended particles and the bottom slope (gravity). Bagnold (1962, 1963) gave e in Eq. (5.40) the value 1; Pantin (1979) arrived at a value of 0.01. Stacey and Bowen (1988) developed a slightly different criterion

$$\frac{w_s \cos \beta}{\bar{u} \, C_D} \leq S_0 \simeq \text{constant},\tag{5.40}$$

where C_D is the drag coefficient [see Eq. (5.20)] and S_0 is a factor smaller than unity. C_D was found to be function of tan β. The criterion indicates that more sediment can remain in suspension with increasing current speed and slope and decreasing particle settling velocity.

When the bottom roughness is higher, it is easier to maintain the sediment in suspension because of increased turbulence, but a change in roughness has only a limited effect: decreasing the roughness length by a factor of 100 changes the minimum velocity necessary to maintain the turbidity current only by a factor of 2. Changes in the particle settling velocity have little effect on the flow itself, but strongly influence the vertical concentration profile. The effect of the presence of particles of different size in the flow is not known. A predominance of fine (mud) particles probably can keep much larger (sand-sized) particles in suspension. The density gradient that will be the result of the vertical concentration distribution will have a stabilizing effect so that high concentrations can develop and be maintained in the bottom layers. Probably after reaching a critical concentration depsosition occurs (Stacey and Bowen 1988).

6 Particle Size

Suspended matter in natural waters – in the sea as well as in estuaries, rivers, and lakes – is mainly present in the form of flocs (or aggregates) which have been observed by many authors in situ in the water or in suspended matter samples collected on filters, or studied with a reversed microscope (Beebe 1934; Sherman 1953; Suzuki and Kato 1953; Nishizawa et al. 1954; Piccard and Dietz 1957; Berthois 1961; Krone 1962; Riley 1963; Riley et al. 1964, 1965; Holmes 1968; Sheldon 1968; Biddle and Miles 1972; Edzwald 1972; Schubel and Kana 1972; Kranck 1973; Paerl 1973 to name the earlier ones). Mostly the flocculated material was observed under a microscope and usually found mixed with single mineral particles. Nishizawa et al. (1954), however, already indicated that flocs in situ in the water are fragile and break up readily during sampling so that suspended flocs have to be studied by in-situ observation. Observations of suspended flocs while diving or from a submersible were recently made by Trent et al. (1978), Shanks and Trent (1980), Silver and Alldredge (1981), Syvitski et al. (1983) and Eisma (1986). Since diving and submersibles have a limited application, camera systems were developed for in-situ photography. A Benthos plankton camera was used by Eisma et al. (1983) and Kranck (1984a). Other camera types were developed by Honjo et al. (1984), Johnson and Wangersky (1985), Wells and Shanks (1987), and Eisma et al. (1990). Most camera systems do not allow in-situ observations of particles smaller than 50–100 μm; the one developed by Eisma et al. (1990) gives results down to about 3 μm. Bale and Morris (1987) used laser diffraction to determine in-situ particle size down to about 1 μm. An optical settling tube giving size distributions from a few microns to several hundred microns was deployed by Spinrad et al. (1989) in the deep ocean. In-situ flocs have a size up to 1–2 mm but large elongated flocs with a length of up to 10 cm have been observed and occasionally, in quiet waters, even much larger flocculated structures.

Many particle size determinations of suspended matter have been made with a Coulter counter (for the measuring techniques see Sect. 6.4), which gives meaningful results that can be related to the distribution of water masses, water depth, degree of resuspension, and river outflow: Sheldon et al. (1972) in the Atlantic and Pacific (surface waters and deep water), Carder and Schlemmer (1973) in the eastern Gulf of Mexico (upwelling and surface current pattern), Plank et al. (1973) in the Panama Basin (abyssal circulation), Kitchen et al. (1978) off Oregon (upwelling), Yamamoto (1979) in the Yellow and East China Seas (river outflow and the Kuroshio), Nelsen (1981) in New York Bight (dispersal of material and water masses), McCave (1983) and Richardson and Gardner (1985) on the Nova Scotia continental rise (near-bottom flow), Eisma et al. (1978a) and Eisma and Kalf (1984) at the Zaire river mouth (river outflow), Eisma and Kalf (1987b) in the North Sea (water masses, water depth). Particles as small as 1.5 μm in diameter can be measured reliably with a Coulter counter. The maximum size observed is ca. 125 μm, but can be larger when particles of planktonic origin are present.

In the interpretation of Coulter counter size distributions, it has generally been disregarded that the in-situ flocs are fragile. Results of Gibbs (1981, 1982a, b), Gibbs and Konwar (1982), and Eisma et al. (1983) indicate that in-situ flocs, although stable in current velocities up to more than $1 \, \text{m} \cdot \text{s}^{-1}$, break up during sampling into particles generally smaller than 100 μm (as normally found after sampling). This also occurs when pumps, pipettes, Niskin bottles, HIAC sensors, or a Coulter counter are used. Only careful sampling by divers allows collecting the flocs intact on a filter or in a tube. From such samples it could be ascertained that the flocs seen in situ are indeed large aggregates of inorganic particles, biogenic particles and particle fragments, and organic matter. Because of the normal practice of sampling and size analysis, the in situ particle size has usually been underestimated (by ca. one order of magnitude, Eisma et al. 1983): the flocs break up so that the Coulter counter size distributions are essentially distributions of floc fragments.

The Coulter counter data, as well as the particle size data obtained after sampling by other methods of size analysis, indicate that flocs from different areas do not break apart into the same sizes but into coarser or finer size distributions that can be related to water masses or other hydrographic features. This indicates that the flocculated suspended material differs systematically in resistance to break-up (floc strength), or in degree of flocculation. Sand grains in suspension, which are hardly cohesive, and have a relatively large mass, are not flocculated. Clay- and silt-sized particles are usually flocculated, but owing to the paucity of data, it is not clear to what extent the fine-grained suspended material is entirely flocculated or whether also single fine-grained mineral particles are present in situ in suspension. Just off a glacier where there is no organic matter, unflocculated fine-grained particles have been observed in suspension (Syvitski and Murray 1981). The fragments that are usually observed are single particles as well as flocs (called microflocs: Eisma et al. 1983; Eisma 1986). They can easily be seen with a microscope (light microscope of SEM). In Makasar Strait at very low suspended matter concentrations ($< 0.3 \, \text{mg} \cdot \text{l}^{-1}$) no microflocs were found, as long as no planktonic diatoms were present, but only single particles (Eisma 1990). This indicates that the particles in situ were not flocculated, or flocculated only very loosely, so that no microflocs remained after sampling. If this can be generalized, the Coulter counter at such low concentrations may give the in situ size distribution, while at higher concentrations it gives predominantly the distribution of floc fragments (see, however, Sect. 6.1.2). The resulting fragments are relatively firm and can be broken up further only by vigorous shaking or ultrasonic treatment.

A third, essentially different way of measuring particle size of suspended matter is by measuring the size distribution of the constituent (unflocculated) particles (mineral grains, biogenic carbonate, and opal) after removal of the organic matter that holds them together. The size distributions thus obtained are of the same nature as the grain size distributions determined in bottom sediments, and can be used for intercomparison.

6.1 Flocculation of Suspended Matter

This section is concerned with the formation of flocs. Flocs have been observed in situ from a few micron upwards and have a rounded or irregular shape (Fig. 6.1; Plate 1). In the ocean they are often of the snowflake type and look like snow when they reflect the light of a lamp while moving through the water, which led the first observers to name them "marine snow" (Carson 1951; Suzuki and Kato 1953). In ocean surface waters also soft gelatinous material as large as 1 cm in diameter, and small organic flakes have been seen (Suzuki and Kato 1953; Nishizawa et al. 1954; Manheim et al. 1970; Jannasch 1973; Silver and Alldredge 1981). Probably more common are long, elongated flocs of up to 9 cm length (Trent et al. 1978) and stringers, or chains of flocs, or particles on a very thin and fragile filament of up to more than 10 cm length (Syvitski et al. 1983; Eisma 1986; Fig. 6.1B). Stringers were observed in Canadian and Dutch coastal waters in quiet flow – in Canada below the wave-influenced surface layer (where only marine snow was seen), in the Dutch tidal inlets during slack tide. Near to the bottom they disappear, probably because of increased turbulence.

The terms "flocculation", "aggregation", "coagulation", "agglomeration", "agglutination", and "conglomeration" have been used to indicate the processes whereby particles are being brought together and remain attached to each other for some time. This terminology can be confusing as to which process or processes is indicated; here for the sake of convenience "flocculation" is used as the general term indicating the process, and "flocs" is used to indicate the result, while "aggregation" is used when the emphasis is on the constituent particles. The other terms are avoided.

The flocculation of suspended particles has a pronounced effect on particle transport and on the transport of particle-associated substances: it results in the formation of large particles out of small ones and in an increase of the particle settling rate which increases as a function of the particle diameter. This can be offset by a lower density when the larger flocs contain relatively more organic matter than the fine ones, or when they enclose more water (or both). Because of a higher settling rate, flocs are more quickly deposited on the bottom and can be accumulated, particularly in deep lakes or in the ocean. Concentration of a large number of fine particles into a few number of large ones gives a greater transparency of the water so that sunlight can penetrate deeper into the water and primary production may be enhanced. Flocculation also reduces the surface

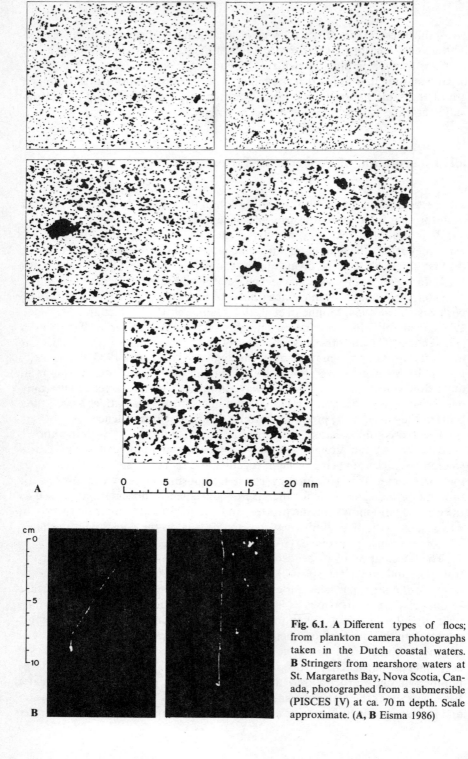

Fig. 6.1. A Different types of flocs; from plankton camera photographs taken in the Dutch coastal waters. **B** Stringers from nearshore waters at St. Margareths Bay, Nova Scotia, Canada, photographed from a submersible (PISCES IV) at ca. 70 m depth. Scale approximate. (**A, B** Eisma 1986)

area of the suspended matter, thus reducing the area available for adsorption of dissolved substances.

For the formation of flocs a considerable number of processes have been indicated:

1. salt flocculation because of higher salinity (van Olphen 1963; Edzwald 1972);
2. adherence of positively charged parts of particles to negatively charged parts of other particles (particularly clay mineral particles: face to face and edge to face structures, sometimes large clusters; Schofield and Samson 1954; O'Brien 1971);
3. concentration of suspended matter increasing the number of interparticle contacts (Krone 1962);
4. turbulence (10^3-10^4 $\mu m \cdot s^{-1}$), bringing particles into contact (Hunt 1980; McCave 1984b);
5. differential settling of larger and smaller particles (10^1-10^4 $\mu m \cdot s^{-1}$) (Hunt 1980; Lal 1980; Hawley 1982; McCave 1984b);
6. Brownian motion (1 $\mu m \cdot s^{-1}$) (Hunt 1980; McCave 1984b);
7. aggregation on bubbles of air or gas (Riley 1963; Baylor and Sutcliffe 1963; Barber 1966; Johnson and Cooke 1980);
8. precipitation, polymerization, adsorbance onto particles and flocculation of dissolved inorganic or organic matter, particularly carbohydrates (Sieburth 1965; Jannasch 1973; Gilmer 1974; Wassmann et al. 1986);
9. active aggregation by organisms: formation of feces, pseudofeces and fecal pellets, eventually followed by decay of the pellets and disaggregation into smaller flocs (Pomeroy and Deibel 1980; Knauer et al. 1982);
10. bacteria and phytoplankton (diatoms, *Phaeocystis*) exuding organic substances (polymers) and producing mucus that acts as glue (Seki 1972; Paerl 1973, 1974; Alldredge 1984; Biddanda 1988);
11. specific sticky organic compounds adsorbed onto the particles such as carbohydrates and fulvic acids (Lal 1980; Eisma 1986);
12. adherence of particles to the remains of phytoplankton and zooplankton (organic material, mucus sheets, discarded feeding webs, etc.; Alldredge 1972, 1976, 1979; Gilmer 1972; Hamner et al. 1975; Silver et al. 1978; Barham 1979; Pomeroy and Deibel 1980; Silver and Alldredge 1981; Knauer et al. 1982);
13. aggregation on the remains of benthic microflora (in salt marshes; Ribelin and Collier 1979);
14. aggregation by mucus of corals and associated organisms (in coral reefs; Johannes 1967; Coles and Stratham 1973).

An important aspect of flocculation is that the in situ particle size is not constant but varies with the conditions in the flow related to the processes indicated above. The response to varying conditions depends on the kinetics of the processes that result in flocculation, as well as of those that result in floc breakup. Particularly in tidal areas, where flow conditions regularly change with

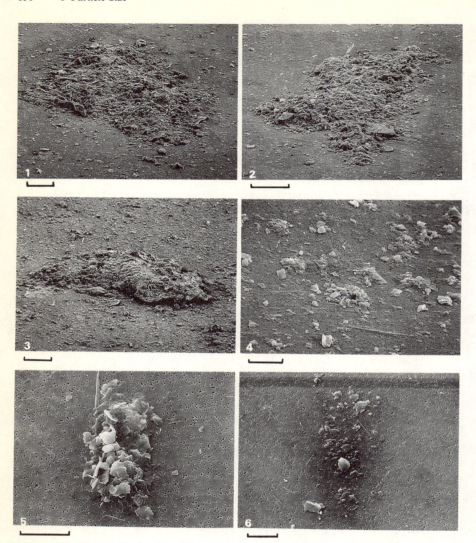

Plate 1A. 1, 2, 3 Unbroken natural flocs sampled in the Ems estuary by a diver. (Eisma et al. 1983). **1** and **2** are more loose and porous, **3** is more densely packed. Scale 100 μm. **4** Broken up flocs after Niskin bottle sampling from the Ems estuary; scale 20 μm. **5** Floc (after Niskin bottle sampling) from Makasar Strait, Indonesia. Fecal pellet; scale 10 μm. **6** Loose floc (after Niskin bottle sampling) from Makasar Strait, Indonesia, with (*dark*) remains of mucus. Scale 10 μm. **5** and **6** both from surface shelf water

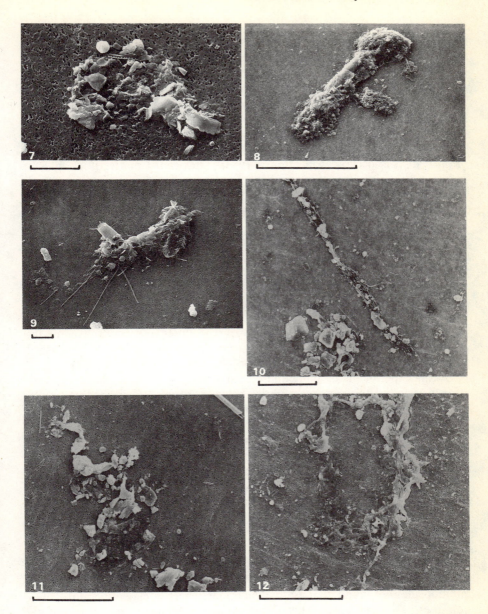

Plate 1B. 7 Loose floc (after Niskin bottle sampling), scale 10 μm. **8** and **9** Fecal pellets: **8** with coccoliths (*right side*), **9** with diatom frustules scale 0.1 mm; (9) scale 10 μm. **7** to **9** Surface shelf water Makasar Strait, Indonesia. **10** Mucus string with attached particles. **11** Loose floc with dark remains of mucus. **12** Detail of mucus string with attached particles. **10** to **12** from the bottom nepheloid layer on the continental rise off Nova Scotia (HEBBLE area). Scale 50 μm. (McCave 1985b)

the tides, probably an equilibrium floc size (or floc size distribution) develops that depends on the flocculation kinetics in relation to the tidal variation, which also includes regular deposition and resuspension of particulate matter.

6.1.1 Salt Flocculation

Suspended particles in natural waters are usually negatively charged (Neihof and Loeb 1972, 1974; Hunter and Liss 1979, 1982) and repulse each other. Assuming the particles to behave like colloids – which is true, e.g., for fine-grained dispersed clays (van Olphen 1963) – suspended matter flocculates at increasing salinity: in fresh-water the particles are kept apart because of their negative charges but at higher salinity these charges become increasingly neutralized by positive charged ions in solution. The positive ions form a cloud around each particle; an electric double layer is present, with the negative inner layer attached to the particle surface and the outer layer consisting of attracted positive ions. Because of the increasing neutralization of the negative inner layer at increasing salinity, the repulsive force diminishes, and when the particles can come very near to each other, the attraction by Van der Waals forces or intermolecular forces (which attenuate rapidly with increasing distance) can become so strong that the particles come together and flocculate. Both the particle charge and the potential between the outer and inner layer of the electrical double layer (the ζ-potential) can be measured in the laboratory. Measurements on colloidal clays support this theory (van Olphen 1963).

In the experiments of Whitehouse et al. (1960) and Migniot (1968) flocculation was found to occur at salinities between 1 and 7%, depending on the mineralogy of the particles (clays). Einstein and Krone (1962) indicated collision of fine particles at salinities above ca. 1% and Drake (1976) at 2%. Salt flocculation of natural suspended particles at the contact of fresh and salt water in river mouths therefore could explain the rapid settling of suspended particles and the high suspended matter concentrations found at low salinities in many estuaries: the flocculated material settles faster and resuspension results in high concentrations. In the laboratory experiments of Whitehouse et al. (1960), it was also shown that different types of clay minerals flocculate in a different way when salt is added to the suspension: flocculation of kaolinite and illite did not continue at salinities above 2%, while for montmorillonite it continued up to 7%. This difference in flocculation, together with large differences in the settling velocities of the flocs that were formed (high for kaolinite and illite and low for montmorillonite), could be used to explain differences in clay mineral distribution in estuaries and on the shelf (Edzwald and O'Melia 1975).

Although the importance of salt flocculation in river mouths and its effect on suspended particles seemed well established, flocculated suspended matter was also found in lakes and in rivers (Sherman 1953; Santema 1953; Meade 1972; Paerl 1973; Holmes 1968; Uiterwijk Winkel 1975; Kelts and Hsü 1978; Eisma et al. 1983; Eisma 1986). In the Rhine river, flocculation could be explained by pollution of the river with salts discharged from the potassium and coal mines (Santema 1953), but it became apparent that also other factors were involved (organisms, natural organic compounds, and cellulose fibers discharged from the paper industry). Measurements of the ζ-potential by Pravdic (1970) and Martin (1971) on natural particles did not show a reversal or neutralization at low salinities and reached this only, if at all, at high salinities. Measurements of the surface charge of natural particles by electrophoresis (Hunter and Liss 1979, 1982) demonstrated that the particles became somewhat less negatively charged at low salinities but never reached zero or became positively charge. Suspended particles, when immersed in natural waters, acquire an organic surface coating that changes the surface characteristics of the particles, inhibits salt flocculation and results in a negative surface charge of the particles also in saline water (Neihof and Loeb 1972, 1974; Narkis and Rebhun 1975; Loeb and Neihof 1975, 1977; Hunter and Liss 1979, 1982; Tipping and Cooke 1982; Gibbs 1983). Experiments by Gibbs (1983) indicated that coated particles flocculate significantly slower at higher salinities than particles with the coatings removed. The effect is larger at low salinities and decreases with increasing salinity. Recently, more detailed work of Loder and Liss (1985) and Garnier et al. (1990) confirmed that the organic matter fully determines the surface characteristics of natural suspended particles. Laboratory experiments with only mineral particles (as done recently, e.g., by Burban et al. 1989) therefore do not realistically reflect conditions in nature as far as the influence of particle surface characteristics and salinity or flocculation are concerned. The measurements of particle surface charge on natural particles, however, do show a charge reduction in saline water, which indicates that some change takes place on the particle surface, although it does not result in a net-positive charge. Weilenmann et al. (1989) showed by comparing data for Lake Zürich and Lake Sempach that calcium ions enhance flocculation and dissolved organic matter retards it: as the chemistry of lake waters varies widely, also the flocculation process can be expected to vary. Particularly the presence of natural polymers on the particle surfaces can influence the degree of flocculation appreciably, as will be discussed below in Section 6.1.4.

Since the settling and concentration of river-supplied suspended matter in estuaries can also be explained by the estuarine circulation or by a flow reduction in general without involving salt-flocculation (see Sect. 7.3.), and since the distribution of clays in estuaries and on the shelf can be related to differences in the origin of the clays and to dispersal and mixing (Meade 1972; Manheim et al. 1972; Feuillet and Fleischer 1980), there remains no clear evidence for salt-flocculation at the contact of fresh and salt water. In-situ floc size measurements made during the past years in estuaries (in fresh-water as well as in saline water) also give no irrefutable indications for a salinity effect (Eisma et al. 1991a). Gibbs et al. (1989), concluded that a salinity effect occurred in the Gironde, but this was based on only one freshwater sample. Other in-situ measurements made in several European estuaries (Ems, Rhine, Scheldt, Gironde) did not indicate any change in size related to salinity (Eisma et al. 1991a). Only in the Rhine were large flocs found at low salinities, but in that area also the flow is reduced and much dredging takes place. An increase in floc size at low salinities in the Zaire river estuary occurs where the river mouth widens and flow velocities decrease (unpublished data NIOZ). In the laboratory saltflocculation of natural suspensions can only be demonstrated at high concentrations ($\sim$ g.l^{-1}), which suggests that the particle concentration and not the salinity is decisive. It follows that there is no good argument for maintaining that suspended matter behaves as a colloid, since colloids are characterized by their flocculation at increasing salt content of the water (van Olphen 1963). Natural suspensions behave more like macromolecular sols, with the difference that the suspended particles settle out when the water is at rest. Salt flocculation of particles smaller than 1 μm (colloid iron hydroxides, humates, and associated substances) has been demonstrated by Sholkovitz and others (Sholkovitz 1976, 1978; Boyle et al. 1977; Sholkovitz et al. 1978; Mayer 1982), but quantitatively this amount is very small in comparison to particulate matter from other sources. (e.g., Wassmann et al. 1986). Also, some material larger than 1 μm may behave in this way but its effect on particle flocculation, if any, is minor. It should be realized that also of the "dissolved" material < 1 μm, only about 10% or less was found to flocculate, while the remainder stays in solution or suspension. Face-to-face and edge-to-face clusters of particles have been observed in saline water off glaciers (Syvitski and Murray 1981) but also in lakes (Kelts and Hsü 1978). Conditions near to a glacier ice-front and in meltwater flows with virtually no organic matter present, probably approximate the laboratory conditions under which salt flocculation has been demonstrated. Syvitski (1989) therefore separates flocculation (= salt flocculation) from agglomeration (= attachment to organic matter), but in nature the processes cannot be separated without more detailed analysis. Flocculation occurs off the ice front at a few ‰ S, where also plankton growth begins and particles may already be coated with organic material. Also in the clusters described by Kelts and Hsü (1978) organic coatings were probably present, as they were formed in the presence of organic matter.

In the size distributions of floc fragments, however, as obtained by pipette analysis and Coulter counter at low salinities, a marked salinity effect can be seen. With only a few exceptions, the particle size obtained in this way becomes finer at low salinities compared to the distributions in the adjacent freshwater (Eisma et al. 1983, 1991; Eisma 1986). This size reduction is associated with a peak in dissolved carbohydrates, which has been observed in a temperate estuary (the Ems) during the winter at low temperatures when also the biological activity is low. In other seasons a decrease in the concentration of carbohydrates in the suspended matter was found, going from freshwater to more saline water (in the Ems and the Gironde). The mobilization of carbohydrates is probably caused by the dissociation of polysaccharides or fulvic acids in the particles which occurs when the water becomes more saline (Leenheer pers. commun.). Thus the flocs lose part of their "glue" and become easier to break up. This increased fragility, however, appears to have no influence on the in-situ floc size.

6.1.2 Particle Collision

The number of collisions between suspended particles that occur depends in the (in-situ) size distribution of the particles and on the movement of the water: Brownian motion, turbulence, or, when the water is quiet, on the differences between the fall velocities of the particles. Whether the particles that collide will stick together depends on the particle surface characteristics. Collision therefore does not necessarily result in flocculation and its efficiency may be as low as 10% or less, as was shown by Edzwald et al. (1974) in experiments with natural sediment particles in artificial seawater. In this section the three physical mechanisms resulting in suspended particle collision are discussed: turbulent shear, Brownian motion, and differential settling.

The frequency of collision because of turbulent shear is given by multiplying the number concentrations of interacting particles $n_i n_j$ with a factor K_S

$$K_S = c \cdot d_{ij}^3 \cdot (\varepsilon/\nu)^{1/2}, \tag{6.1}$$

where $d_{ij} = (d_i + d_j)$ or the sum of the diameters of particles of size d_i and d_j; ε is the turbulent dissipation rate (see Chap. 5.); ν is the kinematic viscosity, and c is a constant. Delichatsios and Probstein (1975) gave for c a value of 0.105.

In the region near the bottom, which covers 10–20% of the water depth, about 80% of the turbulent energy production takes place outside the viscous sublayer. Here the strongest shear and lift forces occur, that control the maximum size of the suspended flocs (Mehta and Partheniades 1975). Only those flocs that are able to withstand the maximum shear in this zone will settle to the bottom. The fragments of the flocs that are broken up in this zone are transported upwards again. Measurements in estuaries, however, have shown that the floc size is larger near to the bottom (at ca. 10–15% of the waterdepth) than near the surface (Eisma et al. 1991a). This goes parallel with a strong increase in suspended matter concentration which reduces the turbulence and increases the frequency of particle collisions. Higher above the bottom, in the core region of the flow, which includes 80–90% of the water depth, the turbulence intensity is less.

High near-bottom suspended matter concentrations were present not in all estuaries that were studied and also in shelf waters both situations can occur: with a higher suspended matter concentration near to bottom and without (see Sect. 7.6). For oceanic suspensions, McCave (1984b) arrived at very low collision frequencies, except for very small particles (ca. 0.5 μm) with very large ones (2–4 mm); Equation (6.1) does not take into account that in turbulent flow particles of different size (mass) have different velocities, but this effect is only important when the particle diameter differs by more than 1 cm (McCave 1984b), i.e., in a suspension containing very large particles as well as very small ones.

The frequency of collision because of Brownian motion is given by multiplying the number of concentrations of interacting particles $n_i n_j$ with a factor K_B:

$$K_B = 2\pi D_{ij} d_{ij}, \tag{6.2}$$

where $D_{ij} = D_i + D_j$ or the sum of the diffusion coefficients for particles with diameters d_i and d_j. The frequency of collision by Brownian motion and by turbulent shear (or the velocity gradient) is the same for particles of ca. 8 μm at a low shear rate (or turbulence intensity) and for particles of ca. 1.5 μm at high shear rates. Above 8 μm collision because of turbulent shear is the most important process, below ca. 1.5 μm the collision because of Brownian motion, while for particles of 1.5 to 8 μm it depends on the intensity of turbulence. Van Leussen (1988) concluded that in estuaries Brownian motion is more important than turbulence only for flocculation of particles smaller than 2 μm. In the viscous sublayer, flow is dominated by molecular viscosity, but is not every-where laminar and fluctuates irregularly, which is related to the degree of turbulence in the layer above. The near-bottom flow is characterized by burst phenomena – fluid ejections and sweeps – where a maximum of turbulence is generated and dissipated. This complicates any prediction on near-bottom particles behavior.

Flocculation by differential settling occurs because large particles settle faster than smaller ones. The frequency of collision is given by multiplying $n_i n_j$ with a factor K_G where

$$K_G = \frac{\pi d_{ij}^2}{4}(w_{sj} - w_{si})(E_{C_{ij}} + E_{D_{ij}}), \tag{6.3}$$

where $\pi d_{ij}^2/4$ is the area of particles with d_i and d_j perpendicular to the direction of fall, and w_{sj} and w_{si} are the fall-velocities of the particles with d_j and d_i. E_{Ci} indicates the ratio of the number of particles the settling particle makes contact with to the number of particles pushed away. E_{Dij} indicates the ratio of the diffusion of small particles onto a large one, to the numbers moving past the large particle because of its settling. McCave (1984b) indicates that smaller particles are most efficiently flocculated onto particles that are a little larger.

The relative importance of the three collision mechanisms is shown in Fig. 6.2 (from van Leussen 1988), following Friedlander (1977) and using density data of McCave (1984b), for particles of 0.5 μm, 5.0 μm, 50 μm, and 500 μm (d_i) colliding with a range of particles from 0.01 μm to 1 cm (d_j). Brownian motion is only of importance for particles smaller than 1–10 μm, depending on the velocity gradient ξ or on the average velocity u at mid-depth (for u = 0.5 cm.s^{-1} the mean value of ξ is 3.5 s^{-1}, for u = 1.5 cm.s^{-1} ξ is 16.5 s^{-1}). Differential settling is the dominant collision mechanism for larger particles.

The number of effective collisions also depends on the particle concentration and on the stickiness of the particle surfaces. For kaolinite particles the collision efficiency is in the order of 10% or less (Edzwald and O'Melia 1975) but when the particles are coated with natural organic polymers it can reach almost 100% (Alldredge and Silver 1988). It can be expected that at very low concentrations no flocculation occurs because the collison frequency is too low. As mentioned above, in the Mahaham river plume, flocculated suspended matter was observed in concentration as low as 0.1 to 0.3 mg.l^{-1}. At lower concentrations flocs were only present in (surface water) samples where diatoms were also present, which probably induced the formation of flocs by producing sticky material (Eisma 1990). This implies that the Coulter counter particle size distributions of most oceanic suspensions (with concentrations < 0.1 mg.l^{-1}) would give in-situ particle size. Settling of flocs from the surface water, however, where concentrations are higher and organisms enhance flocculation, will confuse this. The observations of Silver and Alldredge (1981), Honjo et al. (1984) and McCave (1985b) indicate that such flocs are regularly present down to 5000 m depth (Fig. 6.3), so that also in ocean water with low suspended matter concentrations Coulter counter results are likely to include measurement of floc fragments.

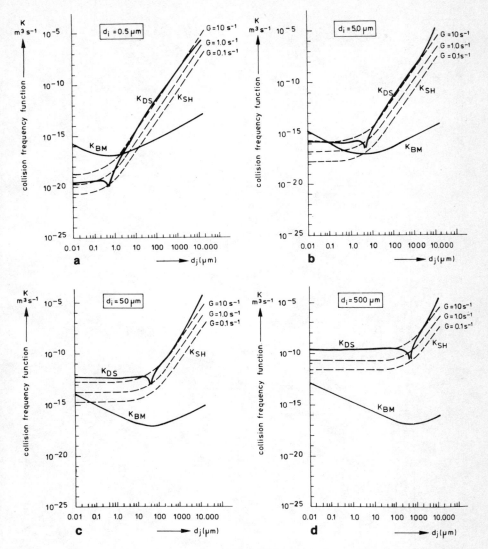

Fig. 6.2a–d. Comparison of collision mechanisms for different particle diameters. BM Brownian motion; DS differential settling; SH turbulent shear; G root mean square velocity gradient. (van Leussen 1988)

6.1.3 Floc Formation by Bubbles

Dissolved organic matter can be converted into particles through adsorption onto gas bubbles in the water and aggregation when the bubbles burst or dissolve (Baylor and Hirschfeld 1962; Sutcliffe et al. 1963; Riley 1970; Johnson

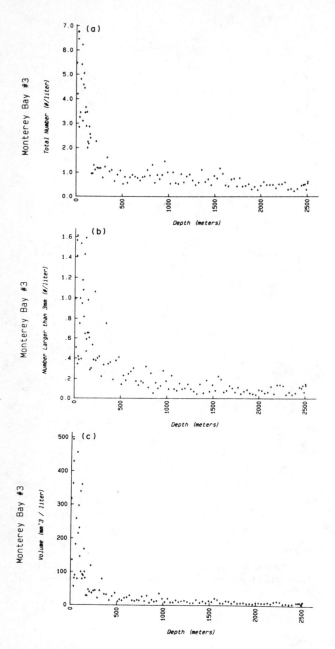

Fig. 6.3. Depth distribution of large amorphous flocs (marine snow) at a station in Monterrey Bay. Volume was calculated for particles > 0.5 mm diameter, assuming a spherical shape. (Honjo et al. 1984)

and Cooke 1980). In-situ observations of bubbles in the sea indicate that most bubbles are smaller than 200 μm (Medwin 1977; Johnson and Cooke 1979; Wu 1988). Experiments by Johnson and Cooke (1980) indicate that particles up to 16 μm in diameter are formed by such bubbles: formation of bubbles always resulted in particle formation. Bubbles also aggregate existing particles, which in experiments resulted in flocs up to 32 μm in diameter. Probably dissolved, colloidal, particulate, organic and inorganic, living, and detrital material is involved in the formation of flocs in this way. The adsorption of small particles stabilizes the bubbles (Johnson and Wangersky 1987). Bubbles are formed by gas escaping from organisms or from bottom sediments, but principally are formed by waves breaking at the surface. They can be moved downwards to a maximum depth of 20 m in the open ocean. It is, however, not yet clear to what extent the formation of flocculated suspended matter through bubbles is significant, relative to floc formation by other processes. Probably bubbles will become important in oceanic surface waters where particle concentrations and biological activity are low.

6.1.4 Coatings

Suspended matter in natural waters is coated with organic matter and some-times also with hydroxides. Coatings are formed very quickly (within an hour after immersion), as was found by Loeb and Neihof (1977), who immersed small platinum plates in natural seawater. The organic coatings retain or inhibit salt flocculation (see Sect. 6.1.1). Coatings can be formed in different ways: by direct adsorption of dissolved compounds onto the particle surface, or by flocculation or polymerization of dissolved compounds in the water followed by adsorption. Coatings can also be produced by microorganisms (bacteria, diatoms, etc.) living associated with particles. The mucus produced by those organisms provides a surface suitable for adsorption (Trent et al. 1978; Alldredge 1979), so that in practice it will often be difficult to know whether a coating was produced by organisms or not. When larger organic particles are formed in the water, they can be incorporated in the flocs, enhancing the adsorption capacity. Small flakes of only organic matter that can easily become attached to larger particles have been observed by Sieburth (1965), Jannasch (1973), and Gilmer (1974). Linley and Field (1982) describe large gelatinous flocs, probably formed by trans-formation of dissolved high molecular organic matter (Polysaccharides) into particulate matter rich in bacteria. The polysaccharides adsorb onto suspended particles, which further increase in size through bacterial growth and further adsorption of dissolved organic matter. One mechanism of floc formation by long-chain molecules or polymers is through the formation of bridges where a polymer chain can attach itself to an adsorption site at the surface of another particle (Gregory 1978). The number of adsorbed long-chain molecules depends on the available adsorption sites, in proportion to the surface area of the particles. Optimum flocculation can already occur at a very low concentration of polymer molecules (≈ 1 mg.l^{-1}), which in nature are formed by micro-organisms. Pavoni et al. (1972) demonstrated that in a kaolinite particle suspension of 300 mg.l^{-1} the turbidity rapidly decreased on adding extracted bacterial extracellular polymers (at pH 7). As seen above (Sect. 6.1.1), the concentration of polysaccharides or humic and fulvic acids in the flocs probably influences the floc strength considerably.

6.1.5 Flocculation by Organisms

Organisms are known to induce or cause flocculation of suspended matter in various ways. Bacteria produce slime films (polysaccharides and other polymers) on submerged surfaces (Riley 1963, 1970; Avnimelech et al. 1982). The films are pale yellow or brownish and amorphous, and are very similar in appearance to the organic matter usually found in flocs. In freshwater, flocculation takes place at very low ion concentrations ($\approx 5 \times 10^{-4}$ M Ca^{2+}) when natural polymers are present. It can also take place at bacterial surfaces (Eppler et al. 1975). Pearl (1973) described from Lake Tahoe close associations of bacteria, cells of fungi, pieces of detritus, and attached algae. Large flocs can be formed actively by many organisms, planktonic as well as benthic, that produce pellets out of suspended particles by clogging them together, usually within a membrane. The pellets are feces or pseudofeces. The former have passed the digestive tract; the latter consist of particles that were not taken in but were rejected and enveloped in mucus on the gills. Both types of pellets have settling velocities that are much larger than those of the constituent particles. Other organisms produce floccu-lated feces, which may grow by flocculating further with other particles. The rate of pellet formation can be very fast: the entire volume of the Dutch Wadden Sea can be pumped through the gills of only two mollusc species (*Cerastoderma edule* and *Mytilus edulis*) in 140 days; ca. 125 000 tons (dry weight) of fecal pellets is produced during that period (Verwey 1952). As the total mud deposition in the Dutch Wadden Sea is in the order of 1×10^6 t.y^{-1}, the yearly production by these molluscs alone involved ca. 30% of the total amount that is deposited. During this process, suspended matter is removed from suspension, but can be returned through resuspension after the pellet membrane is broken or decomp-osed. Pellet formation results in more transparent water. Measurements on oysters have shown that the water can be cleared to almost 100% of trans-parency within 15 h through the formation of pellets, whereas only to 70% transmission when particles are allowed to settle during the same period of time (Meade 1972).

The formation of "marine snow" occurs in two ways (Alldredge and Silver 1988):

– it can be newly (actively) produced by mucus-producing organisms in the form of pellets, fecal material, and feeding structures and by clogging of mucus (polymer)-producing bacteria and algae;

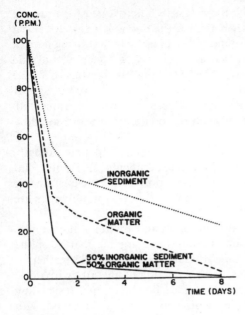

Fig. 6.4. Relation between concentration and time for settling suspensions consisting of inorganic (single mineral) particles, of organic matter, or of flocs consisting of 50% organic matter and 50% mineral particles (Kranck 1984a)

– flocculation of suspended particles can be enhanced by mucus-producing organisms that enhance the stickiness of the particle surfaces.

Bacteria are omnipresent, and are known to grow very quickly on particulate material (Tito de Morais 1983; Johnson et al. 1986). Flocculation during coccolith and diatom blooms has been observed in situ (Cadée 1985; Kranck and Milligan 1988; Riebesell 1991a, b). Large flocs of up to 5 cm in diameter are formed following the decline of the bloom, and result in the diatoms settling to deep water. Also coccolithophorids can act as nuclei for flocculation. Flocculation proceeds as a successional flocculation of selected diatom species, not by flocculation of the entire algal community. The flocculation during a diatom bloom was modeled by Jackson (1990) by considering the kinetics of both the flocculation process and the algal growth. Important parameters were found to be the fluid shear, the particle (algal) concentration, and the size and stickiness of the algae. The flocculation of mineral particles with organic matter greatly enhances their settling velocity, as experiments by Kranck (1984a) have shown (Fig. 6.4).

6.1.6 Floc Breakup

Flocs may break up, or may be eroded particle by particle through surface shear (Parker et al. 1972). Both particle breakup and surface erosion have been observed, in laminar as well as in turbulent flow, in experiments by Pandya and Spielman (1982). Tambo and Hozumi (1979) related the maximum floc size to the turbulent dissipation rate ε and the Kolmogorov microscale of turbulence (Fig. 6.5). When the floc diameter is larger than λ, the floc will break up. For a smaller floc diameter viscous effects dominate. Van Leussen (1988), using data of Nakagawa and Nezu (1975), estimated that at a velocity of $0.5\,\mathrm{cm.s^{-1}}$ at one tenth of the water depth from the bottom, floc size would be between 600 and 800 μm, which is indeed about the maximum size of flocs measured in situ at that depth in the estuaries of the Ems, Rhine, Schelde, and Gironde (Eisma et al. 1991a). This means that stable flocs are only subject to viscous or transition flow shear.

Particle breakup due to turbulent shear, is also related to floc strength, which for the clay-aluminum flocs used in the experiments of Tambo and Hozumi (1979) is not very high. A coating of organic polymers increases the floc strength (as well as the flocculation efficiency) and for that reason polymers are used in waste water treatment.

Most natural flocs contain a visible amount of organic matter, but also flocs that seem to consist only of mineral particles can have an organic coating on the constituent grains. As well as by the coatings, these flocs may be held together by van der Waals forces, electrostatic forces, or by a coating of (metal) hydroxides. When a shear stress occurs along the particle surface, floc breakup is resisted by a yield stress, which in laboratory experiments was shown to decrease with increasing floc size (Krone 1963). For this reason the dependence of the floc size

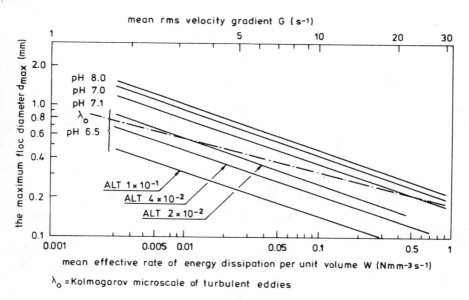

Fig. 6.5. Relation between maximum floc diameter and the mean effective energy dissipation of clay aluminum flocs. (van Leussen 1988; after Tambo and Hozumi 1979)

on the velocity gradient (or the turbulent kinetic energy dissipation) would not be very different for flocs larger or smaller than λ.

Flocs may also decrease in size by gradual breakup through surface erosion of the floc by turbulent drag, as was suggested by Argaman and Kaufman (1970) and Parker et al. (1972). The rate at which this takes place is proportional to the floc surface area and the surface shearing stress, as well as to the properties of the fluid and the floc combining inertial and viscous effects. The maximum floc size is then in the order of mm, which is not very different from the size of the Kolmogorov microscale. Experiments by Tomi and Bagster (1978) and Tambo and Hozumi (1979) are consistent with this (Fig. 6.5).

Collision of particles (flocs) can result in floc breakup, as well as in flocculation. A general formula for the rate of change of a particle size distribution with time is given by Burban et al. (1989) based on a paper by Lick and Lick (1988):

$$\frac{dn_k}{dt} = \frac{1}{2} \sum_{i+j=k} A_{ij} \beta_{ij} n_i n_j - n_k \sum_{i=1}^{\infty} A_{ik} \beta_{ik} n_i - B_k n_k +$$

$$+ \sum_{j=k+1}^{\infty} \gamma_{jk} B_j n_j - n_k \sum_{i=1}^{\infty} C_{ik} \beta_{ik} n_i + \sum_{j=k+1}^{\infty} \gamma_{jk} n_j \sum_{i=1}^{\infty} c_{ij} \beta_{ij} n_i. \qquad (6.4)$$

Here n_k is the number of particles per unit volume in size range k. The first two terms on the right hand side refer to the formation of flocs of size k by collisons between particles of size i and j, and to the loss of flocs of size k by collisons with all other particles. β_{ij} refers to the collison frequency function for collisons between particle i and j because of turbulent shear, differential settling and Brownian motion as discussed in Section 6.1.2. The third term represents the loss of flocs of size k by breakup because of shear, the fourth term represents the gain of flocs of size k because of breakup of flocs of size j > k due to shear, the fifth term represents the loss of flocs of size k due to collisons with all other particles, and the last term represents the gain of flocs of size k after collisions between all particles i and j, where j > k. The coefficient B_k is a function of the shear, the floc diameter, and the effective density of the floc; γ_{jk} is the probability that a particle of size k is formed after the breakup of a particle of size j; C_{ik} is the probability of breakup of a particle of size k after collision with a particle of size i.

If it is assumed that particle breakup does not occur, or has no effect at all, with only flocculation being of importance, the (median) floc size is proportional to the square of the particle concentration because particle collisions increase with concentration. That this probably occurs is indicated by the larger floc size measured in situ in several estuaries in the bottom water where concentrations are higher (Eisma et al. 1991a). Floc breakup, because of fluid shear, is proportional to the concentration and floc breakup because of collisions is proportional to the square of the concentration. Thus, if fluid shear is the dominant process in floc breakup, the median diameter would also increase with concentration, and if collisions are the dominant mechanisms the median diameter would remain approximately the same at increasing concentration; but if it is assumed that not two-particle collisions but three-particle collisions are the dominant process in floc breakup, the median floc diameter decreases with particle concentration, whereas we would not easily observe any change when floc breakup and reflocculation are rapid processes. Burban et al. (1989) found in laboratory experiments a marked decrease in the floc size with increasing concentration (at steady state), which indicates floc breakup by three-particle collisions as the dominant process. Since floc breakup also depends on the floc strength which is related to the presence of organic matter acting as "glue", the effect would probably have been smaller if the experiments had been done with natural particles. At present, the influence of flocculation as well as of floc breakup on in-situ floc size distributions is largely unknown, apart from the general fact that suspended matter is flocculated and therefore flocculation dominates over breakup.

As a measure of floc strength the maximum diameter in relation to the turbulence intensity (or the velocity gradient) is used following the relation:

$$\alpha_{max} = a \cdot \varepsilon^{-m}, \tag{6.5}$$

where ε is the energy dissipation rate, a is a constant depending on the shear strength of the floc and m depends both on floc strength and the type of breakup mechanism. Also a and m are very dependent on whether $d_f < \lambda$ or not (van Leussen 1988). Hannah et al. (1967) used the floc breakup by shear in a Coulter counter as a measure for floc strength. Direct measurements of floc strength only exist for cohesive sediment suspensions of high concentration ($> 5 \, g \cdot l^{-1}$). Krone (1963) found that at a given floc density several values could be found for the plastic viscosity, which indicated that the same suspension can flocculate in several ways. Van Leussen (1988) stressed the fact that the history of the particles in the flocs is of great importance for their flocculation behavior.

In flocs formed by organic matter, organic filaments are important that bridge the gaps between the constituent particles. Number and strength of these filament determine the floc strength. Thus, floc strength is related to the fraction of the particle surface covered by adsorbed polymers (Healy and La Mer 1962). In general, flocs held together by polymers are stronger than those held together by electrostatic or van der Waals forces only (Kitchener 1972). In a strong jet stream, clay-Fe^{3+} flocs broke down to 100–200 μm at most, whereas flocs held together by polymers were still present as large as 600 μm. No estimates or measurements exist for natural flocs; attention has been given almost exclusively to synthetic organic polymers used in industry.

6.1.7 Floc Structure: Floc Types

Direct observations of natural floc structure are limited. A few undisturbed flocs were collected by divers (Trent et al. 1978; Eisma et al. 1983) and an even lower number was studied from a submersible (Syvitski et al. 1983; Eisma 1986). Most observations are on floc fragments by light microscope or SEM. In all, the following types have been found:

– Loose, apparently unstructured mixtures of organic matter and mineral particles including biogenic particles such as diatom frustules and frustule fragments, and biogenic carbonate; found in rivers, coastal and shelf waters.
– More or less layered dense mixtures of organic matter and mineral particles and biogenic particles; more densely packed than the former type; found in rivers, coastal and shelf waters.
– Densely packed layered mineral material (clays?) with no visible admixture of organic matter; found in estuaries and coastal waters.
– Fluffy flocs of organic matter with mineral and biogenic particles; found in the deep ocean.
– Long (< 10 cm) chains of organic matter, with mineral and biogenic particles; coastal and surface ocean waters.
– Stringers: long threads (up to > 10 cm) usually with a large floc at the lower end and small flocs or particles attached to the thread; nearshore/estuarine and shelf waters.
– Flocs consisting almost exclusively of cocoliths or diatoms
– Larvacean houses with attached particles (ocean waters)
– Flocs consisting mainly or entirely of individual mineral grains that touch each other over a small area (edge-to-face and face-to-face contact); there is no visible organic matter; freshwater lakes; coastal water near glacier outflow.
– Fecal aggregates and pellets: flocs of regular shape (oval-shaped, cylinders), but not necessarily with a regular, internal structure; coastal, shelf, ocean waters.

Krone (1963) in laboratory experiments with natural muds found an ordered floc structure and "orders" of aggregation could be distinguished. Minerals glued together in a cluster with a uniform porosity were called zero-order aggregates; flocculation of those zero-order aggregates gave first-order aggregates, flocculation of first-order aggregates gave second-order aggregates, etc. Up to sixth-order aggregates could be found. The different orders of aggregation were related to discontinuities in the relation between floc size and velocity gradients. With increasing aggregate order the floc density decreases from 1.16–1.27 to 1.056–1.079 (seawater is 1.025). A similar order of floc formation – in four orders – was postulated by Michaels and Bolger (1962), Firth and Hunter (1976), van de Ven and Hunter (1977), and François and van Haute (1985) for suspensions of kaolinite, colloids, and hydroxide flocs. The microflocs, produced by sampling in-situ (macro)flocs, and which are more stable than the in-situ flocs they were part of (Eisma et al. 1983; Eisma 1986), can be regarded as "first-order flocs" that are flocculated into second-order flocs.

Plate-shaped minerals (clay minerals like kaolinite and mica) and other mineral grains with pronounced edges (such as the calcium carbonate particles formed by precipitation in lakes; Kelts and Hsü 1978) on flocculation show edge-to-edge and edge-to-face contacts that can form a firm flocculated structure. Starting with particles of 1 μm or smaller, units of 10–20 μm can be formed, that can flocculate further into flocs of up to 200 μm size (McDowell and O'Connor 1977). On deposition of such flocs, a cell-like sediment structure is formed, probably induced by further flocculation just before deposition. When deposition continues on top of the cell-like sediment, the water is gradually squeezed out of the hollows and the cell-like structure collapses into a firm consolidated mud layer.

6.1.8 Final Remarks on Flocculation

The general dynamics of flocculation have been presented by Friedlander (1977) and Jeffrey (1982), who gave a general equation for particle flocculation in any flow system. In the two-dimensional and time-averaged form as given by McCave (1984b) for oceanic suspensions it includes the particle distribution by volume, Brownian diffusion, turbulent diffusivity, horizontal advection, the particle settling velocity distribution, and gains and losses of particles because of flocculation and breakup. Terms can be added for near-bottom conditions and shallow water. A more recent equation relating changes in floc size to the rates of flocculation and floc breakup is the one from Lick and Lick (1988) given above in Section 6.1.6.

The general balance equation for particle flocculation in a flow system can only be solved for special situations which include an equilibrium or steady state achieved during a (relatively) long period without additional supply or loss through sedimentation, or a quasi-steady state of production of fine particles balanced by removal through sedimentation of large ones, or by assuming a situation (determined by Brownian motion, turbulent shear, and differential settling) with a constant flux of material from fine particles to large particles. Such special situations are not realistic for particles in natural waters.

A more simplified model is given by O'Melia (1985), whereby the water column is divided into compartments and 25 size fractions are considered:

$$\frac{dn_{k,I}}{dt} = \frac{1}{2} \sum_{i+j \to k} \alpha_I \lambda(i,j)_I n_{i,I} n_{j,I} - n_{k,I} \sum_{i=1}^{\infty} \alpha_i \lambda(i,k)_I +$$

$$+ \frac{W_{k,H}}{Z_I} n_{k,H} - \frac{W_{k,I}}{Z_I} + \frac{Q_0}{Z_I} n_{k,0} - \frac{Q_1}{Z_I} n_{k,I}, \tag{6.6}$$

where:

$n_{k,I}$ = the number concentration of particles of size k in lake compartment I;

α = the collision efficiency factor reflecting the stability of the particles and the surface chemistry of the system;

$\lambda(i,j)$ = a collision frequency function that depends on physical modes of interparticle contact;

$W_{k,H}$ = the settling velocity of particles of size k located directly above
        box I;
Z_I = is the depth of box I;
$W_{k,I}$ = the rate of production or destruction of particles of size k in
        box I;
Q_0 and Q_1 = the areal hydraulic loadings into and out of compartment I;
$N_{k,0}$ = the number of concentration of particles of size k in the water
        flowing into compartment I.

$\lambda(i, j)_t$ is the total (by addition) of the rate coefficients for collision through
Brownian diffusion (B), turbulent fluid shear (SH) and differential settling (DS):

$$\lambda(i, j)^B = \frac{2kT}{3\mu} \frac{(d_i + d_j)^2}{d_i d_j}$$

$$\lambda(i, j)^{SH} = \frac{1}{6}(d_i + d_j)^3 G$$

$$\lambda(i, j)^{DS} = \frac{\pi(\rho_p - \rho)}{72\mu}(d_i + d_j)^3 |d_i - d_j|,$$

where d_i and d_j are the diameters of colliding particles i and j, k is the
Boltzmann's constant, τ is the absolute temperature, μ is the water viscosity, G is
the mean velocity gradient in the water, ρ_p is the particle density, and ρ is the
water density.

This model was given for lakes, and the boxes have horizontal boundaries,
but it is also applicable to situations where the water column is not layered (as in
a river) and the boxes have vertical boundaries, or in stratified seas, where the
water column should be divided vertically into two boxes. Where horizontal
flow is important, there should also be vertical divisions. The model does not
specifically consider the effects of floc formation by organisms but $W_{k,I}$ can be
made to include such particle production (as well as breakup). All quantities
used in the model can be measured except the collision efficiency, the in-situ
particle settling velocity and the in-situ particle density. The collision efficiency,
which is related to the particle surface characteristics and reflects the particle
stability, has to be estimated from laboratory experiments; It is generally
considered to be less than 10% (Ali et al. 1984) but when polymers act as glue an
efficiency of up to 100% is possible. The in-situ density can be estimated when
the particle composition and its water content are known. Estimation of the
in-situ settling velocity of the particles will be discussed in Section 6.2.

Algal cells also can flocculate and loss of algal cells by flocculation can be of
the same order as losses by zooplankton grazing (Weilenmann et al. 1989).
Actual flocculation of algal cells was observed in nature by Smetacek (1985),
Cadée (1985), Alldredge and Gotschalk (1988a), Kranck and Milligan (1988),
and Weilenmann et al. (1989), and probably is responsible for the formation of
algal mats on the ocean floor soon after algal blooms in the surface water (Billet
et al. 1983; Lochte and Turley 1988). The relation between algal growth and

flocculation was modeled by Jackson (1990), based on exponental algal growth and a flocculation model similar to Eq. (6.6) but leaving out the effect of Brownian motion. Flocculation was found to dominate when the single cell concentration reached a critical value, which is in the order of 1000 cells . cm^{-3}. This is of the same order as concentrations during algal blooms in high productivity regions. At lower concentrations, flocculation occurs but is less rapid. Also when zooplankton consumes a substantial number of cells, flocculation is reduced (but fecal pellet production increases).

6.2 Floc Density and Settling Velocity

Flocculation results in the formation of larger particles which settle faster than the original constituent particles: the settling rate increases as a function of the particle diameter. This can be offset by a lower density of the flocs as compared to the original constituent particles when the larger flocs have a higher content of organic matter, e.g., by the inclusion of organic particles, and/or more water.

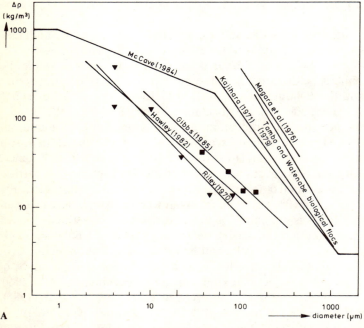

Fig. 6.6. A Differential density Δp as a function of floc diameter. (van Leussen 1988). **B** Relation between median settling velocity and concentration for muds of various estuaries. The *numbers* indicate the value of n in the formula $W_s \propto C^n$. (Dyer 1989)

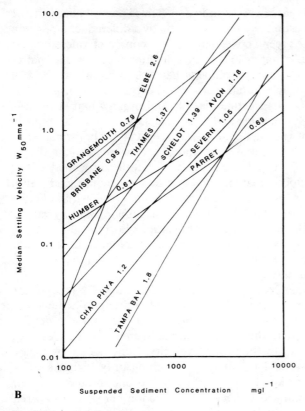

B   Suspended Sediment Concentration   mgl^{-1}

Fig. 6.6. (continued)

Krone (1963, 1978), Riley (1970), Kajihara (1971), Magara et al. (1976), Tambo and Watanabe (1979), Hawley (1982), McCave (1984b), and Gibbs (1985) have indicated that the density of flocs decreases with increasing size (Fig. 6.6). Density determinations, however, were based on (1) floc size and Stokes Law, with some assumptions on the effect of floc shape, or (2) settling in a sucrose solution: the density of the floc is equal to the density of the solution when the floc does not sink or rise. All determinations have been made on floc fragments and it remains uncertain whether such density estimates indicate the density of flocs in situ. The general relation between floc density (ρ_f) in relation to the density of the waters (ρ_w) and floc size (d_f) is given by:

$$\Delta p = \rho_f - \rho_w = A \cdot d_f^{-m}, \tag{6.7}$$

where A and m are empirical constants. The values found for m range from -0.90 to -1.65. McCave (1984b) proposed two curves, one with m $= -0.42$ for $d_f < 50 \, \mu m$ and one with m $= -1.3$ for $d_f > 50 \, \mu m$ (with $\Delta \rho = 1.0$ for $d_f < 1 \, \mu m$ and $\Delta p = 0.003$ for $d_f > 1200 \, \mu m$). These figures indicate that for a certain diameter d there is a large variation in $\Delta\rho$. This is partly caused by the different ways the density is estimated, but also partly by a different composition of the flocculated material. As seen above, natural flocs consist of different types of constituent grains, organic matter and enclosed water. The organic matter content of suspended matter can vary from less than 1% to more than 90% of the total weight. Mineral particles (quartz, feldspars, clays) have density of ca. 2.65, calcium carbonate 2.7, opal frustules ca. 1.8, and organic matter (mainly carbohydrates, humic and fulvic acids and fatty acids) ca. 1.3. The amount of enclosed water is usually not known, but an average content can be estimated by a Coulter counter, when the total weight and the organic content of the suspended matter is known. It should also be realized that the suspended matter may be mixture of already flocculated material of different origin, e.g., material eroded from different types of soils or supplied from areas with different weathering conditions or vegetation. Therefore flocs of similar size may vary in composition, provided the flocs do not easily breakup. This may also happen when there is a sudden influx of newly formed particulate material, as will occur during plankton blooms. Up to now, however, no detailed analysis has been made of the composition of individual flocs.

Particle settling velocity, as seen in Chapter 5, follows Stokes' law up to the point where Reynolds number becomes larger than 0.1, although the difference with Stokes' law remains small up to Re $\approx$ 5. For larger particles of a density considerable larger than water, the particle surface roughness besides size, shape, and density contributes to the settling velocity. Experiments by Chase (1979) indicate for lacustrine and marine flocs a significantly higher settling velocity than according to Stokes law. This was most pronounced for smaller flocs. It was attributed to nonlinear viscous drag reduction but, as van Leussen (1988) has pointed out, the physical mechanisms for this reduced settling velocity are not well understood.

From laboratory measurements it was found that the settling velocity w_s is related to the particle concentration C following

$$w_s = k \cdot C^m, \tag{6.8}$$

where k and m are constants mainly depending on the type of particle and on turbulence; for m values near to 1 were found (Krone 1962; Migniot 1968). In situ measurements, usually made with an in-situ settling tube (see Sect. 6.4), gave results as summarized in Fig. 6.6, based on data from NEDECO (1965), Thorn and Parsons (1980), Burt (1986), Puls and Kuehl (1986), and Puls et al. (1988). The values for m in these measurements ranged from 1.0 to 2.2. The relation of the settling velocity with concentration changes at $C \approx 5\,g.l^{-1}$ and becomes lower because of hindered settling (Fig. 5.34). Settling velocities of large oceanic flocs (marine snow) are in the order of $1-400\,m.day^{-1}$ (mostly $50-200\,m.day^{-1}$; Alldredge and Silver 1988) which is $0.01-4.7\,mm.s^{-1}$ ($0.6-2.4\,mm.s^{-1}$). McCave (1985b) gives ca. $0.05\,mm.s^{-1}$ for loose mucus-bound oceanic flocs. Measurements on estuarine flocs by Kineke et al. (1989) gave $0.2-5.7\,mm.s^{-1}$, those by van Leussen and Cornelisse (1992) in the Ems river estuary $1-5$ mm.s^{-1}, and those by Sternberg et al. (1986) in San Francisco Bay $0.7-5.5$ mm.s^{-1} (av. 2.8). Carson et al. (1988) give 0.5 mm.s^{-1} for small flocs in a saltmarsh and $0.5-5$ mm.s^{-1} for fecal pellets. Fecal pellets of copepods and euphausids have settling velocities of $0.2-9$ mm.s^{-1}, those of barnacles $3.5-22$ mm.s^{-1} (Komar et al. 1981). Amos and Mosher (1985) found in the Bay of Fundy in-situ floc settling velocities of $1-5$ mm.s^{-1}, where in settling tubes velocities of $0-2$ mm.s^{-1} were measured. Ross (1988, in Mehta 1989) found that the settling velocities measured in estuaries were about one order of magnitude higher than found in laboratory tests where the same mud was used. This was considered to be due to the high shear rate in estuaries, which would result in larger and stronger flocs. In general, there is a relation between the settling velocity and the particle (floc) concentration: $W_s = C^{1.6}$, which was already found by Krone (1962).

The concentration distribution with depth of the suspended matter while settling out, differs with the initial concentration (Mehta 1989). At low concentrations (≤ 1 g.l^{-1}) log-linear distributions with depth are found with the largest concentration decrease at the surface because of differential settling (Fig. 6.7). The settled mud is concentrated in a thin unconsolidated layer at the bottom. At a high initial concentration (in the experiments 5.5 g.l^{-1}), three layers develop separated by clearly marked interfaces: a surface layer with a log-linear concentration gradient with depth, a middle layer where the concentrations remains constant with depth, and a lower layer at the bottom with high concentration. The upper interface gradually moves downward; the lower interface marks the beginning of hindered settling. The middle layer with constant settling suggests the formation of flocs of approximately the same size that start to settle out once they are formed. The thickness of the lowest layer depends on the total amount of suspended matter in the water column.

The measurements and estimates mentioned above were made in quiet (nonflowing) water. In flowing water, the particle settling is determined by the flow characteristics versus the settling properties of the particles. This is complicated by possible flocculation or breakup of the particles, which will result in different settling velocities as compared with still water. Above a critical shear stress no settling takes place. At lower shear stress the rate of deposition depends on the shear stress and the initial concentration. In steady flow settling is also steady. Krone (1962) considered the probability of particle deposition from turbulent flow $p_i = 1 - (\tau_b/\tau_{ci})$, where τ_b is the (time-mean) shear stress and τ_{ci} the critical shear stress for deposition. If the properties of the settling particles in a suspension are uniform, the concentration will decrease exponentially with time (or distance). In nature, however, the flocs will show a size distribution which implies considerable differences in settling velocity. Therefore the decrease in concentration will deviate from the exponential decrease and will be much slower (Mehta 1989).

Settling flocs in turbulent flow have to pass the more turbulent near-bottom layer and if they pass this without breakup and reentrainment into the main flow, settle into the viscous sublayer where the velocity gradient is about

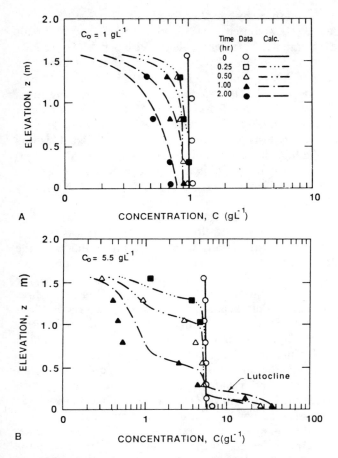

Fig. 6.7A,B. Suspension concentration profiles in a settling column at different times for Tampa Bay mud. **A** Initial concentration 1 g.l^{-1}. **B** Initial concentration 5.5 g.l^{-1}. (Mehta 1989; after Ross 1988)

constant and the shear is controlled by the viscosity of the water. At a rough bottom, the entrapment of flocs in the sublayer will be irregular, but as soon as enough flocs have been deposited to reduce the bottom roughness, the thickness of the sublayer (which is in the order of 0.5–1.0 mm) increases two to five times (Gust 1976) and entrapment of the flocs is enhanced. A very rough bottom, like gravel or coarse sand, may allow some entrapment of flocs between the grains or stones where the sublayer is thicker, but may not result in the development of a mud deposit, so that "muddy gravel" or "muddy coarse sand" is found. This indicates that there is a critical relation between flow characteristics, bottom roughness, suspended matter concentration, and average particle settling rate, which determines whether deposition occurs, whether the formation of a mud deposit is stopped in an early stage, or whether it can continue. The mechanisms of a net entrapment of suspended matter in the viscous sublayer have been discussed by Einstein and Krone 1962, McCave 1970, McCave and Swift (1976), and Hunt (1986).

There are indications that the deposition of suspended flocs is determined primarily by the turbulent shear stress at the bottom and is influenced by flocculation processes and settling rates. Conditions within the viscous sublayer are difficult to verify and a further complication is that the effects of bed roughness, the dampening influence of the suspended flocs on the turbulent velocity fluctuations near the bottom, and the degree of floc breakup near the bed are insufficiently known.

6.3 Grain Size in Relation to Particle Size: Size Sorting

6.3.1 Size Spectra in the Ocean

Particle size spectra in ocean waters, determined by a coulter counter as well as with a scanning electron microscope, can be fitted to a power law distribution:

$$N = kD^{-\beta}, \tag{6.9}$$

where N is the cumulative number of particles, D is the particle diameter, k is a coefficient related to the total concentration of suspended matter, and β is an

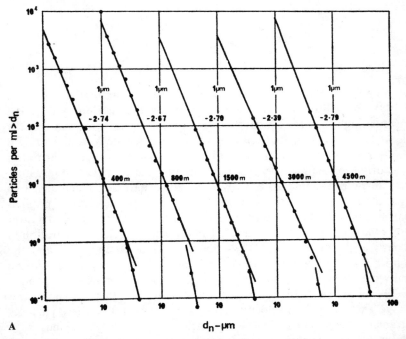

A

$d_n - \mu m$

Fig. 6.8. A, B. Size distributions of suspended particles determined with a Coulter counter. **A** ocean surface water; **B** deep ocean water. (McCave 1975)

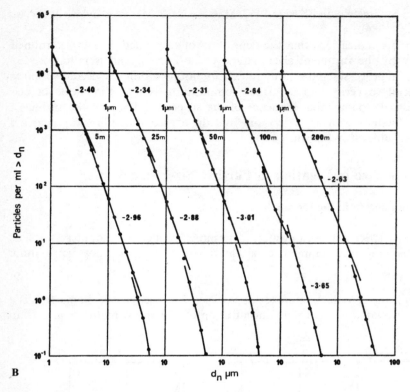

Fig 6.8. (continued)

exponent ranging from 2 to 5 (McCave 1975, 1984b; Lambert et al. 1981; O'Melia 1985; Fig. 6.8). For open ocean suspensions, the distributions are flat and follow a straight line for which $\beta = 3$, but in the surface waters and in the nepheloid layer the size distributions are peaked and deviate from the straight line. The interpretation of such curves is difficult, not only because of the mixed nature and origin of the particles, but also because the particle size measurements may include floc fragments as well as single particles. Generally, two particle populations have been distinguished in the ocean: a common population of particles with particle sizes that are smaller than about 5 μm, and a rarer population of large particles > 50 μm (Lal 1977). The particle concentration in the fine particle population is in the order of 0.02 to 0.05 mg.l^{-1} and the particles have a settling velocity of less than 1 m.day^{-1}. They are transported horizontally over large distances and dominate a standing stock of suspended particles. The larger particles (> 50 μm) are mainly flocs (marine snow) and have relatively rapid settling velocities (50–200 m.day^{-1}). For these particles vertical transport dominates. This two-particle-population concept has been very useful in geochemistry to explain the behavior of trace elements and trace compounds in the ocean. The size distributions that fit (Eq. 6.9) have been considered to describe the population of fine particles because the large particles would be too rare to show up in the majority of samples, but the relation between the settling particle flux (mainly the large flocs), the fine particle population and the behavior of trace elements is not straightforward. The vertical flux of trace elements (and of radioisotopes) is at least one order of magnitude larger than that based on the sinking rate of the fine particles, but several orders of magnitude smaller than the downward flux of large particles (Lal 1980). Scavenging of small particles by large flocs that settle down may explain this (Lal 1980; McCave 1984b), but the scavenging is probably also reversible, so that trace elements are not transported directly to the ocean floor but in steps (Chester 1990). Lavelle et al. (1991) found that scavenging of fine particles by large ones (in Puget Sound, which is 200 m deep) occurs with a time scale of 4 to 6 days over a wide range of large floc settling speeds. But also deflocculation of large flocs occurs on a time scale of 1 to 4 days at a large floc settling velocity of $100 \, m.day^{-1}$. Fine particles in the surface water have a residence time of 11 to 16 days. The separation into large and fine particle population, however, may be arbitrary: there may be a continuous size spectrum of very fine to very large particles (flocs) as recent in-situ measurements in the deep ocean indicate (Spinrad et al. 1989).

6.3.2 Size Spectra of Constituent Grains

The size of mineral grains, biogenic carbonate, and opal particles, that constitute the natural flocs together with organic matter, is smaller than the in-situ floc size and usually also smaller than the Coulter counter size, the pipette-settling-tube size, or the size found after ultrasonic treatment. Mostly fine grains of a few micron in diameter dominate (according to their number). Kranck (1973, 1975) found for suspensions in the open sea a direct relation between grain size mode and floc mode (both in μm), the grains being the constituent grains and the floc size being measured by a Coulter counter. Taking the floc size, determined in this way, to be floc fragment size, this suggests a relation between grain size and floc fragility. The relation was found to be different in estuaries (Fig. 6.9), which indicates that in estuaries, where concentrations are higher, and where probably deflocculated material is present because of higher turbulence and resuspension, flocs are of a different nature. For in-situ flocs, measured with a benthos camera, again a relation was found (Kranck pers. comm.): the maximum floc size is equivalent to the size of the largest un-flocculated grains, but the mechanism behind this is not clear.

Kranck (1980, 1984b, 1986) and Kranck and Milligan (1985) developed a two-component model of settling, i.e., a combination of unsorted floc-deposited material being deposited alternately or intermittently with single grains, or well-sorted coarser material following Stokes' law. The log-grain concentration in logarithmically equally sized classes $\{dC[d(\log D)]^{-1}$ versus $\log D\}$ gives a

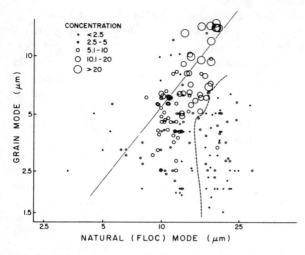

Fig. 6.9. Relation between grain size mode and floc size mode as determined with a Coulter counter. *Size of circles* indicates particle concentration (in ppm). *Solid line* regression line from Kranck (1975) for open marine suspensions. *Samples to the right of the broken line* had a salinity of less than 2%. (Kranck 1981)

characteristic curve (Fig. 6.10), which can be approximated by

$$C = QD^m, \qquad (6.10)$$

where C is the particle concentration (in ppm), D is the particle diameter, Q is the intersect at 1 μm, and m is an empirical constant whose value varies with the sediment source and is close to zero. This can be related to the modification of unsorted sediment by currents (Kranck 1986) and a value K_D can be calculated from the relation

$$C = QD^m e^{-W_s K_D}, \qquad (6.11)$$

where W_s is the particle settling rate and $K_D = (t - t_0) \cdot d^{-1}$ where d is the water depth and $t - t_0$ is the time during which the sediment is modified. K_D (in s $\cdot \mu m^{-1}$; not to be confused with the K_d values, which in geochemistry are used to describe the partitioning of an elements between the dissolved and the particulate form) describes the point of exponential falloff in grain concentration at the coarser end of the distribution. Different curves that follow Eq. (6.10) indicate different ways of settling that are characterized by changes in Q and K_D: as single grains (no change in Q, a positive change in K_D), compound settling as grains and flocs (a negative change in Q and a positive change in K_D), as flocs or by hindered settling (a negative change in Q and no change in K_D), whereby hindered settling takes place only at high concentrations. Both settling in the form of flocs and by hindered settling (probably also as flocs) result in bottom sediment with the same particle size distribution as the original material; settling as grains or in a combination of grains and flocs results in size sorting.

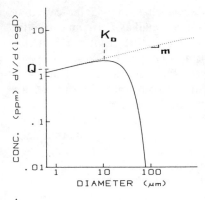

A

Fig. 6.10. A Example of a grain size distribution of suspended matter or bottom sediment relating log concentration (in ppm) to the log diameter (in μm). Q is the intersection of the graph with the Y-axis: m describes the slope of the source distribution; K_D determines the exponential fall-off in grain concentration at the coarse end. (Kranck 1986). **B** Examples of grain size distributions as shown in **A** for suspended matter at different localities with a varying percentage of flocs. Grain size determined with a Coulter counter. (Kranck pers. comm.)

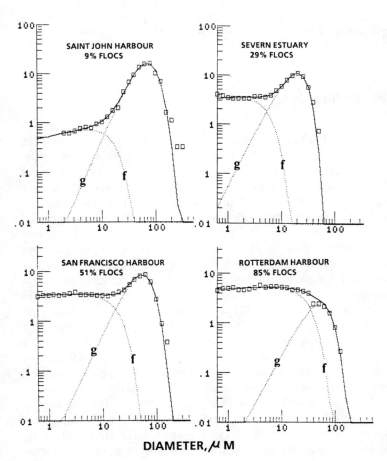

B

The results of Kranck (1984a) and Kranck and Milligan (1985) indicate that not only sand-sized particles but also silt particles as small as 5–10 μm can settle as single particles. This may also be indicated by the sorting of silt-size particles on Chinese tidal flats (Ren 1986; Wang and Eisma 1990) but on West European tidal flats often a constant ratio is found between the amounts of particles < 2 and 2–16 μm (summarized in Verger 1988). A similar constant ratio for particles < 100 μm in rivers and estuaries is shown by some of the data of Kranck and Milligan (1983, 1985). Biddle and Miles (1972) describe estuarine flocs containing silt/sand grains up to ca. 100 μm. The constancy in ratio suggests that flocs containing clay + silt + fine sand particles in the same ratios settle out.

Apparently, silt-sized material can be transported and settle out as part of flocs or as single particles. Data published by Kranck (1980, 1984b) suggest that the fractions of ca. 10–100 μm can settle out as single particles in the ocean and on the shelf, whereas in the estuaries that were studied, it was flocculated. This would correspond to the observations mentioned in Section 6.1.2 that at very low concentrations (< 0.3 mg.l^{-1}) in oceanic waters particles are probably present as single particles because the collision frequency is too low.

6.4 Measuring Particle Size

The size of suspended particles has been determined mainly by Coulter counter, light microscope and scanning electron microscope, and by settling (pipet analysis). The Coulter counter, originally designed to count blood cells (Coulter 1956; Allen 1966), is easiest to apply; it gives size distributions by particle volume. The particles are sucked through a small aperture which is in an electric field maintained by electrodes at both sides of the aperture. The interference of a particle with this field when it passes through the aperture can be measured and is related to the diameter of the particle. Particle sizes of 10–50% of the diameter of the aperture can be measured reliably. A broad range of particle sizes can be measured by using apertures of different size. The aperture is at the lower end of a glass tube through which the water is sucked. The range of sizes measured in natural suspension is from ca. 1 to 125 μm. When plankton is present, the maximum size can be (much) larger. In order to do the measurements, the particles have to remain suspended for some time, which is maintained by a stirrer. Heavy and large particles, such as sand grains, nevertheless will sink to the bottom so that the measurements are biased to finer and lighter particles. When measuring sand grains, a fluid of higher density is used to keep them in suspension. McCave and Jarvis (1973) used a mixture of glycerine and water.

Sizing and counting by microscope is laborious and time-consuming. Not many analyses have been made in this way. The size distribution is given by numbers of particles per size fraction. A modern extension is to take microphotographs of the particles and determine the size distribution from the photographs by scanning them in an image analysis system. Pipette analysis until recently was possible only for high concentration suspensions and was used mainly for analyzing artificial suspensions made out of bottom sediments or soils from which the organic matter and often also the calcium carbonate had been removed. The particles settle in quiet water and, after homogenizing the suspension, it is left standing for some time. After a certain interval of time, particles with a certain settling velocity (or size) are not present any more at a certain distance y from the top. The time interval and the depth y are related to a certain particle size d, using Stokes' law [see (5.32) on p. 118]. The water pipetted off at the water depth y after a calculated time interval therefore is taken to contain only particles with a size smaller than d. Repeating this for different time intervals and depths (while each time homogenizing the suspension again) gives a particle size distribution: the amounts are determined by filtering over a preweighed filter a fixed volume of water pipetted off at each pipetting, and weighing the filter again after drying. From these weights a particle size distribution (by weight) is calculated. Pipette analysis tends to be inaccurate because of turbulence developing in the tube and because of concentrations have to be low to prevent hindered settling, but not so low that a 20-ml sample does not give sufficient resolution. Using low-weight membrane filters (Nuclepore) and a microbalance, very small amounts of particles can be weighed accurately so that pipette analysis is possible at concentrations of $5-10$ mg.l^{-1}. For determining the concentration of particles smaller than $1-2$ μm, the method is not accurate unless special precautions are taken to prevent temperature changes and gradients developing in the tube.

As was seen above, particle size measured after sampling and using a Coulter counter, pipette analysis, or a microscope, gives the size of floc fragments. In-situ laser diffraction, or photography of suspended matter followed by image analysis of the photographs allows determination of the in-situ size from ca. 1.3 μm upwards (Bale and Morris 1987; Eisma et al. 1990), or, depending on the system, upwards of $50-100$ μm or more. This gives a distribution by number. Particle shape influences the results of all methods, particle settling velocity influences only those of the pipette analysis and those of the Coulter counter when particle settling velocities are too high to keep the particles in suspension.

For in-situ size analysis using Stokes' law and for determination of the in-situ settling velocity, various types of in-situ settling tubes have been developed that can be lowered into the water. The Owen tube (Owen 1971) is lowered open at both ends and in a horizontal position. At the required depth both ends are closed. At the surface, on deck, the tube is turned into a vertical position and settling starts. The sediment is collected at various intervals as it settles on the bottom. Because the settling of fine particles takes a long time, a tube was developed that allows samples to be withdrawn from apertures at the side of the tube (van Rijn 1985). The settling rate in the tube is strongly influenced by the dimensions of the tube and a minimum width is needed to obtain reliable results. The conditions in the tube are different from those in the water: turbulence decays in the tube which may result in additional flocculation, and there will also be differential settling. Vibrations, shock waves, and turbulence induced by handling the tube may also influence the results by causing floc break-up and secondary circulations in the tube. Van Leussen and Cornelisse (1992) used a videocamera to observe the behavior of flocs in a tube and to estimate their settling velocity.

Note added in proof: Particle size analysis is extensively treated in a recent book edited by Syvitsky (1991).

7 Transport Systems and Fluxes of Suspended Matter

In the previous chapters, the sources, composition, and concentration distribution of suspended matter were discussed, as well as the mechanisms of particle transport and flocculation. In this chapter, the dispersal or concentration processes in the different aquatic environments are described. By necessity this chapter will have to be more qualitative and descriptive than quantitative. Lack of sufficient knowledge of in-situ particle size and settling velocity, of the interaction between waves and currents and its effect on particle behavior, of the bottom boundary layer and the behavior of freshly deposited mud, and of the difficulties of a three-dimensional approach, as discussed in Chapters 5 and 6, limit any quantitative treatment. Therefore the following summary of our knowledge on suspended matter in the various environments is primarily a description of the processes that are involved, together with an evaluation – as far as possible – of their relative importance.

7.1 In Rivers

The multitude of different types of rivers and streams is classified according to discharge, length, basin area, and sediment load (bed load, suspended load, and wash load), and by various interrelationships such as between discharge and solid load (as opposed to dissolved load), solid load concentration, seasonal variations, and basin relief. All these factors influence the flux of suspended matter through a river. To this should be added the type of rock or soil that is eroded and the changes brought about by man in a basin as well as in the river itself. They affect the supply of sediment to the river as well as the flux through the river: climate, rock type, basin relief and man's activities largely determine the present river sediment flux. This is (partly) shown in Table 7.1 (after Meybeck 1977) which gives the relation of the particle transport with surface runoff and relief. Although a relationship clearly exists, it should, however, be realized that the Colorado river and the Nile are strongly influenced by dams, while the Yellow river and many of its tributaries receive very much sediment from a loess area of medium to low relief.

Most of the eroded material is supplied to a river through the small streams near to the watersheds, which together supply the main river channel. Supply of

Table 7.1. Relation of particle transport (Ts = bed load + suspended matter in $t \cdot km^{-2} \cdot y^{-1}$) with surface runoff (q in $l \cdot s^{-1} \cdot km^{-2}$) and relief (qualitative). (After Meybeck 1977)

	q > 15 High runoff	5 < q < 15 Medium runoff	q < 5 Low runoff
High relief	Ts > 100 (Magdalena, alpine rivers)	Ts > 150 (Ganges, Blue Nile)	Ts > 250 (Colorado)
Medium relief	10 < Ts < 100 (Amazon)	20 < Ts < 150 (Yukon, Danube, Parana)	50 < Ts < 250 (Nile)
Low relief	Ts < 10 (Quebec)	Ts < 20 (Yenissei, Zaire)	Ts < 50 (Dniepr, Murray)

suspended matter through channel erosion is important on a seasonal time scale, but over longer periods is usually negligible, except where bank erosion is strong, landslides occur and waste is discharged in particulate form. The importance of the small tributaries is shown in Table 7.2 (after Leopold 1962), which gives the number and length of river channels of various sizes in the USA. The small tributaries of only a few km in length and with a drainage area of only a few km^2 not only are the most numerous, but also cover the largest part of a river basin. The erosion in the vicinity of the small tributaries therefore to a large extent determines the amounts and the composition of the sediment that is transported. An extreme example is the Amazon, where more than 80%, probably more than 95%, of the suspended load comes from the Andes mountains where most of the small tributaries are located, the relief is steep, and physical weathering is strong (Meade in press). The amount of sediment delivered to a river can in most river basins be estimated with reasonable

Table 7.2. Number and length of river channels of various sizes in the United States. (After Leopold 1960)

Stream[a] order	Number	Average length (miles)	Total length (miles)	Mean drainage area (miles²)
1	1570000	1	1570000	1
2	350000	2.3	810000	4.7
3	80000	5.3	420000	23
4	18000	12	220000	109
5	4200	28	116000	518
6	950	64	61000	2460
7	200	147	30000	11700
8	41	338	14000	55600
9	8	777	6200	264000
10	1	1800	1800	1250000

[a] Stream order 1 is a channel without tributaries; order 2 is a channel with only order 1 tributaries between the upstream junction of two order 1 channels and the downstream junction with another order 2 channel, etc.

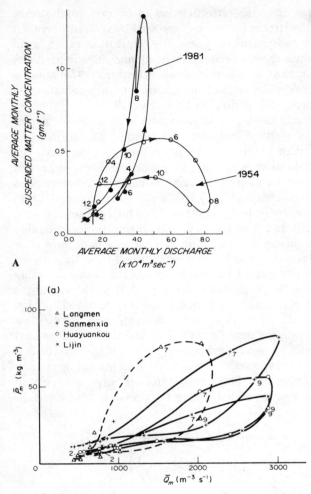

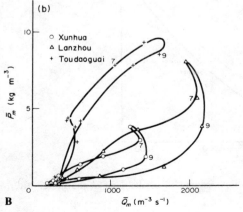

Fig. 7.1. A Relation between average monthly discharge and average monthly suspended matter concentration in the Chang Jiang at Datong near to the river mouth in 1954 and 1981. Between 1954 and 1981 deforestation in the upper river basin in Sichuan increased to 60% of the total area of this province and in 1981 the rainfall intensity was very high. (Shi et al. 1985). **B** Relation between mean monthly water discharge ($\bar{Q}_m$) and mean monthly suspended sediment concentration ($\bar{P}_m$) at gauging stations in the Yellow River below the loess plateau (**a**) and above (**b**). (Ren and Shi 1986)

accuracy from the drainage area, the ratio of relief to average total stream length, and the density of the stream pattern as expressed by the ratio between the number of successively higher stream orders (the bifurcation ratio; Roehl 1962). On the watersheds, sheet erosion predominates; channel erosion produces at most 34% of the total amount of sediment. Rendon-Herrero (1974) found a relation between the production of washload and the amount of runoff in excess of base flow during or shortly after periods of heavy rainfall.

Of great importance is the timing of erosion events as a result of precipitation and ice- or snow-melting, irregular and episodic rainstorms, and exceptionally heavy rains such as occur only a few times per century. The seasonal and regional variations in water and suspended sediment supply in many rivers result in a hysteresis loop in the relation between the suspended sediment concentration and discharge (Figs. 7.1; 7.2). A similar hysteresis effect can also occur during or after individual rains (Fig. 7.3). During short rains the effect is hardly visible but it can become very clear during prolonged rainy periods. Hysteresis does not always occur, as the Tanana river data show (Fig. 7.4). The hysteresis indicates a decrease in the availability of erodible material ("exhaustion"). During the dry period before the rain, subaereal weathering produced particulate material that was hardly transported. It is taken up at the beginning of the rain that produces runoff, and is quickly removed in high concentrations, after which the suspended matter concentrations in the flow decrease. In a river with strong seasonal flow, the material that was deposited during falling water is picked up again during rising water, (Fig. 7.5) which is enhanced when the deposit was exposed to subaereal weathering during the low level stage of the river. Ephemeral flows can transport large amounts of particulate material that had accumulated during the previous dry period. In

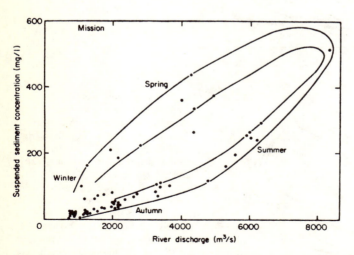

Fig. 7.2. Relation between average monthly discharge and suspended load in the Fraser River, 1966–1971. (Milliman 1980)

semi-arid areas where there is no dense vegetation to protect the soils (and where ephemeral flows are most likely to occur), this can result in hyperconcentrations of suspended matter, in the order of 30 to 40% by weight (Beverage and Culbertson 1964). In less ephemeral streams during a rare storm such as occurs only once every 50 years, up to 10 times the normal annual load can be removed (Finlayson 1978). Deforestation or any other large-scale removal of vegetation (e.g., by plowing grassland) can have a similar effect: rapid removal of loose soil resulting in high suspended matter concentrations followed by a decrease and a

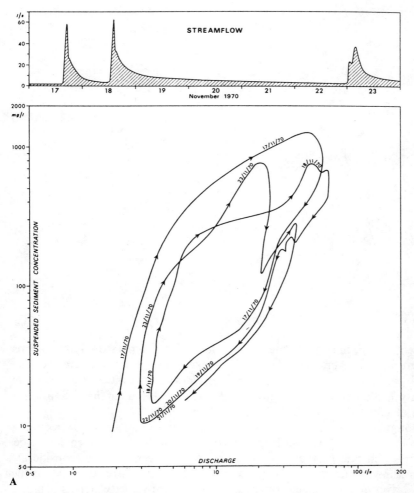

Fig. 7.3. A The relation between suspended sediment concentration and discharge in a small catchment area in Exeter, U.K., during a sequence of three storm runoff events. (Walling 1974). **B** Water-sediment discharge relations: effect of hysteresis for the San Juan River, effect of a progressive lag for the Big Horn River. (Guy 1964)

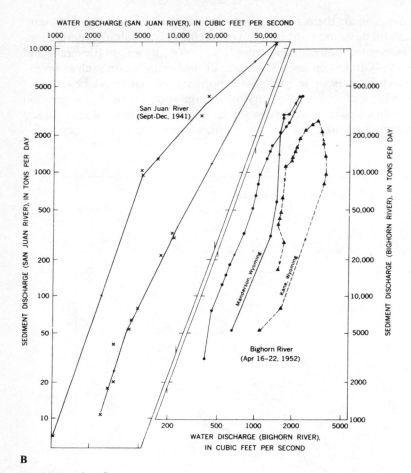

B

Fig. 7.3. (continued)

permanent change in river regime because much less water is stored than before. The decrease, however, does not need to take place in areas where erodible material continues to be produced in considerable quantities (e.g., in loose sediment). In that case there is only accelerated erosion after the vegetation is reduced or gone (Douglas 1967; Milliman et al. 1987).

Although the contribution of suspended matter from channel erosion is small and rivers tend to establish some degree of equilibrium between discharge, river slope, and bed particle size, processes in the river itself can consideraly influence the suspended matter concentration. In a downstream direction the slope of a river decreases, while (in most rivers) current velocity increases in that direction or remains constant. Also water depth and the width of the river increase downstream: the water depth increases faster than the river slope decreases, and the width increases faster than the depth (Leopold 1960). High

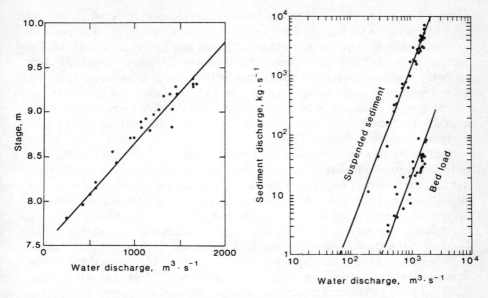

Fig. 7.4. Relation between water discharge and stage and sediment discharge for Tanana River, Alaska. (Nordin 1985)

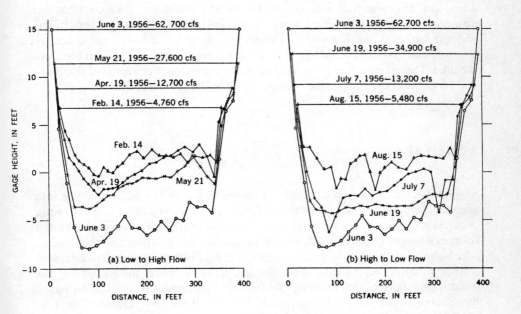

Fig. 7.5. Scour followed by infill during the passage of flood in the snowmelt season, Colorado River, 1956. (Leopold 1960)

and low flow are partly compensated by erosion or deposition (scour and fill). In spite of this apparent regularity a large part of the eroded material is stored in the river basins by deposition in stream channels and valleys (as already seen in Sect. 2.1). In the USA more than 90% is stored (Holeman 1980), which is probably related to increased sediment supply because of increased land use. Such deposits are partly being eroded again and have become a local sediment source. Storage is enhanced by the construction of dams and reservoirs but this is partly compensated by erosion of the streambed downstream of the dam (Faktorovitch 1967; Trimble 1977).

Secondary flow in rivers results in the formation of flow cells with a flow component perpendicular to the main direction of flow. Meanders are formed, but also the flow may meander within a more or less straight channel with bars alternating on both sides of the channel. This is not known to influence suspended matter transport very much, but this becomes different when a floodplain is flooded and the river water spreads out laterally (Barishnikov 1967; Hughes 1980). The flooding results in a reduction in flow velocity and in sediment load as the water depth becomes less and the effect of bed friction on the flow increases, partly because of the decrease in water depth but also because the flood plain usually has a rather coarse roughness and a complex relief. The effect on suspended matter transport can be variable: when the plain is flooded, material may be picked up but when flow is reduced, and during falling water, suspended matter is deposited, which can proceed rapidly (van Heerden et al. 1983: Atchafalaya river flood plain). In many plains the flooding is reduced by dikes, and sediment that was lost on the flood plain before dike construction now remains between the dikes, which results in building up the river bed. In this way the bed may reach a level far above the flood plain level, with extensive flooding and disastrous consequences when the river during high flood breaches the dike. This process has been rapid with, as a consequence, repeated floodings and large-scale changes in the course of rivers like the Yellow River.

7.2 In Lakes

Suspended matter in lakes has not been studied very much, but taking into account research on the formation of bottom deposits allows the following conclusions (Sly 1978). Coarse material remains mostly nearshore near to the source of supply (rivers, runoff, landslides, erosion of the shore and the shallow lake bottom). Transport to deeper levels in the lake is through density underflow (turbidity currents), slidings, and slumps. This is much influenced by the underwater topography. Dispersal by turbidity currents is over large distances, by slides or slumps over relatively short distances (Frey 1963; Dussart 1966). As on the continental slope and shelf, turbidity currents excavate canyons in the lake floor which often meander, with the meanders increasing in size with distance from the source. Because of the Coriolis force, they tend to move to the right (on the northern hemisphere). Turbidity currents have been observed in many lakes and reservoirs (lake Geneva, Lake Constance, Dussart 1966; Walensee, Lambert et al. 1976; Lake Mead, Lake Mendota, Frey 1963; Lake Brienz, Sturm 1975). Turbidity currents can also be formed when surface waves resuspend large amounts of bottom sediment, producing turbid water of high density (Frey 1963).

The particle flux through the water in suspension can be quite complicated because of the interaction of wave-forced currents and waves, upwelling, and turnover when the surface water cools off. In the bottom sediments, a general relationship exists between particle size and water depth. As is evident from the discussion on the origin and concentration of suspended matter in lakes (Sects. 2.2 and 3.2), there is a close relation between the particulate matter and the geochemical processes in lakes.

The easiest and most used way to determine suspended matter concentrations in lakes is by measuring the (relative) transparency, either horizontally – with a beam-transmissometer (Petterson 1934) or a nephelometer – or vertically by measuring the penetration of sunlight (for which the Secchi disc was already used around 1870 by Forel in Lake Geneva; see Sect. 3.4.4). The transparency measured by beam transmission often shows a fine-layered structure that is related to the density stratification. The surface water is mixed by the wind, which results in a surface mixed layer (epilimnion) with uniform transparency (which in very shallow lakes can reach down to the bottom). Below the epilimnion, in the thermocline and in the hypolimnion below the thermocline, the transparency profiles show layers of several cm to several m thickness that are internally homogeneous. Near to the bottom there is usually a marked decrease in transparency (Whitney 1937). The transparency is related to particle size and concentration of the suspended matter, including living as well as nonliving material, and to the colour of the water.

The dispersal of suspended matter, supplied from the shore, goes primarily through the surface water, from where it gradually settles out. Often the turbid inflow is markedly separated from the clear lake water by a sharp front (bataillère in French, Brech in German). The settling of very fine particles is influenced by the thermocline as the settling velocity is reduced where the viscosity of the water increases with depth. This was experimentally shown by Kindle (1927, in Welch 1935); it also results in size sorting. Very fine particles and nannoplankton have been regarded as nonsettling suspended matter (Welch 1935) and near to colloids in size. As there is almost always some water movement, either because of atmospheric circulation or because of standing waves (seiches) which easily develop in lakes, some fine-grained particles normally will remain in suspension.

Changes in the turbidity of a lake can have several causes simultaneously, such as overturning of the water resulting in turbid deeper water reaching the surface, high plankton growth, inflow of suspended matter from a river and from runoff, and supply or increased resuspension of bottom sediment by the wind.

Suspended particle loss with depth may occur because of mineralization of organic matter and dissolution of calcite and opal, but this depends very much on the geochemical conditions in the bottom water. In aerobic water organisms will consume at least part of the organic matter that settles down from the surface, and carbonate can dissolve because of a shift in its solubility at lower temperatures. Also opal particles may dissolve where the dissolved silica has not yet reached saturation. Some deep freshwater lakes have saline bottom waters. This occurs in a number of former fjords (listed in Pickrill et al. 1981) that were cut off from the sea by isostatic uplift after the melting of the Pleistocene ice cover, or that were cut off by a sedimentary barrier with the salt water remaining inside. In Lake McKerrow (New Zealand), which is separated from the sea by a spit, the saline bottom water is periodically replenished during high tides in combination with low lake levels (Pickrill et al. 1981). The distribution of temperature, salinity and suspended matter concentrations is given in Fig. 7.6. River inflow occurs at Upper Hollyford, outflow and episodic tidal inflow at the mouth but settling seems to be slowed down by the pycnocline. Most of the dispersal is through the epilimnion, where the residence time of the water is about 127 days, whereas in the hypolimnion it is 5.8 years. Suspended matter concentrations near to the bottom are greater and extend high up along the slope to near Lower Hollyford, which suggests supply and/or resuspension by the salt water inflow. This implies that the sampling was done not long after such an influx, but there may have been resuspension in shallow water and the formation of one or more episodic turbidity currents. The concentration change at the thermocline is accompanied by a change in particle composition: terrigenous mineral particles dominate in the epilimnion, organic particles in the hypolimnion (Fig. 7.7). In both the epilimnion and the hypolimnion the suspended matter seems flocculated.

In Loch Earn, as in Lake McKerrow, the suspended sediment dispersal appears complex when studied carefully (Duck 1987). The supply is from six streams and from primary production. The latter results in organic matter contents in the suspended matter of more than 90% during the summer months. The suspended matter concentrations vary with the thermal structure of the lake. Low, nearly uniform concentrations ($\sim$ 1–2 mg.l^{-1}) occur throughout the lake during the winter. During spring to autumn the water is thermally stratified: in the thermocline concentrations are five to ten times higher (up to 25 mg.l^{-1}) than in the epilimnion and the hypolimnion, where the concentrations remain mostly in the same range as during the winter. This maximum in the thermocline is believed to be the result of interflows whose depth is regulated by the relative densities of the inflowing water and the lake water. Such interflows have also been observed in Lake Brienz (Nydegger 1967) and in Lake Geneva and Lake Constance (Wagner and Wagner 1978; Dominik et al. 1983) at 10 to 30 m depth, and were formed by river inflow continuing in the lake at an equilibrium depth determined by the temperature and the concentrations of dissolved and particulate matter (Fig. 7.8). Also slowing down of particles settling from the surface probably contributes to the maximum. Observations of

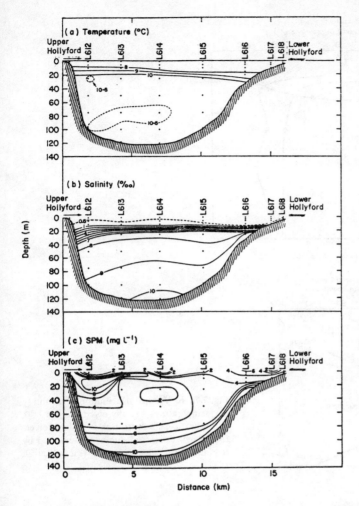

Fig. 7.6a–c. Longitudinal profiles of temperature, salinity, and suspended matter concentration through Lake McKerrow, New Zealand. (Pickrill et al. 1981)

other interflows or near-surface flows indicate a deflection because of the Coriolis force. In Bow lake (Canada) interflow velocities at ca. 9 m depth, measured with drogues, were found to be ca. $2 \, cm \cdot s^{-1}$ (0.6–$4.4 \, cm \cdot s^{-1}$) with the direction always down-lake, regardless of the wind direction. Fine particles will be carried by this current through the entire lake in a few days (Smith and Syvitski 1982). In Loch Earn other inflow occurs along the surface as well as below the thermocline, but to a much smaller extent than through the thermocline. There are no indications for turbidity currents but slides of bottom sediment occur along the side slopes of the lake, which may explain locally high concentrations of suspended matter near the bottom. In the surface water

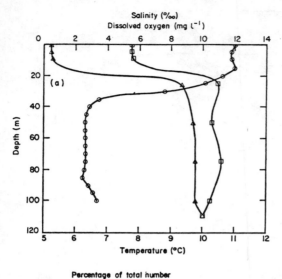

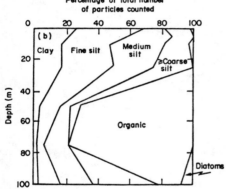

Fig. 7.7a, b. Vertical distribution of **a** salinity (Δ), temperature (□), dissolved oxygen (○), and **b** particle concentration in Lake McKerrow. Particle size and number determined by scanning electron microscope. (Pickrill et al. 1981)

parallel foam streaks occur at 15 to 25 m distance from each other when wind speeds exceed 6 to 7 m.s.$^{-1}$. Within the foam streaks, suspended matter concentrations are higher than between the streaks. This is explained by the development of Langmuir circulations (Langmuir 1938): foam is concentrated in the convergence zones where the water motion is downward and suspended matter concentrations are higher, whereas lower concentrations occur in between the streaks, where there is upwelling.

The suspended matter in Loch Earn is flocculated and seldom occurs as single particles. The flocs consist of inorganic (mineral) and biogenic material and are much less present in the streams that flow into the lake than in the lake itself. In the streams they can be totally absent. There are generally two floc types: one that is spherical or barrel-shaped and consists of densely face-to-face packed plates, and a (more frequent) loosely bound type of less regular shape.

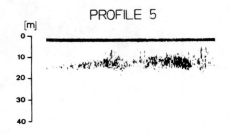

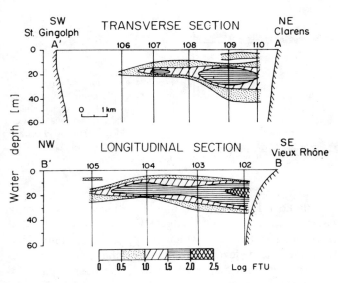

Fig. 7.8. Turbidity (in log FTU units) along two transects across the Rhône River plume in Lake Geneva and a 100 kHz echogram following approximately the SW–NE transect. (Dominik et al. 1983)

The first type is rather small and contains no diatom debris, the other type is generally larger and contains (pennate) diatoms. Floc sizes range from a few to several hundred micron.

In very shallow lakes, such as Lake Arari in Brazil, which is 2.5 to 4 m deep (see Sect. 3.1.2) wind effects dominate. This is also the case in Lake Loosdrecht, a very shallow but rather large lake (28 km^2) in the Netherlands. It has a mean depth of 1.9 m and resuspension by wind occurs every 2 days, even in summer, affecting more than 50% of the lake area (Gons et al. in press). This, coupled to strong wind-forcing, results in rapid dispersal and mixing of the particulate matter, which is important in spring after the lake has been covered by ice. Concentrations of suspended matter in the lake may reach 200 mg.l^{-1} during storms. The particulate matter is highly organic with peat fragments in the coarser size fractions ($> 33\ \mu$m) and algal remains dominating the fine size fractions ($< 15\ \mu$m).

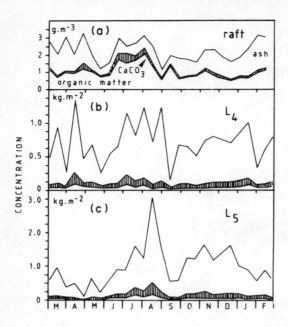

Fig. 7.9a–c. Seasonal changes in the total dry weight composition of suspended matter (a) and particulate matter at the sediment-water interface at two locations (b and c) in Lake Vechten (water depth 11 m). (Verdouw et al. 1987)

In a somewhat less shallow Dutch lake (Lake Vechten, ca. 11 m deep), particle settling dominates from May until mid-August. During this period there is a strong temperature gradient in the water, which disappears at the end of the summer. When thermal homogeneity develops because of enhanced turbulence, particulate matter is resuspended and dispersed towards the deeper parts. A turbid layer of ca. 1 m thickness is formed near to the bottom where concentrations increase two to three times (Gons and Kromkamp 1984; Verdouw et al. 1987). The particulate material is 5 to 20% organic matter, 10 to 15% calcite, and the remainder mineral particles (Fig. 7.9). Calcite production takes place in June and July and is related to primary production. Production is highest at ca. 3 m depth where locally the calcite content of the particulate matter may reach 41% (Gons 1982).

7.3 Transport and Fluxes in Estuaries

As was seen in Section 2.3, estuaries receive particulate matter from rivers, from the coastal sea and from a number of other, mostly much smaller sources (Fig. 2.5). Removal of particulate matter from estuaries usually is in several directions: downward (deposition) or sideward (towards the shore, a floodplain, or tidal flats) and then again downward, outward into the coastal sea, or up-river by tidal dispersal. There can also be some loss of particles to the atmosphere but quantitatively this is negligible as compared to the other forms of removal. In many estuaries, large amounts of material are removed by dredging. In the Rhine mouth, where the annual river supply is ca. 3.5 million tons sediment (fine sand and mud) and annual net outflow into the coastal sea is 6 million tons, ca. 10.5 million tons are yearly removed in this way.

The landward limit of an estuary is taken either at the point where the river water meets more saline water, or at the landward limit of tidal influence. The seaward limit most often lies at the coastline, but off larger rivers the estuarine area may include a large part of the adjacent sea. In a steady state, the total supply of sediment to the estuary, given in mass/time units, should be balanced by the total that is removed. The time during which the particles remain in the estuary is the residence time, and can be calculated from the total quantity of particulate matter being present in suspension or moving along the bottom, divided by either the supply rate or the removal rate. It can also be defined as the average time needed for particulate matter to move from the point of supply to the point of removal. Particulate matter in estuaries varies in size, density (composition), shape, and surface characteristics, so that there are large variations in the way particulate material moves through estuaries. Different types of material (sand, clay, organic matter) will have different residence times as well as a different mass balance. Within the estuary the transport and deposition of especially the finer-sized particles can be very much influenced by aggregation or deaggregation, which changes the transport characteristics of the particles so that particulate matter may be removed in a different state than in which it was supplied. Moreover, the supply from one source will be removed in more than one way (e.g., partly by deposition and partly by outflow) and material removed in one way will have been supplied from various sources. Usually, however, material from different sources cannot be clearly distinguished on the basis of size and composition so that only a gross average residence time or mass balance is calculated involving material of different origin and characteristics. Even the obvious distinction between the mass balance of particulate organic and inorganic matter is not always made.

7.3.1 Constraints on Particle Transport Through Estuaries

Usually a division is made between fully mixed estuaries, partly mixed estuaries, and salt-wedge estuaries (Bowden 1980; Postma 1980; Fig. 3.9). A classification diagram based on the ratio of the difference in salinity (% S) between a) the surface water and the bottom water to the salinity at mean depth (S_0), and b) the ratio of the mean surface velocity to the mean flow due to river discharge, is given in Fig. 7.10 (from Hansen and Rattray 1966). Because of the variations in river flow and because the inner parts of an estuary are usually less well mixed than the outer parts, estuaria appear in this diagram as a line. The classifications shown in Figs. 3.9 and 7.10, however, are not quite sufficient because the sediment distribution mechanisms, and the resulting fluxes, also depend on the general geomorphology of the estuary, the flow and sediment load characteristics of the river, the tidal characteristics, the wind pattern (resulting in waves,

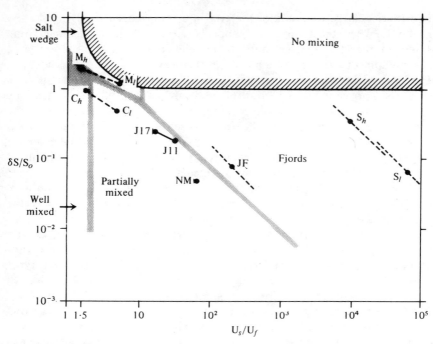

Fig. 7.10. Estuarine classification diagram (Hansen and Rattray 1966). δS difference in salinity between surface and bottom; S_o depth of mean salinity; U_s mean surface salinity; U_f mean flow due to river discharge. M Mississippi River; C Columbia River estuary; J James River estuary; NM Narrows of the Mersey River estuary; JF Juan de Fuca Strait; S Silver Bay. Subscripts h and l refer to high and low river discharge. *Numbers* at the James River indicate miles from the mouth. (Dyer 1986)

currents and atmospheric particle input), the effects of the Holocene sea level rise, the tectonics of the area (uplift, subsidence), and the effects of estuarine organisms on the transport, deposition, and erosion of sediment.

River flow, tides and wind effects (in some areas also evaporation), the bottom topography and the general configuration of the estuary determine the estuarine circulation. River flow and wind effects (and evaporation) are related to the climatic conditions as are also local conditions, such as river sediment supply, the presence or absence of a large flood plain, of numerous distributory channels or only one narrow channel, and the presence of large amounts of sand or mud. Where a lagoon, large tidal flats or a flood plain are present, net sediment outflow to the coastal sea may be zero but it may be relatively large where the estuary is confined to a narrow channel. Also in a narrow channel, however, the presence of bends and shoreline irregularities causes lateral transport and lateral accumulation of sediment. River sediment input, the tides, and wave action are the main factors determining the geomorphology of an estuary (Fairbridge 1980). Deltas are predominantly formed by rivers with a high sediment load along coasts with a low tidal range and limited wave action (Mississipi, Orinoco, Nile, Rhône, Danube, Fraser, Yukon). On such coasts the estuaries are usually of the salt-wedge type and only the rivers with low sediment loads have open estuaries (Texas bays). Where the tidal range is high, the V-shaped drowned-valley type of estuary usually is preserved because more sediment is swept out of the estuary towards the coastal sea. Such estuaries are often partly filled up with sediment (e.g., the Loire, Long Island Sound; Barbaroux et al. 1974; Bokuniewics et al. 1976; Fig. 7.11) and usually are of the partly or fully mixed type. All estuaries have been formed during the Holocene sea level rise and are up to 10000 years old, having started their existence at a much lower sea level. Where sediment supply has more than kept pace with sea level rise, the estuary was completely filled up and further extension in the form of a delta became possible after the sea level rise leveled off. Fjords typically indicate conditions whereby the low sediment supply has not been able to fill up the deep valleys. Most estuaries are in an intermediate stage. The Thames estuary was already an estuary ca. 8900 years ago and has remained an estuary up to the present (Greensmith and Tucker 1973). Subsidence will retard or prevent filling up of an estuary, whereas uplift will enhance this because of reduced water depths and stronger erosion of the adjacent land. In tectonically active regions, local or regional movements can completely overshadow the effects of tides and waves. In Cook Inlet, with very active tectonics, where earthquakes can alter large areas of the coastline in a few seconds, deltas are formed which vary in shape and extension with the rate of river sediment supply and wave action (Hayes and Michel 1982). This happens in spite of a tidal range of 4–6 m and currents up to $7 \text{ m} \cdot \text{s}^{-1}$.

Another constraint on the particle flux through estuaries is the river regime. The relation between river discharge and sediment load is complex (see Sect. 2.1) and predominantly determined by the conditions in the drainage basin. Also for a single river the relation is not straightforward and seasonal effects are not simply related to discharge because during the different seasons different parts of

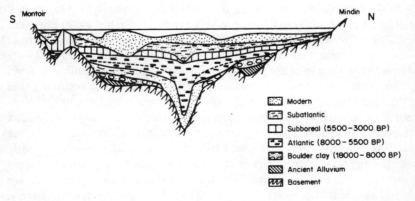

Fig. 7.11. Deposition in the Loire estuary during the Quaternary. (After Barbaroux et al. 1974)

the drainage basin may contribute to the sediment load. Also many rivers show a hysteresis curve: at the same discharge the sediment load is higher during rising water than during falling water, which is related to the availability of sediment for transport. Some rivers do not show this effect and some rivers may even show a reverse effect (Chang Jiang; Sect. 3.1). Some rivers have several flood peaks within a few months; during the later floods the suspended load is much lower than during the first one (for the Rhine: Eisma et al. 1982). Moreover, many rivers are regulated in some way by dams, weirs, or canalization upstream of the estuary, or receive increased quantities of sediment because of agricultural practices, deforestation, or urban developments resulting in soil erosion, or through mining activities.

The impact of estuarine organisms on sediment transport, deposition, and erosion in estuaries is not well understood. Organisms can increase particle settling and impede erosion by clogging particles together, but can also increase the erodability of bottom deposits by increasing the water content or the bottom roughness (Nowell et al. 1981; Jumars et al. 1981). The influence of variable weather conditions on the estuarine circulation is equally not well known. Winds induce surface waves as well as variations in water level and wind-generated currents. In the Potomac estuary, fluctuations in the circulation pattern of 2.5 days are attributable to the wind, which is able to reverse the expected normal flow pattern with seaward surface flow and landward bottom flow (Elliot 1978). Reversals occur ca. 20% of the time, the "normal" circulation 43%, total inflow along the surface and the bottom about 20%, and in the remaining time total outflow or more complex circulation patterns are present. In shallow microtidal estuaries, such as are present along the Texas Gulf coast, where the tides are further reduced by the presence of bay mouth barriers, the wind is the dominating process that regulates sediment transport through controlling dispersal patterns and resuspension of bottom sediment (Shideler 1984). Seawater piled up along the coast by onshore wind results in a strong influx of seawater into the estuary and the freshwater is temporarily stored in the inner estuary. Interaction of estuarine flow and wind effects in the coastal sea resulting in storage followed by increased outflow has been described for several estuaries (Louisiana and Texas Lagoons: Kjerve 1975; Smith 1977; Patuxent estuary: Elliot 1976; Narragansett Bay: Weisberg 1976; Weisberg and Sturges 1976; a ria in NW Spain: Heath 1973; fjords: Pickard and Rodgers 1959; Farmer and Osborn 1976). Estuaries are protected against large waves coming in from the sea. Therefore resuspension by waves of bottom material is restricted and sediment tends to accumulate, but in shallow areas wave action can be sufficient to prevent the building up of bottom deposits. In this way, a large tidal influx can be maintained because a large volume of water can continue to enter during the flood tide (Postma 1980). In the Rhine estuary, mud is predominantly deposited during the winter, which is attributed to the effect of winter storms stirring up large quantities of mud in the nearshore coastal sea that are subsequently deposited in quiet areas in the estuary (chiefly harbor basins; Salomons and Eysink 1981; Eisma et al. 1982). Exceptionally severe storms and

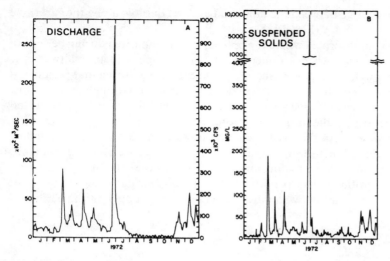

Fig. 7.12A, B. Changes in discharge and suspended sediment concentration in the Susquehanna River as a result of flooding caused by the hurricane Agnes in June, 1972. (After Hayes 1978)

heavy river floods, occurring only rarely, may change more within a few days than happens otherwise during many years. This is well documented for the effects of hurricane-generated processes – strong currents, wave action and flooding resulting from heavy rainfall – on Chesapeake Bay and other bays and inlets along the US east coast and the Gulf of Mexico in 1972 (Schubel 1974; Hayes 1978; Fig. 7.12).

7.3.2 Saltwedge Estuaries

Saltwedge estuaries occur where vertical mixing by the tides is insignificant relative to river outflow so that a surface layer of freshwater flows out over salt bottom water separated by a halocline. This is normally the case in vertically tideless estuaries like those along the Baltic and most of the Mediterranean, in microtidal estuaries where the tidal range is small [< 2 m in the classification of Hayes 1975), and in estuaries with a larger tidal range where the tidal influx is small relative to river discharge (Orinoco, Fraser river; Eisma et al. 1978b; Kostaschuk et al. 1989). In a saltwedge estuary the suspended matter is mostly of river origin: it is deposited at the head of the estuary or carried seaward in the surface water. Also the bottom load is deposited at the head of the estuary and gradually fills it up (Meade 1972). An exception is the Orinoco estuary: 60 to 70% of the sediment is supplied from the Amazon along the north coast of South America (Eisma et al. 1978b). In the estuary a fluid mud is formed, which a stable feature; the vertical density distribution in the estuary shows that the highest (sediment + water) density occurs in the fluid mud. The inflowing salt water forms a wedge between the fluid mud and the outflow along the surface of low-density river water (Fig. 7.13).

Usually a bar is formed where the bottom load is deposited during periods of high river discharge when sediment transport is high and the saltwedge is pushed seaward. Horizontal density gradients are strong during that period and usually occur at the bar, where fresh and salt water meet. During periods of low river discharge the saltwedge penetrates inward over the bar into the river and horizontal density gradients are less strong. Also there is usually some entrainment of salt water with the freshwater that flows along the surface, but inflow from the sea along the bottom is weak so that even where sediment is available, there is not much sediment transported from the coastal sea into the estuary. Higher concentrations of suspended matter in the inner parts of the estuary at the contact of fresh and salt water are caused by a large supply during periods of

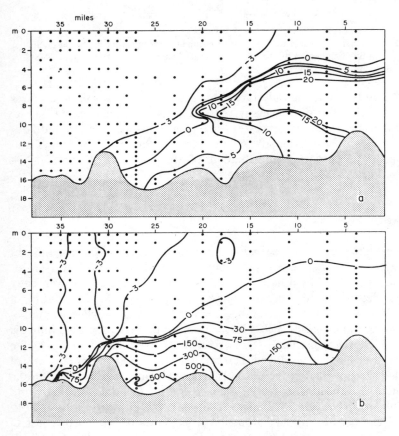

Fig. 7.13a, b. Distribution of water density (σ_t) (**a**) and density of the water-suspended matter mixture (**b**) along a longitudinal profile through the Boca Grande outflow channel of the Orinoco River. (Eisma et al. 1978b)

high discharge, which settles where the flow slows down. When high amounts are supplied simultaneously a (temporary) bottom nepheloid layer can be formed that flows down from the river mouth over the shelf into the adjacent deep sea. This occurs regularly off the Rhône river mouth (Aloisi et al. 1982; Fig. 7.14). and is probably of a nature similar to the density flows in lakes that are caused by a high river supply of sediment.

In the Mississippi river mouth (South Pass), which is microtidal, the mouth bar consists of fine sand and silt (Fig. 7.15). The inertia of the outflowing jet stream is balanced there by friction so that the less dense water spreads laterally and forms a gradually thinning surface layer (Wright and Coleman 1974). Because of the reduction in velocity, this leads to deposition of sediment on the mouth bar. The position of the saltwedge varies with the river discharge. During river floods the outflowing river water pushes the wedge against the bar (Fig. 7.15A). The high outflow velocities and the shallow depths over the bar result in very intense shear at the bar and strong turbulent mixing. There is much resuspension of bottom material and a very turbid surface layer is formed that moves outward into the sea over an inflowing salt water layer with much lower suspended matter concentrations. Further seaward, the surface flow slows down

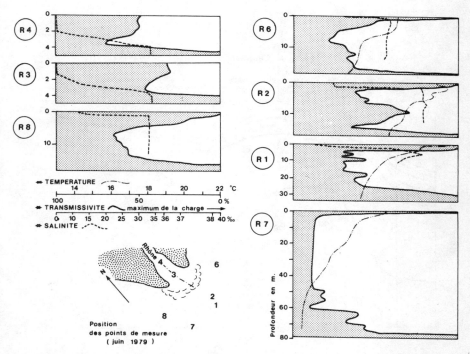

Fig. 7.14. Relation between turbidity, water characteristics, and water depth in the channel and mouth of the Rhône River. Depth in meters, transparency at a scale of 100 to 0% (maximum turbidity), salinity in % S. (Aloisi et al. 1982)

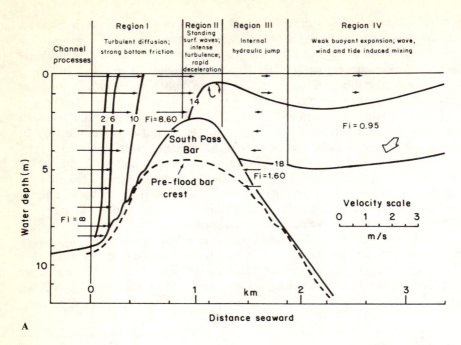

A

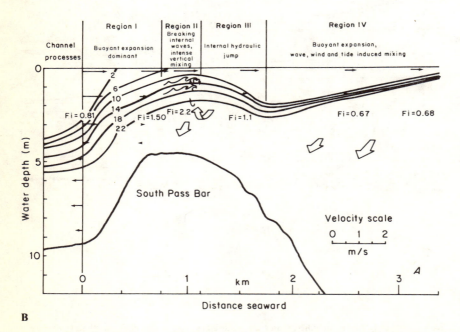

B

Fig. 7.15A, B. Density and currents cross-section through South Pass of the Mississippi River mouths during flood-stage (**A**) and during low stage (**B**) (Wright and Coleman 1974)

and most of the suspended material is deposited. During low river discharge the saltwedge extends further inward but the lateral expansion and thinning of the outflow over the bar results in internal Froude numbers of ca. 1 and the formation of internal waves with strong vertical mixing in the surface water (Fig. 7.15B). The tidal currents, which are strong, although the tidal range is small, because of the large volumes of water that are involved, move the saltwedge in and outwards.

Stable fluid mud below a saltwedge, as in the Orinoco river mouth (Fig. 7.13), also occurs in other estuaries. Kendrick and Derbyshire (1983) describe such highly turbid layers from several (tropical) estuaries and indicate that they can be transported by weak currents without significant vertical mixing.

7.3.3 Tidally Mixed Estuaries

These include partially and fully mixed, mesotidal, and macrotidal estuaries where the tidal range is more than 2 m, as well as microtidal estuaries with a tidal range less than 2 m, where river flow is relatively weak. In the tidally mixed estuaries the residual flow along the bottom is stronger and mixing in such estuaries is schematically shown in Fig. 7.16 (from Dyer 1979). At the head of the estuary there is an area of low velocities where the residual bottom currents – seaward in the river and landward in the estuary – converge. Here sediment is trapped, resulting in a turbidity maximum. Local resuspension during full tide or spring tide may contribute considerably to the maximum as, e.g., in the Tamar and Gironde estuaries (Morris et al. 1982; Allen et al. 1977). Seasonal changes in flow characteristics can cause a turbidity maximum to be a seasonal phenomenon with the turbidity maximum being washed out of the

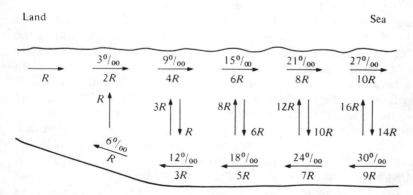

Fig. 7.16. Schematic representation of mixing in a partially mixed estuary. Volume exchange in units of river flow. (Dyer 1979)

estuary when river flow is high, and being built up when river flow is low (e.g., Ems river; Postma 1981). In the Thames, Gironde, Seine, and Severn estuaries, fluid mud is predominantly formed during neap tides because then the currents are decreasing while the duration of the slack tide period increases and increasingly less material is kept in suspension or is resuspended. The deposited material had been brought in suspension during the previous spring tide when currents were strong and resuspension of bottom material resulted in a turbidity maximum and probably a complete resuspension of the fluid mud deposited during the previous neap tide (Inglis and Allen 1957; Allen et al. 1977; Avoine et al. 1981; Kirby and Parker 1977, 1983). The Chang Jiang estuary is well mixed during spring tides, with high current velocities in both the ebb and the flood directions and suspended matter being kept almost continuously in suspension. Around neap tides, the estuary is partially mixed to stratified, with low current velocities but with a higher net flow and less suspended matter kept in suspension. During neap tides, the estuary is a temporary suspended matter sink, but a turbidity maximum is not or only weakly developed (Milliman et al. 1984). This is probably caused by a large particle size of the suspended matter, but Su and Wang (1986) have shown that there is a great spatial and temporal variability in suspended matter transport in the estuary so that the absence (or presence) of a turbidity maximum may well be a temporary effect, related to the amount of discharge.

The different parts of even a small estuary may have a very different suspended matter regime, as was found by Uncles et al. (1985) in the Tamar estuary. The inner part of the estuary shows maximum current velocities and a maximum of resuspension of bottom material. During spring tide there is a net inward transport of sediment, which contributes to the turbidity maximum. In the outer estuary, during both spring tide and neap tide there is net outflow of suspended matter into the coastal sea.

The maximum suspended matter concentrations during a tidal cycle do not coincide with the maxima of flow velocity but occur somewhat later, giving a tidal hysteresis effect (Fig. 7.17). This can be related to lag effects in settling and resuspension (Postma 1980; Dyer 1986) or to the fact that turbulence intensity – and hence the suspended matter concentration – is higher during the decelerating phase than during the accelerating phase (Gordon 1975; Anwar 1975; Bohlen 1976; Anwar and Atkins 1980; Vittori 1989). Lag effects are generated because the suspended sediment, when it settles out because of decreasing flow velocities, needs some time to settle. During accelerating it will take some time before the resuspended sediment is moved upward in the flow. Lag effects can also occur because it takes more energy to resuspend a particle than to keep it in suspension. Particles will therefore be resuspended at higher flow velocities than those at which they were deposited. For fine-grained material that has just been deposited and is still very fluid, this difference is not very great, but after some consolidation it will have a marked effect, as it also has for sand grains. Experiments by Creutzberg and Postma (1979) indicated that ca. 1 h of consolidation had already a significant effect. Kostaschuk et al. (1989)

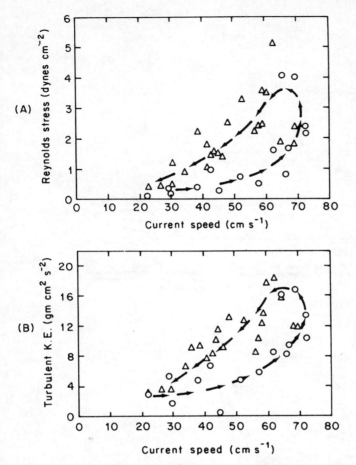

Fig. 7.17A, B. Hysteresis of **A** Reynolds stress and **B** turbulent kinetic energy, as a function of flow velocity. *Open circles* accelerating flood tide; *triangles* decelerating flood tide. (Gordon 1975)

concluded that in the Fraser river estuary the tidal hysteresis is due to settling lag associated with settling of resuspended particles, which is generated by enhanced shear velocities during decelerating flow.

A turbidity maximum can develop independently from the density circulation in estuaries where the tidal flow dominates and a tidal asymmetry creates a sediment trap in the upper estuary (Gironde and Aulne river estuaries; Allen et al. 1980; Fig. 7.18). Tidal asymmetry develops in many macrotidal estuaries. Because of friction, the crest of the tidal wave travels faster than the trough: the flood currents have a higher velocity than the ebb currents but are of shorter duration. More inward in the estuary, the asymmetry becomes more pronounced and a tidal bore may develop. In mesotidal or microtidal estuaries (i.e., where the tidal range is less than 4 m) friction is of much less importance and

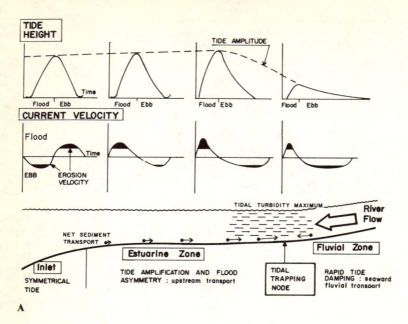

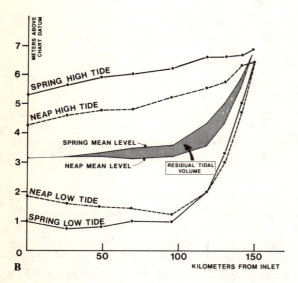

Fig. 7.18. A Schematic model of net tidal transport and trapping of suspended sediment in a macrotidal estuary with little or no density circulation. **B** Longitudinal variation of water level for high tide and low tide in the Gironde estuary during spring and neap tides. **C** Longitudinal variation of the tidal range in the Gironde estuary during low river flow, and total tidal power dissipated on the bottom during a tidal cycle. **D** Variation of total suspended sediment in the Aulne estuary during a spring tide, and variation of bottom current shear above a critical erosion value of 3 dyn . cm^{-2}, integrated over the entire estuary. (**A–D** Allen et al. 1980)

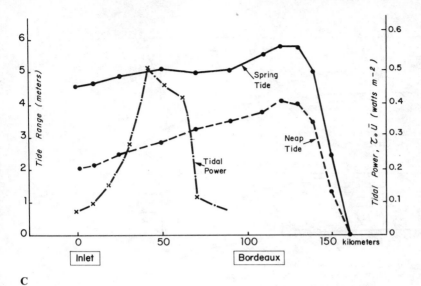

C

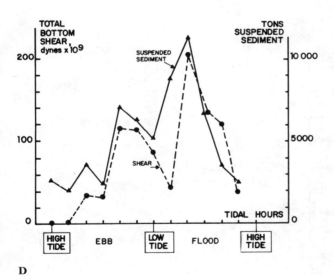

D

Fig. 7.18. (continued)

tidal asymmetry is usually weakly developed, but in many estuaries the tidal range increases because of convergence and decreases again further inward where friction becomes important. In such estuaries the tidal amplitude attains a maximum within the estuary, which also contributes to the inward accumulation of sediment. Because sediment is transported only when the current

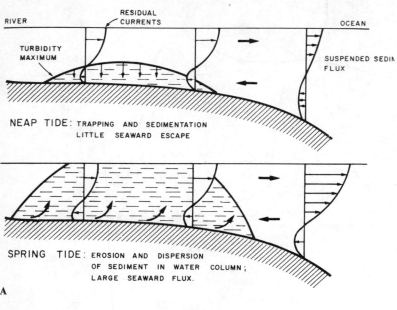

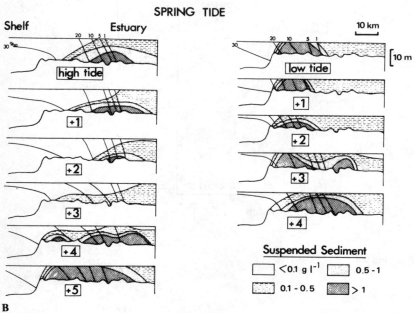

Fig. 7.19. A Schematic diagram showing seaward advection of suspended matter during spring tide along the surface and net inward drift during neap tide along the bottom in the Gironde. (Castaing and Allen 1981). **B** Evolution of the turbidity maximum in the Seine estuary during a (Semi-diurnal) tidal cycle, at spring tide and low river inflow. *Numbers* indicate hours after high or low water at Le Havre. (Avoine et al. 1981)

velocity (or the bottom shear) is above a threshold value, tidal asymmetry has a strong effect on the transport of suspended matter: the period during which the currents are above the threshold value is longer during the flood, and the currents are stronger, which favors transport in the flood direction. Where there is no tidal asymmetry, the difference in water depth at high tide (on the flood plain or the flats) and at low tide (in the channel) plays a major role in the settling and accumulation of suspended material. The flooding of flats during flood tide itself can induce a tidal asymmetry of a different kind than discussed above: during flood tide the water is first confined to the channels but when the flats are flooded the crosssectional area is increased and currents become slower. The reverse happens during ebb tide so that around high tide there is a much longer period of low current velocities than around low tide, which favors deposition on the flats.

Seaward escape of suspended matter is related to the position of the turbidity maximum in the estuary (which is related to river flow and the tides) and to the degree of resuspension of bottom sediment (which is related to the spring-neap tidal cycle). In the Gironde and in the Aulne river estuary, maximum escape occurs during spring low tides combined with high river flow (Castaing and Allen 1981). At this time the turbidity maximum is located downstream in the

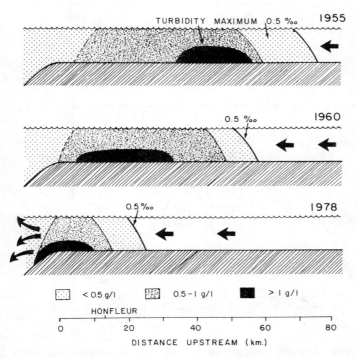

Fig. 7.20. Schematic diagram demonstrating seaward migration of turbidity maximum and salt intrusion (at 0.5%) in the Seine estuary between 1955 and 1978 (Avoine et al. 1981)

estuary and the increased surface residual flow transports large amounts of suspended sediment into the coastal sea (Fig. 7.19). The seaward flux is then two orders of magnitude larger than during low riverflow and neap tide. In the Aulne river estuary this is enhanced by canalization and dike construction, resulting in a seaward migration of the turbidity maximum (Avoine et al. 1981; Fig. 7.20).

Two other tidal phenomena influence the suspended sediment transport in tidally mixed estuaries. Drag coefficients, estimated from maximum bed shear stress values, were found to vary widely in estuaries, with much higher values during the ebb than during the flood, as was found in the Tees, Severn, Conway, and Thames estuaries (Lewis and Lewis 1987). This was mainly related to the slope of the estuary bed, more than to the surface slope or to longitudinal pressure gradients. A steeper slope causes a thicker boundary layer and higher drag during ebb tide.

Co-oscillation of the tides within an estuary was described for the Gironde by Castaing (1989). Short period currents are related to wavelengths that correspond to those of the semi-diurnal tides in the estuary. At the nodal points the bottom sediments are stable, at the antinodes the bottom sediment is regularly resuspended. The unstable areas are 18 to 26 km long and alternate with stable areas of less than 5 km length.

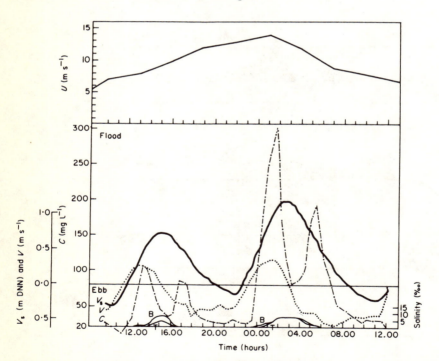

Fig. 7.21. Dynamic conditions just inside the mouth of the Varde Å estuary, on 16–17 April 1977. U Wind speed; V_s water level; V mean velocity; C depth-integrated suspended matter concentration; B bottom water salinity; T surface water salinity (in % S). (Bartholdy 1984)

Secondary maxima can develop in long estuaries along the bottom (James River, Rappahannock, Thames; Officer and Nichols 1980) or in the water column and not connected with the channel floor (St. Lawrence; d'Anglejean and Ingram 1976). A peculiar situation develops in the Varde Å, a small river discharging into a tidal flat area which is part of the Wadden Sea (the Ho-Bugt). In the Wadden Sea suspended matter is concentrated along the inner margins (see Sect. 7.4), so that with the flood turbid water flows into the estuary. Around

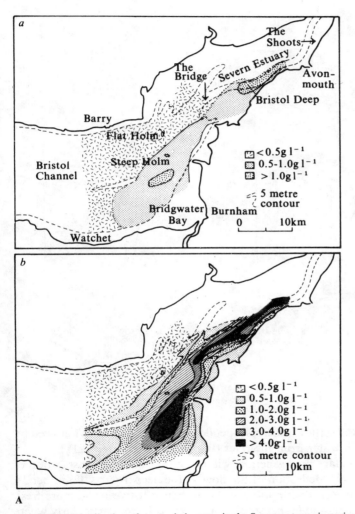

Fig. 7.22. A Distribution of suspended matter in the Severn estuary in spring in the surface water (**a**) and in the bottom water (**b**). (Kirby and Parker 1982). **B** Changes in the suspended matter concentration distribution with the tide in a cross-section of the Severn estuary. **a–d** Ebb tide sequence; **e–g** flood tide sequence. (**A, B** Kirby and Parker 1982)

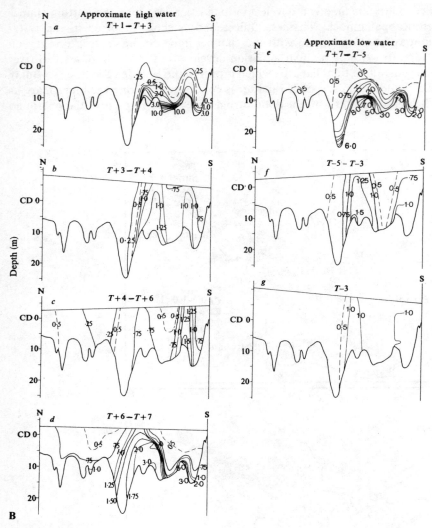

B

Fig. 7.22. (continued)

high tide this material settles, but is resuspended by the ebb, which is followed by inflow of clean fresh water from the river (Bartholdy 1984; Fig. 7.21).

Transverse secondary flow generally will occur (1) where the crosssection of the estuary varies, with deflection of the flow to the deeper parts, (2) in bends, (3) because of an uneven distribution of the exchange between the more fresh and more saline water, and (4) because of the Coriolis force. Particularly where the vertical variations in salinity are small, as is the case in well-mixed estuaries, transverse flow is important. Transverse flow can be related to the spring-neap tidal cycle with different water masses occurring on both sides of the estuary. In the Gironde this is related to the inflow of different water masses from the Dordogne and Garonne rivers (Allen et al. 1977). In the Severn, two water masses with different suspended matter concentrations develop, separated by a steep change in concentration ("front") which is independent of the salinity distribution (Kirby and Parker 1983; Fig. 7.22). It is probably maintained because the front marks a sharp change in density. Longitudinal density fronts have also been described in the Wadden Sea and from York River estuary (Postma 1954, Huzzey and Brubaker 1988). They are generated by differential advection between the channel and the shoals during a tidal cycle and are short-lived (in the order of hours).

7.3.4 Estimates of Particle Flux Through Estuaries from Numerical Models

Modeling of estuarine circulation and mixing is based on the freshwater or salt budget, on tidal mixing, on wind effects, and on a consideration of advection and diffusion in the estuary. An advection-diffusion equation in one form or another forms the basis of numerical models. Where there is a strong density gradient, as is the case in many saltwedge estuaries, a two-layer model gives a more appropriate representation, and in most estuaries, even in relatively simple channels, lateral flow across the estuary has to be taken into account. A full discussion of the circulation and mixing in estuaries is given by Officer (1976) and Bowden (1980).

Modeling of sediment transport in estuaries is more complicated because of the nonconservative behavior of particles in estuaries, the different transport modes — bed load and suspended load — and the cohesiveness of fine-grained material, resulting in a different mechanism of erosion, transport, and deposition as compared with non-cohesive particles that remain constant in size, shape, and density. By reducing a three-dimensional configuration to two dimensions, or to one, and by introducing values for particle settling and erosion (resuspension) rates, and taking into account the tides, the salinity distribution, river flow, and the estuarine geometry, models of estuaries have been made that give a reasonably correct indication of the areas of erosion and deposition and of the concentration distribution of suspended matter, including a turbidity maximum (e.g., Fig. 7.23; Odd and Owen 1972; Ariathurai and Krone 1976a, b; Schubel and Carter 1976; Kuo et al. 1978; Festa and Hansen 1976, 1978; Officer 1980; McAnally et al. 1980; Officer and Nichols 1980; de Grandpré and du Penhoat 1978, in Allen et al. 1980; Galappatti and Vreugdenhil 1985; Johnson et al. 1986; Borah and Balloffet 1986; Vongvisessomjai and Pongpiridom 1986; King et al. 1986; Teisson and Latteux 1986; Wolanski et al. 1988; Odd 1988; Le Hir et al. 1989; Falconer and Owens 1989. One-dimensional models were used to simulate suspended matter concentration distributions along longitudinal or cross-sections, but their application is limited because the vertical as well as the horizontal components are lacking, which are essential to understand suspended

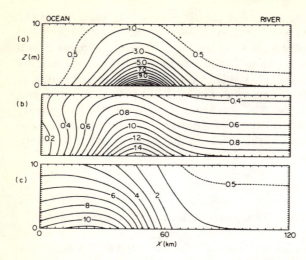

Fig. 7.23a–c. Steady-state numerical model of a turbidity maximum in a partially mixed estuary.
a River to ocean source ratio 10; particle settling velocity 3×10^{-3} cm.s^{-1}. **b** River to ocean source
ration 10; particle settling velocity 1×10^{-3} cm.s^{-1}. **c** River to ocean source ratio 0.1; particle
settling velocity 1×10^{-3} cm.s^{-1}. (Festa and Hansen 1978)

sediment fluxes. This also applies to most two-dimensional models, but in
estuaries with a well-defined flow which is largely limited to a main channel, the
lateral flux can be neglected. A three-dimensional approach is approximated by
using a two- or multi-layered model.

Flux calculations from numerical models, however, are liable to include
considerable errors because several aspects of the sediment transport process are
insufficiently known, such as the resuspension of sediment, the conditions in the
bottom boundary layer, particle aggregation or flocculation, in-situ settling
velocities of sediment particles, and nonlinear tidal effects. Budgets calculated
from models can also be biased because of preconceived ideas on the sediment
sources that have to be considered. Especially because knowledge on particle
aggregation and de-aggregation processes and near-bottom and bedload transport conditions is insufficient, total sediment budget estimates based on models
of estuarine circulation and mixing have only been made in a tentative way and
need confirmation from independent approaches, including confirmation of the
sediment sources that are involved, the transport directions, and the sediment
fluxes. Such confirmation can come from the distribution of natural tracers, and
from estimating the sediment fluxes across transects through the estuary or
across the estuary boundaries. In only a few cases up to now could the flux
calculated from a numerical model be checked with measured transport rates, as
was done by Bokuniewics (1980) for sand transport in the eastern part of Long
Island Sound.

7.3.5 Tracer Studies in Estuaries

Naturally or artificially introduced tracers give an indication of the percentage contribution of specific sediment sources to the estuarine sediment. From such data an idea can be obtained of the transport and mixing of particulate matter in an estuary. Tracers do not give a transport rate but when for one of the sediment sources an input rate is known (e.g., for a river), the input rates from the other sources can be estimated. Natural tracers usually will be useful: they cover an entire estuary, whereas the use of artificial tracers is limited by the amount of tracer that can be introduced and usually gives only an indication of local conditions. As natural tracers, various characteristics of natural sediments have been used with variable success: the mineralogy, the elemental composition, the isotopic composition of certain sediment components (carbonate, organic matter, clay minerals, quartz, carbonate-free sediment) and the nature of particles (dead remains of marine or freshwater organisms). Examples of the use of various tracers are given in Table 7.1 (chiefly after Salomons and Förstner 1984). The mineralogy of the fine size fraction that is transported in suspension is characterized by clay minerals and variable amounts of quartz, calcite, etc. Clay minerals have been used to indicate provenance in the estuaries along the U.S. east coast (Hathaway 1972) and the westward dispersal of Amazon mud into the Orinoco delta (Eisma et al. 1978b). Although clay mineral analysis is semi-quantitative, a reasonable estimate of the contribution from a specific sediment source can be made when its clay mineral composition is clearly defined. For sand sized sediment heavy mineral analysis is a well known and much used method to determine provenance (Milner 1962), but rather complicated calculations are necessary to estimate from such data the relative contribution of several sources in a mixture (Eisma 1968). There is also the possibility of density sorting because the heavy minerals have a higher density than the bulk of the sand. This does not apply to the use of carbonate particles as tracers (calcite/dolomite ratio) which were used by Salomons (1975) to estimate the influx of marine sediment into the Ems and Rhine-Meuse estuaries. Also other light minerals (types of quartz and feldspars) have been used, e.g., in the Gulf of Paria (van Andel and Postma 1954) and in Yaquina Bay (Byrne and Kulm 1967), but these tend to be less specific.

Chemical analysis gives more precise data for the elemental composition than is possible at present for the mineralogical composition, but since the elements are contained within particles, their concentrations in sediments can be subject to rather larger variations because of particle size and sorting effects. Differences in chemical and mineralogical composition of sediments ultimately are caused by differences in rock composition and weathering processes in the source area. Only in a few cases – such as the boron content in clays, the composition of carbonates, and the nature of organic compounds – are the compositional differences related to a freshwater or marine origin. The isotopic composition of carbonates and organic matter (δC^{13} and δO^{18}) is often a better indicator than their chemistry because marine carbonates and organic matter contain more of the heavier isotopes. In a number of estuaries this difference has been successfully used to distinguish river supply from marine supply (Table 7.3) but can be complicated by the supply of eroded old peat or marine carbonate of marine origin by rivers. Also combinations of methods can be useful (Sr-content and isotopic analysis of carbonates and clay minerals in the Rhine and Ems estuaries, Salomons 1975; Salomons et al. 1975; C/N ratios in organic matter, lignin concentrations, and isotopic composition of organic matter in the St. Lawrence estuary Tan and Strain 1979; Pocklington and Leonard 1979).

The use of radioactive tracers is usually not allowed, but clay marked with ^{46}Sc was used in 1973 in the Gironde to follow the dispersal of mud (Allen et al. 1974). Discharges of ^{137}Cs and ^{134}Cs from La Hague, France, which had become attached to suspended particles, were used by Jeandel et al. (1981) for the same purpose: in the Seine estuary suspended matter coming from the sea was found to penetrate inward as far as the contact with freshwater. Brydsten and Jansson (1989) used suspended matter labeled with ^{137}Cs from the Tsjerno-byl fallout to follow the dispersal of river-supplied suspended matter in the estuary of the Őre river in Sweden. This is a nontidal estuary that flows into the very brackish Botnian Gulf. During the period of high discharge, the net

Table 7.3. Compositional particle characteristics used as tracers for sediment dispersed in estuarine areas

Tracer	Application	Reference
Mineralogy		
Calcite/dolomite ratio	Rhine, Ems estuaries	Salomons (1975)
Heavy minerals	Yaquina Bay, Oregon	Byrne and Kulm (1967)
Clay mineralogy	Estuaries U.S. East Coast	Hathaway (1972)
	Orinoco delta	Eisma et al (1978b)
	James River	Nichols (1972)
	Garolim Bay, Korea	Song et al. (1983)
Elemental composition		
Sr-content of carbonate fraction	Rhine, Ems estuaries	Salomons (1975)
Fe/Ti and Fe/Zn ratios	Chesapeake Bay	Eaton et al. (1980)
B, Cr, Co, Ni, V	Marine and freshwater muds	Shimp et al. (1969)
Ca, Eu. La, Ta, Th, Yb	Schelde estuary	Salomons et al. (1978)
Silicones	New York Bight	Pellenbarg (1979)
Isotopic composition		
Organic matter	Schelde, Rhine, Ems est.	Salomons and Mook (1981)
	St. Lawrence estuary,	Tan and Strain (1979)
	estuaries Gulf of Mexico	Shultz and Calder (1976)
	Gironde	Fontugne and Jouanneau (1981)
Carbonates	Schelde, Rhine, Ems	Salomons (1975)
	estuaries	Salomons and Mook (1977)
		Salomons and Eysink (1981)
Clay minerals	Rhine estuary	Salomons et al. (1975)

suspended water flow was outward but preceded by deposition in the estuary and resuspension by wind-forced currents and surface waves. During the period of low discharge, there was further dispersal of earlier deposited suspended matter by surface waves towards the deeper parts of the estuary. Particularly during storms a significant redistribution took place.

Tracers are only useful when they are specific for a source and show conservative behavior in the estuary. Minerals and organic matter used as tracers should be resistant against decomposition and weathering; tracer elements should not be mobilized from the sediment particles or be deposited from solution; differential settling or sorting, changing the percentage composition, should be minimal. The tracers indicated in Table 7.3 to a large extent satisfy these requirements. Clay minerals have been considered unsuitable as tracers because of flocculation effects and differential settling (Whitehouse et al. 1960; Edzwald and O'Melia 1975), or because of diagenetic processes, as was postulated for, e.g., the estuaries along the U.S. east coast (Powers 1954; Nelson 1960). Diagenetic processes in estuaries changing clay minerals have not often been observed and in any case have small effects (Eisma et al. 1978b; in the Orinoco estuary; Feuillet and Fleischer 1980, in James River estuary). Clay minerals can change in estuaries because of uptake of potassium from the seawater (Weaver 1958) or by the formation of interlayer hydroxide sheets of aluminum and iron (Eisma et al. 1978b), which can be easily recognized. Flocculation effects and differential settling, which have been demonstrated in laboratory tests (Whitehouse et al. 1960), have not generally been found to influence clay mineral distributions in estuaries. Gibbs (1977b) found some evidence for selective transport of clay minerals from the Amazon along a distance of ca. 1400 km along the north coast of South America, but this compositional change is small as compared with the compositional difference between Amazon mud and Orinoco mud, so that the clay mineral composition could be used as an indicator for sediment sources in the Orinoco estuary (Eisma et al. 1978b). In James River, Feullet and Fleischer (1980) found that the estuarine circulation, causing inland transport of marine sediment, could account entirely for the observed clay mineral gradients. However, Gallenne (1974) found a strong increase in montmorillonite content in the turbidity maximum of the Loire estuary, probably because of a lower settling velocity. This increase is reflected in the clay mineral composition of the mud banks in the estuary, so that there remains some caution as to the use of clay minerals as estuarine tracers.

7.3.6 Sediment Balance Estimates

Natural sediment tracers offer the possibility to estimate the relative contribution of different sediment sources in an estuary. The output from an estuary through removal by transport either into the sea or up-river canals can be traced in this way, but not the removal by deposition because the deposition material is of the same composition as the material that is being transported. Therefore a more general approach has to be adopted whereby the mass balance of estuarine sediments is estimated from the fluxes in the input and output directions or from the fluxes through transects in the estuary.

This is also the only approach left when suitable natural tracers are not present. Resultant fluxes through transects were estimated, e.g., by Dyer (1978) in the Thames estuary and the Gironde. Apart from the errors introduced by inadequate sampling along transects, the error in the residual transport is large because the transport is calculated from small differences in rather large numbers. This can be demonstrated for the Dutch Wadden Sea where ca. 30 $\times 10^6$ ton (dry weight) of suspended matter is yearly moved in and out through the tidal inlets. Only a few percent of this amount is deposited in the Wadden Sea. Even in the Dollard, a tidal embayment between Holland and Germany, where the deposition rates are rather high (0.7×10^6 tons $. y^{-1}$), only ca. 3% of the total amount of sediment that is yearly moved in suspension is deposited (Dahl and Heckenroth 1978). In these cases, the percentages for deposition were not obtained from calculating the difference between inflow and outflow, but from independent estimates of the deposition rate. To calculate reliably the difference between inflow and outflow would require very long and time-consuming series of measurements along standard transects.

Some of the difficulties involved can be demonstrated with Fig. 7.24. (Dyer 1986). Inward of the turbidity maximum in Chesapeake Bay (Schubel 1969), ebb currents fully dominate. The flood currents are too weak to resuspend bottom sediment. Seaward of the turbidity maximum, the classic estuarine circulation is present with seaward residual flow along the surface and inward residual flow along the bottom. Maximum concentration occur both during ebb and flood. Balancing in- and outflow of suspended matter in the estuary involves concentration or loss of suspended matter in the turbidity maximum, which varies with the tides. This variation can be large: in the Aulne estuary the total amount of material in suspension varied during one tide between 2000 and 12000 tons (Allen et al. 1980). The variation becomes much larger when those during the spring-neap tide cycle and the seasonal variations relative to river discharge are included. To obtain a reasonable balance in such estuaries, large series of data are necessary. Schubel (1974) showed how during a tropical storm in 1972, the sediment concentration in the Susquehanna river increased 40 times and the sediment discharge was the equivalent of 30 to 50 years of normal supply.

Where suspended sediment transport is less variable, a reasonable series of measurements can suffice. At the Chao Phya river mouth near Bangkok, Thailand, a good relation was found between the tidal level and the transport rates through several transects, and between the wind force and the deviations from the average transport rates (NEDECO 1965). From these data it was possible to estimate that during the wet season the resultant transport through the estuary, through the offshore channel, and over the large shoals that lie off the river mouth, is outward (Fig. 7.25), whereas during the dry season, when river discharge is low and salt water intrudes inward, sediment is chiefly moved

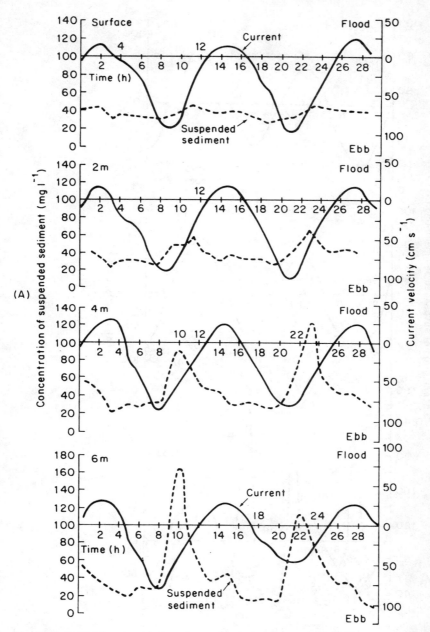

Fig. 7.24A, B. Tidal cycle variation of current velocity and suspended sediment concentration in Chesapeake Bay. **A** At the landward end of the turbidity maximum. **B** On the seaward side of the turbidity maximum. (Dyer 1986; after Schubel 1969)

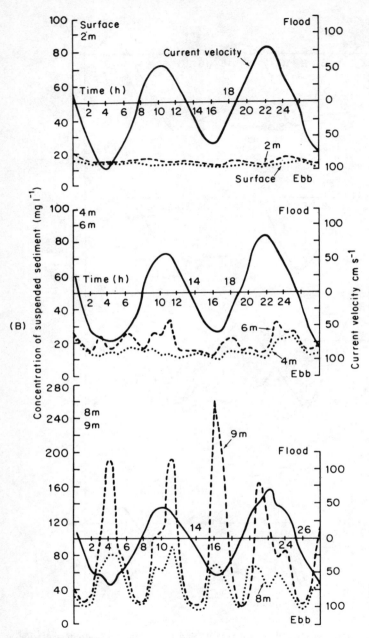

Fig. 7.24. (continued)

into the opposite direction, into the river mouth and to the eastern parts of the shoals. For Chesapeake Bay, (Schubel and Carter 1976) arrived at a sediment budget by estimating mean fluxes. Of a total of 1.89×10^6 tons $\cdot$ y^{-1}, 57% came

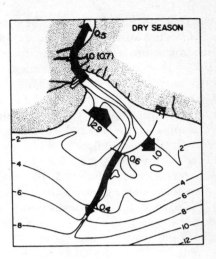

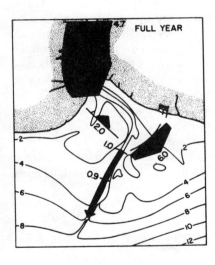

Fig. 7.25. Sediment flow at the Chao Phya river mouth. *Numbers* indicate transport in 10^6 t.y^{-1} (NEDECO 1965)

from the Susquehanna river, 32% from shore erosion, and the remainder from the ocean. Ca. 92% of this total is deposited in the bay.

Including data on deposition rates makes a sediment mass balance more complete and provides an often essential check on the flux balance, as can be seen from estimates on sediment supply in Delaware Bay (Table 2.5; after Wicker 1973 and Meade 1982): although on the average ca. 6.2 million tons of sediment are yearly deposited in the navigation channels, a sediment supply of only 5.3 million tons can be accounted for. A similar problem was encountered by Bokuniewicz et al. (1976) and Bokuniewicz and Gordon (1980), who estimated that in Long Island Sound for the past 8000 years, when the Sound was connected with the coastal sea, the present sediment input from the rivers can easily account for most of the sedimentation. Some additional supply may have come from shore erosion and ice rafting, but the most likely source for the excess supply is inward transport of sediment from the continental shelf. Krone (1979), in the San Francisco Bay system, included the results of soundings made since 1879 (indicating sedimentation and erosion) and dredging data to estimate an average budget (Fig. 7.26). The calculated outflow came to 47% of the average river supply (clay + silt and some fine sand), whereas Conomos and Peterson (1976) on the basis of flow data and suspended matter concentrations, came to only 6%, with large-scale loss of sediment to the ocean only at very high river discharges occurring once in 5–10 years. In the Gironde estuary, sedimentation and erosion volumes were estimated from soundings made between 1900 and 1973. The average yearly deposition of silt and clay amounts to 1.6 million tons whereas the average river supply (Garonne and Dordogne) is 2.2 million tons, indicating that on the average ca. 70% of the supply is trapped in the estuary and only ca. 30% is moved outwards into the coastal sea (Allen et al. 1976). For the estuaries along the U.S. east coast, Meade (1982) has shown that suspended sediment is stored in them that has been supplied from the river or from the near shore sea. Relative to the northern estuaries, however, the southern estuaries have a much higher storage efficiency (expressed as the percentage of suspended matter deposited in the estuary relative to the total of supply and production in the estuary itself). Particularly the northern estuaries store suspended sediment supplied from the coastal sea (Nichols 1986).This is probably because the flushing of the northern estuaries is slow relative to the southern estuaries, and their volume is large. Thus the northern estuaries have conditions that are more favorable for settling of suspended matter and retaining it in the form of a bottom deposit, but it should be kept in mind that the flushing velocity (defined as river discharge divided by the cross-sectional area; Dyer 1986) or flushing

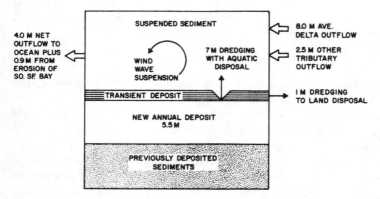

Fig. 7.26. Suspended sediment budget for San Francisco Bay. $M = 10^6$ t . y^{-1}. (Krone 1979)

time (volume of the estuary divided by river discharge) relates to the water and not to the suspended sediment.

Organic matter follows a different budget as compared with the inorganic fraction because of organic production in the estuary as well as decomposition and consumption. When the supply of degradable organic matter involves a sizeable fraction of the total supply of suspended matter, a large error can be introduced when this is not taken into account. The data of Biggs (1970) for the upper and middle Chesapeake Bay show how the budget for the organic matter in suspension differs in that area from the budget for the inorganic suspended matter (Table 2.4).

In most river mouths estuarine mixing is confined to a limited, predominantly inshore area, but at some very large rivers most or all of the mixing and the dispersal of river supplied sediment takes place in the coastal sea. For the Amazon, a mass balance should include not only the river supply (ca. 11–13 million $t \cdot y^{-1}$, Meade et al. 1985) but also the westward removal by longshore transport towards the Orinoco Delta (100–200 million $t \cdot y^{-1}$; Allersma 1968), the deposition on the shelf north of Venezuela, and the deposition on the continental shelf and slope off the Amazon river mouth (Fig. 7.42, p. 231). It can be calculated from the content of terrigenic material in ocean sediments and from estimates for eolian supply of dust to the ocean that only 5–8% of the total supply of sediment from rivers to the sea reaches the deep ocean (Postma 1980; Eisma 1981). The bulk is deposited in estuaries, in the coastal seas and, locally, on the outer shelf, the continental slope, and on submarine fans. Postma (1980) estimates that about half of the total supply is deposited in estuaries.

The only sediment budgets known to the author that include sand as well as suspended matter are the ones made for Long Island Sound (Bokuniewicz and Gordon 1980; Bokuniewicz 1980) and for the Rhine estuary (the Rotterdam Waterway, Fig. 7.27; V and M Report 1979, in Eisma et al. 1982). In Long Island Sound all river inputs are trapped in silt-mud deposits (with an unknown amount of exchange between the Sound and the coastal sea), whereas sand is dispersed from submerged glacial deposits. The silt-mud balance is based on a consideration of river inputs and deposition rates in the Sound, the sand balance on flow characteristics, sedimentation rates, and assumptions on the sand transport velocity and sand diffusion coefficients. By assuming the sedimentary processes to proceed at a constant rate and adding some simplifications, the sand-mud transition zone in the Sound could be adequately described. The calculated sand flux was found to agree reasonably with the average sand flux measured in the eastern part of the Sound.

The sediment budget for the Rhine mouth is based on data for inflow from the two river branches into the estuary, estimates for the supply of (fine) sand from the river, data on the composition and the quantities of sediment dredged from the harbor basins, and the outflow of suspended matter into the coastal sea. Supply from the coastal sea to the estuary makes up the difference, but is not based on measurements because supply from the sea is largest during storms

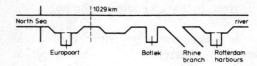

Yearly mud-transport (10⁶ ton)

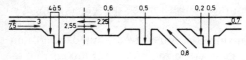

Yearly sand-transport (10⁶ ton)

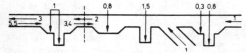

Yearly total sediment transport (10⁶ ton)

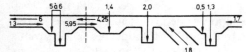

Fig. 7.27. Sediment budget for the Rhine estuary. (Eisma et al. 1982)

and probably takes place mainly by flow very near to the bottom. Since, moreover, the dredging data are not always reliable because of the varying water content of the dredged deposits, this budget still includes uncertainties with regard to removal of sediment to, and supply from the coastal sea, as is also the case with the silt-mud budget for Long Island Sound.

Estimates of the residence time of sediment in an estuary are few, and complicated by the deposition of sediment in the estuary: some sediment will be moved again shortly after deposition, e.g., at the next tide or the next flood, whereas other sediment may remain deposited for thousands of years. Jouanneau (1982) calculated that in the Gironde some particulate matter coming from the Garonne and the Dordogne may reach the coastal sea within 3–5 days, but that another part of the sediment supply may stay in the estuary almost indefinitely. From his data on the total amount of sediment in suspension in the Gironde, it can be estimated that the suspended matter supplied to the estuary by both rivers, has an average residence time of 2–2.5 years in suspension before it is either deposited or removed seaward.

7.4 Intertidal Areas

The intertidal areas – tidal flats – are located between the levels of spring low tide and spring high tide but include also channels with a depth below the level of spring low tide. Based on the tidal range, tidal areas have been described as microtidal (tidal range 0–2 m), mesotidal (tidal range 2–4 m), or macrotidal (tidal range more than 4 m; Hayes 1975). Ehlers (1988) modified this classification, based on observations made in the North Sea: a microtidal area has a tidal range less than 1.4 m, and a mesotidal area between 1.4 and 2.9. Barrier islands with tidal deltas and inlets develop particularly in macrotidal areas, tidal flats and salt marshes in meso- and macrotidal areas. The intertidal areas are characterized by shallow depths, strong cooling and heating compared to the adjacent sea, periods during which the flats fall dry, and rather small salinity gradients that increase when a river flows out through the flats. There are tidal flats with hardly any influence from a river, (like parts of the Waddensea) flats located in river mouths and along estuaries (Severn, Scheldt, Chang Jiang) and flats with an intermediate degree of influence from river outflow. Flats in the tropics are usually covered with mangrove, at higher latitudes they are bare of plants or covered with seagrass or *Spartina*. Flats are often bordered on the landward side by salt marshes that are located above the high tide level and are only flooded during storms or very high spring tides. Many salt marshes have been reclaimed. The dominant agents transporting suspended matter in intertidal areas are the tidal currents, surface waves, and wind forced currents.

Tidal areas are strongly influenced by the deformation of the tidal wave because of the inward shallowing water depth, the bottom topography, and the configuration of the coast. This deformation may already have started in the coastal sea and is then enhanced in the intertidal area. It results in an asymmetry of the tidal velocity curve which has been observed in many tidal areas (Postma 1961, 1967, 1980; Pethick 1980; Uncles 1981, and Lincoln and Fitzgerald 1988 among others). Dronkers (1986a, b) distinguished two types of intertidal areas: (1) where the relative increase in wet cross-section with increasing water level is smaller than the relative increase of the storage area, and (2) where the relative increase in wet cross-section is greater. In the first case, the tidal channels are deep and the tidal flats high relative to mean sea level. This favors a slack water period that is longer before the flood than before the ebb (and export of sediment from the area). In the second case, the channels are shallow and the flats are low relative to mean sealevel. This favors a long slack water period before the ebb and import of suspended matter. In regularly shaped basins and in the absence of river flow, distortion of the tidal wave results in strong maximum flood currents and weaker maximum ebb currents, and a shorter period of low current velocities around low tide than around high tide. This favors inward transport. Also in tidal basins co-oscillating with the tides in the adjacent sea, this type of asymmetry develops (as in the Gironde; Castaing 1989). In small basins where the water depth in the inlets is less than the tidal amplitude, the flood currents dominate while the ebb currents are strongly retarded. This results in a strong inward transport of sediment, as is the case in some tidal basins along the U.S. east coast (Lincoln and Fitzgerald 1988).

The difference in duration between successive slack water periods primarily affects the transport of fine material in suspension (flocs), whereas the differences between the maximum current velocities during flood and ebb also affect the transport of coarser material (bed load, temporarily suspended sand) as well as the total quantities of the material transported in suspension: depending on the availability of fine material, the amounts transported in suspension are a function of the second to third power of the mean velocity. Large flats induce a maximum ebb current: because the tidal wave propagates faster in the channel than on the flats, the decrease of the water level during ebb takes place later on the flats than in the channel, which gives a strong current in the direction of the channel. This current is particularly strong during the late ebb and forms small gullies that often meander. In mangrove swamps the vegetation retards the flow, enhances deposition and hampers resuspension of deposited fine sediment. This also occurs with seagrass and particularly with *Spartina*, which is why it is often planted for coastal protection and the formation of new land. In mangroves, during periods of strong currents flowing between the trees, resuspension of bottom sediment can create a fluid mud layer (Wolanski and Ridd 1986).

The influence of waves is very pronounced in intertidal areas because of the shallow water depth, even when the waves remain small relative to those in the neighboring sea because of the shelter provided by barrier islands or by land. Particularly the superposition of waves and tidal currents can result in high transport rates where transport by currents or waves alone would not amount to very much. In the Wadden Sea high suspended matter concentrations were found to be correlated with strong winds (Kamps 1962). This is also the case at the mouth of the Eden River, which was studied by Jarvis and Riley (1987).

On these flats there is an overall flood dominance and transport by the flood is enhanced by waves as the suspended matter concentrations increase approximately quadratically with wave height. Storms usually have a strong erosive effect on tidal flats, although this is not so much the case at the Eden River mouth, which is rather sheltered. Observations in the Wadden Sea show that mud deposited during the summer is almost entirely removed during the winter storms. This is partly due to an increased flow in the ebb direction because the water is piled up on the flats during the storm and flows back (Eisma et al. 1989; unpublished data NIOZ). Enhanced ebb flow and high suspended matter concentrations were also observed on the tidal flats in Korea during summer storms (taifoons; Wells et al. 1990). To maintain the relatively high elevation of the flats, the sediment has to be returned during normal conditions. Sedimentation resuming directly after a taifoon was observed on Chinese tidal flats, which mostly have no channels but only flat slopes (Ren 1986; Wang and Eisma 1988, 1990). The sediment is brought up-slope by the flood which is much stronger, but of shorter duration, than the ebb. During the flood a turbulent and turbid front is formed which moves up-slope and is caused by the friction that the accelerating wave experiences while moving up-slope in very shallow water depth. During the ebb, flow velocities increase down-slope but since acceleration takes some time, part of the suspended matter brought up by the flood and deposited around high tide remains behind.

In intertidal areas, generally an inward concentration of suspended matter can be observed, which is reflected in the bottom sediments that become increasingly finer towards the inner margins of the tidal areas (Fig. 7.28). Several mechanisms contribute to this:

- The asymmetry of the tidal wave resulting in a relatively strong flood and a relatively long slack period around high tide combined with scour lag (particles are resuspended at higher velocities than those at which they were deposited) and settling lag (particles need some time to settle when the current cannot keep them in suspension). The result is a net landward migration of the particles which accumulate where the maximum tidal currents equal their threshold velocity (van Straaten and Kuenen 1958; Postma 1961; Fig. 7.29).
- The low water depths around high tide when the water is over the flats and the larger water depth around low tide when the water is in the channels.
- Fixation of settled fine material by diatoms or higher plants that cover the flats although strong currents can result in resuspension even in dense mangroves. In the Wadden Sea, fixation by benthic diatoms is not strong enough to prevent erosion by winter storms.

Pejrup (1988) found that during the summer the net flux of suspended matter over a tidal flat in the Danish part of the Wadden Sea was seaward when concentrations were low and inward when concentrations were high. The overall net transport was inward, which corresponded to measured rates of deposition of suspended matter on the inner flats. The net inward flux of suspended matter over the flats was opposed by a net outflow through the channel. Presumably, pronounced scour and settling lag effects take place only when the concentration of suspended matter is high. This was attributed to an increased flocculation because of increased concentrations which resulted in sufficiently large settling velocities for the suspended matter to be deposited during the slack tide period. The above considerations indicate that the concentration and deposition of suspended matter in intertidal areas is a delicate balance between slow net inward transport, erosion, and net rapid outward transport during storms, consolidation of the deposited mud during the period of predominant deposition, the degree of protection against erosion offered by organisms on the flats, and the force needed for resuspension. Taking these various factors into account (Dronkers 1986a,b) could estimate in the Ameland area (where the net flux is inward) and in the Oosterschelde (where the net flux is outward) the residual flux as well as its direction to within ca. one order of magnitude. (Pejrup 1986), in a simple model, could describe about 80% of the variance of the suspended matter concentrations at the observation site on a tidal flat by the wind speed and direction, the salinity, and the tidal current velocity.

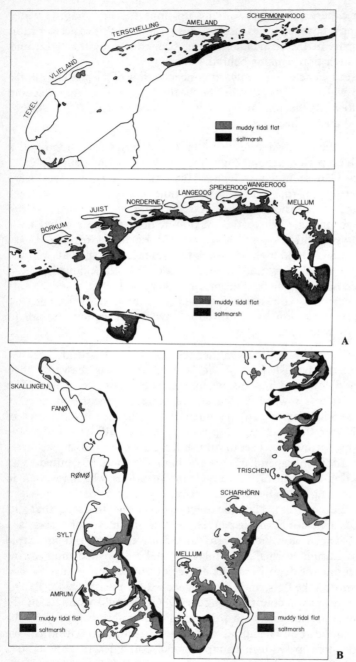

Fig. 7.28A, B. Distribution of muddy tidal flats and salt marshes in the Waddensea. (Eisma and Irion 1988; after Abrahamse et al. 1976). Salt marshes also include present land reclamation areas. *White parts in the Waddensea* sandy bottom sediment

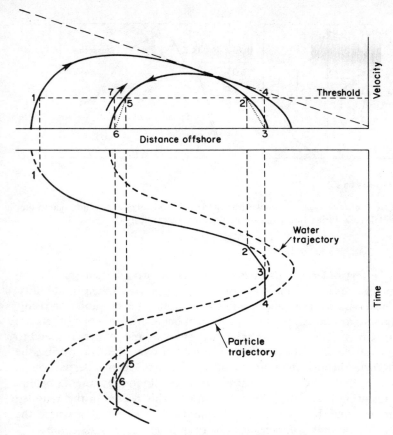

Fig. 7.29. Schematical representation of the tidal transport of suspended matter shorewards by an asymmetric tidal current. (Dyer 1986; after van Straaten and Kuenen 1958 and Postma 1961)

Salt marshes, located along the inner margins of intertidal areas above high tide level and only occasionally flooded, nevertheless receive suspended matter that settles out. Particularly material resuspended during storms is brought to salt marshes. Although during storms waves are large over the marshes (Fig. 7.30), they are damped so that suspended matter can settle and be left behind when the water returns to its normal level. In the Gradyb area in the Danish Waddensea, ca. 50% of the mud deposited in that area is deposited on the salt marshes, the other ca. 50% is deposited on the muddy tidal flats (Bartholdy and Pfeiffer Madsen 1985). This is also related to the relative extent of the salt marshes, which shows large regional variations: in the Dutch Wadden Sea salt marshes cover only ca. 3% of the total intertidal area and mud flats ca. 16%, whereas salt marshes along the Maryland and South Carolina coasts in the USA cover almost the entire intertidal area. Conditions on the salt

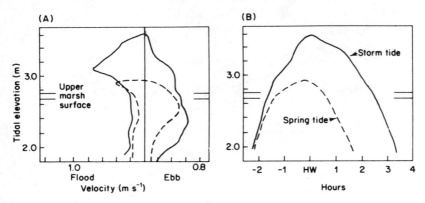

Fig. 7.30A, B. Curves of velocity against tidal elevation (**A**) and curves of tidal elevation against time (**B**) for an intertidal creek during storm and spring tides. (Healey et al. 1981)

marshes may be very different: channels that meander through them may be ebb or flood channels, suspended matter supply may differ very much in quantity, storms may have less or more effect, and the area may be subsiding or rising. Some salt marshes are therefore in the process of being eroded while others are being built up. Channels in tidal flat areas, salt marshes and mangroves tend to be ebb-dominated because the water that comes in over the flats during the flood partly returns through the channels during the ebb. Meanwhile, part of the suspended matter that is brought inwards during the flood may remain behind on the flats, resulting in long-term net deposition. This occurs in the Wadden Sea (van Straaten and Kuenen 1958; Postma 1961; Eisma 1981), where the sedimentation on the flats compensates for a sea level rise of 1 to 2 mm . y^{-1}. Along the US Atlantic coast, however, there are tidal marsh areas that regularly lose sediment (in the order of 1 to 2 kg . m^{-2} . y^{-1}) besides those that trap sediment (e.g., Chesapeake Bay and salt marshes in South Carolina, that keep pace with a relative sea level rise of 1.4 to 4 mm . y^{-1}; Court Stevenson et al. 1988; Wolaver et al. 1988). In lagoons in the USA, deposition generally compensates for the relative sea level rise (Nichols 1989). The differences between the various tidal areas arise from differences in the degree of vegetation, the effects of storms, the occurrence of seasonal changes in sea level, the degree of infill changing (gradually) the tidal characteristics, and, probably most important, from differences in sediment supply.

Most of the estimates of net export or import are based on estimates of net deposition rates. Estimates based on flux measurements in channels tend to be misleading, depending on the extent to which storm conditions, spring and neap tides, and transport over the flats can be included in the measurements. In a New Zealand mangrove, export or import of suspended matter was found to depend on the method of estimation (Woodroffe 1985a, b). Some channels clearly show dominant ebb transport, as do the small channels in the Waddensea and those in the South Carolina tidal marshes (van Straaten 1954; Boon 1978; Ward 1981). Flow in such channels shows an asymmetry in the tidal velocity curve that is opposite to the one found by Postma (1961) for the Ameland channel in the Wadden Sea (Fig. 7.31). Lateral effects should also be taken into account: the flow in tidal areas is complex because of nonsymmetric mixing and dispersal processes (Dronkers and Zimmerman 1982). Gao et al. (1990) describe an estuarine bay on the Chinese coast where the time-averaged flood and ebb flow shows net ebb surplus, but where the net transport of suspended matter is inward. This is caused by mixing in the bay mouth, where the bay water flowing out during ebb mixes with coastal water with a higher suspended matter concentration. The mixture goes inward with the flood and the suspended matter in the bay is concentrated in the bottom water, which during spring tides has a net inward flow.

Small features in the suspended matter concentration distribution in tidal channels are related to the development of longitudinal density fronts, lateral flow from the flats, and the formation of billows by the strong turbulence of the incoming or outgoing tide (van Straaten 1952: Dutch Wadden Sea). High suspended matter concentrations can develop in channels in Korean tidal flats (Adams et al. 1990; Wells et al. 1990). Here at maximum ebb, a lutocline is formed with concentrations of about $1 \text{ g} . 1^{-1}$ in the bottom water and less than $100 \text{ mg} . 1^{-1}$ in the overlying water. Internal waves develop as well as internal surges and short frequency instabilities. These cause fluctuations in turbidity that show up as variations in reflection at the water surface. Amos (1987) in Chignecto Bay, which is part of the Bay of Fundy, Canada, describes turbid fronts, ribbons, plumes, and boils. The turbid fronts are related to strong gradients in density (and probably contribute to these gradients). Ribbons are elongated areas with a constant concentration which is higher than in the

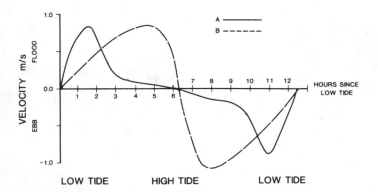

Fig. 7.31. Current velocity during a tidal cycle in a tidal channel in the Wadden Sea (*A*) and in a South Carolina marsh (*B*). Wadden Sea data from Postma (1967). (Ward 1981)

surrounding areas. Plumes show lateral diffusion and boils occur in areas of strong wave turbulence.

7.5 Fjords

Fjords are long elongated valleys filled with sea water; they have a direct connection with the open sea and are characterized by a sill at the seaward end, although there are also a few fjords without a sill, such as, e.g., Coronation Fjord on Baffin Island, Canada. They were excavated by glaciers during the Pleistocene out of previously existing valleys and therefore occur only at higher latitudes (Norway, Greenland, northern Canada, Southern Chile, New Zealand). The sill marks the position where the erosive force of the ice was reduced. Over the sill, water flows outward along the surface and inward along the bottom. In some fjords the deeper water is renewed only very slowly. When this occurs in combination with a large supply of organic matter (usually from plankton production in the surface water in the fjord), anaerobic conditions can develop in the bottom water, when the supply of dissolved oxygen is not sufficient to oxydize all the organic material and oxygen becomes depleted.

The suspended matter distribution and processes of sediment transport and deposition in fjords have been discussed at length by Syvitski et al. (1987) and Syvitski (1989). Suspended matter comes from the glacier(s) that end in the fjord and from streams that flow into it, from landslides and shore erosion, from resuspension of older deposits by surface waves and tides, from the adjacent shelf by inflow over the sill, from the atmosphere, and from primary and secondary production in the fjord (Fig. 7.32; Syvitski et al. 1987). Where a glacier flows into a fjord, sediment is supplied by ice-front melting, by glacio-fluvial discharge and melting of floating ice. Dispersal is normally through the surface

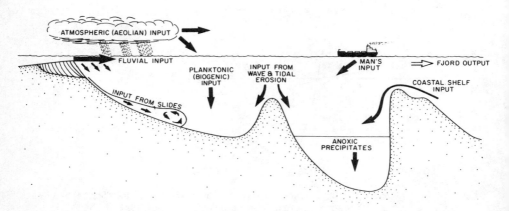

Fig. 7.32. Sediment input into a nonglaciated fjord. (Syvitski et al. 1987)

water, but submarine discharge gives a vertically rising buoyant jet. Most of the sediment is deposited near to the ice front, but by floating ice, sediment dispersal is wider and material from floating ice may be deposited over the entire fjord. Calving icebergs produce solitary waves that may deposit material above the high tide level. Further away from an ice front or river mouth suspended sediment is dispersed through a surface plume, which is very susceptible to the wind and is influenced by the Coriolis force and centrifugal forces. The surface layer of low salinity water is only a few meters thick and separated from the more saline deeper water by a strong halocline. In the deeper water below the halocline, suspended matter is dispersed by settling, as well as by currents, bottom mass movements, and turbidity currents that develop along the often

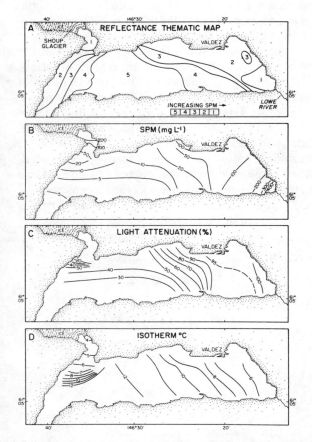

Fig. 7.33A–D. Distribution of reflectance measured by satellite indicating relative suspended matter concentrations in the near-surface water, and of the suspended matter concentration, light attenuation, and temperature in the surface water of Port Valdez on Aug. 2, 1972. (Syvitski et al. 1987; after 1979)

steep slopes along the fjord. In some fjords, where sediment supply from other sources is small, there is a significant contribution of wind-blown sand and silt. Besides mineral particles and organic material supplied from land (the particles are often coated with hydrous oxides, oxides, organic matter, or carbonate), there is an often large particle supply from organic production (organic matter, diatom frustules, coccoliths) in the water.

The concentration distribution of suspended matter in fjords, or the distribution of light scattering, reflects the largely seasonal supply. Concentrations decrease rapidly with distance from the source by settling (Fig. 7.33). Often a turbid bottom layer is present, caused by waves or by bottom currents (renewal of bottom water), or by resuspension during downslope movement of bottom material, or stirred up by bottom fauna. Settling from the surface is rather rapid: fluctuations in the surface water concentrations are within days reflected in the bottom water concentrations, as was measured in Howe Sound (B.C., Canada) in ca. 100 m water depth (Syvitski and Murray 1981). Storm waves can re-suspend bottom sediment at 10 to 20 m depth, but under favorable conditions down to ca. 100 m depth. Storms usually result also in an increased particle supply because of an increase in runoff and shore erosion. Short-term variations in the dispersal of suspended matter in the surface water – in the order of minutes – are related to fluctuations in force and direction of the outflow plume and the position of short-lived fronts. Longer-term variations – in the order of 10 min to several hours – reflect variations in river discharge, ice melting, tides, and wind fluctuations. Variations in the order of days reflect large-scale events that are meteorologically controlled. As vertical mixing is usually very limited, dispersal of suspended matter through the surface waters can cover large distances. The suspended matter distribution can become very complex where river discharge is through more than one outlet or where suspended matter is supplied from more than one source in different quantities.

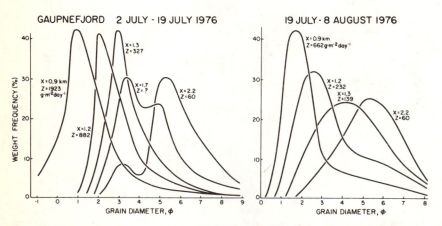

Fig. 7.34. Grain size distributions of sediment deposited in anchored sediment traps in Gaupne fjord. (After Relling and Nordseth 1979).

The particulate matter enters a fjord mainly as single particles of mineral of biogenic origin. Clay particles may already have been eroded in aggregated form and supplied as such to the fjord. Flocculation begins at salinitites of 3 to 5% S, where besides dissolved salts also organic matter is present from biological production. Some dissolved organic matter (probably present as colloids or particles smaller than 0.1 μm) also flocculates (Wassmann et al. 1986). Stringers appear in deeper water and increase in size with increasing water depth as well as in a seaward direction. They become coated with small particles and can form

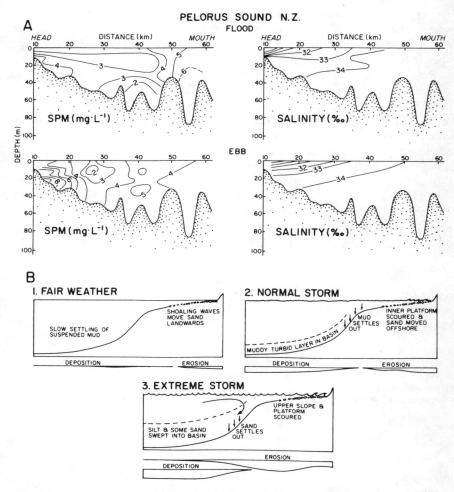

Fig. 7.35. A Generation and mobility of turbidity maxima in Pelorus Sound (New Zealand; after Carter 1976). B Suspended sediment behavior in wave-dominated fjords with low sediment concentrations: 1 during fair weather, 2 normal windy weather, and 3 during exceptional storms. (After Piper et al. 1983)

delicate webs. The grain size distributions of particles collected in sediment traps in Gaupnefjord (Norway) shows a marked size sorting: the coarsest material is deposited nearest to the source and the finer material at some distance (Fig. 7.34). Near to the source the material is considered to settle as single particles, and further away as flocs.

Hyperpycnal flow, or density underflow, occurs only when the concentration of suspended matter is high enough to overcome the density of seawater (ca. 38 g.l^{-1} in a freshwater suspension, when the seawater salinity is 30% S and the temperature 3 °C). Concentrations in meltwater streams are normally less than 2 g.l^{-1} but may be as high as 10 to 39 g.l^{-1} (Gilbert 1983). Hyperpycnal flow, although common in lakes, is therefore rare in fjords. At present, the most likely places for hyperpycnal flow to occur are the discharge outflow points of mine tailings which are brought into fjords as a slurry though pipes that end at some depth below the surface water.

Resuspension can occur through several processes (tidal forcing, wave erosion), resulting in a near-bottom turbid layer, but has not often been observed directly. Examples are the resuspension observed during ebb tide in Pelorus Sound (New Zealand; Fig. 7.35) and the resuspension observed in sediment traps in two shallow Norwegian fjords (Wassmann 1985). Most indications for resuspension come from the distribution of bottom sediments around sills and in relation to water depth.

Plankton production produces high particle concentrations in surface waters, which is mostly measured by the degree of light transmission. Spring blooms usually develop, and where they occur before the stratification has been formed in the water, nutrients will not be limiting and the particle concentration is determined by grazing and particle settling. At high latitudes there tends to be

Table 7.4. Annual supply and loss of POC in the upper 40 m of Nordåsvannet (in g.m^{-2}.y^{-1}). (Wassmann et al. 1986)

	g.m^{-2}–-1	Total supply (%)
Supply		
Primary production	190	68
River discharge: Particulate	45	16
Flocculation of DOC	8	3
Sewage	35	13
Total	278	100
Loss		
Sedimentation	81	29
Export	23	8
Mineralization	174	62
	278	100

a time lag between the development of plankton populations and of grazers, which results in a comparatively high flux of organic matter to the bottom. In fjords with a very shallow sill depth (called "polls" in Norwegian), grazing may be reduced because of a lack of large copepods, for whose development this environment is unfavorable. Fecal pellets produced by zooplankton contribute substantially to the downward flux of biogenic as well as mineral material because the enveloping membrane is degraded at a slower rate at lower ambient temperatures (Honjo and Roman 1978). This contributes to a rapid transfer of material from the surface waters to the bottom. A suspended sediment balance made by Wassmann et al. (1986) for a norwegian fjord (Nordåsvannet; Table 7.4)

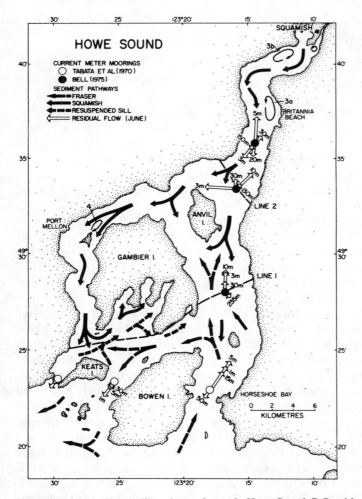

Fig. 7.36. Major sediment dispersion pathways in Howe Sound, B.C., with residual currents from current moorings. Sources: Fraser and Squamish rivers, resuspension on the sills. (Syvitski and Macdonald 1982)

shows a strong dominance of particulate carbon supply from primary production over supply from other sources.

Although in many fjords the suspended sediment dispersal, as suggested by the suspended matter distribution and the distribution of fine-grained bottom sediments, seems to be rather straightforward, fjords can be complex environments. An example is Howe Sound (Syvitski and Macdonald 1982). Figure 7.36 shows the sediment dispersal pathways in the surface water during the maximum runoff period with some additional data for deeper levels at current meter stations. The particles sinking from the surface water are assumed to give a vertical residual downward flux.

7.6 The Continental Shelf

The continental shelf, 100 to 200 m deep (average 132 m) and characteristically less than 100 km wide, is usually divided into an inner part (inner shelf) and an outer part (outer shelf): the inner shelf is directly affected by coastal processes, the outer shelf hardly or not at all, but this division is often not relevant where the shelf is small or where rivers, river plumes, or ice cross the shelf. On the landward side, the shelf is bordered by beaches, flats, inlets, river mouths, rocky shores, and, in the tropics and subtropics, coral reefs and mangrove swamps. The shelf may include valleys or troughs with water depths exceeding 200 m, or elevations and islands.

At the ocean side, the shelf is bordered by the shelf edge, where the continental slope begins. The average continental slope has a dip of 6° as compared to 0.1° for the average shelf. Canyons are a conspicuous feature at the edge of many shelves. They cut in the shelf and slope with steep V-shaped profiles and sometimes near-perpendicular or overhanging walls. At the upper canyon end there are often several tributaries. Most canyons occur on the outer shelf but some begin more inward and may be located at or even in a river mouth (as at the Ganges-Brahmaputra and Zaire Rivers). The outer shelf is part of the ocean margin, which includes also the continental slope and, where present, the continental rise. The latter is a thick sedimentary deposit formed at the base of the continental slope and gently dipping towards the ocean floor. Where a large part of the shelf is situated below 200 m water depth (off California, Barents Sea) it is called a continental borderland. For a more detailed description of the geology of continental shelves, the reader is referred to the reviews of Shepard (1973), Kennett (1982), and Eisma (1987).

To determine the sources of the suspended material, the fluxes from these sources, the concentration distributions, and tracer characteristics are used in the same or a similar way as discussed in Section 7.3.5 (e.g., Biscaye and Olsen 1976; Holmes 1982), to which can be added the estimates of suspended sediment dispersal rates on the Washington shelf based on the dispersal of Mount St. Helens ash (Ridge and Carson 1987). Most of the sources that supply suspended matter to the shelf are located along the coast and on the inner shelf: river mouths, glaciers, coastal erosion (mainly erosion of cliffs and coastal deposits), and resuspension by waves and strong (tidal) currents. The latter is favored by the shallow depth and the deformation of the tidal wave on the inner shelf that results in high tidal ranges and strong tidal currents. Therefore the highest suspended matter concentrations – up to $100 \text{ mg} . l^{-1}$; locally in the order of $g . l^{-1}$ – occur on the inner shelf along the coast, and, except off large rivers, decrease with distance from the coast to less than $5 \text{ mg} . l^{-1}$. This has been observed in the North Sea (summarized in Eisma and Kalf 1987; Fig. 7.37), along the Alaska Coast (Feely et al. 1979), on the Aquitania shelf (Castaing et al. 1982), along the U.S. east coast and the Gulf of Mexico (Manheim et al. 1970, 1972), along Southern California (Gunnerson and Emery 1962; Drake 1976), the north coast of South America west of the Amazon River mouth (NEDECO 1968; van Andel and Postma 1954; Nittrouer and DeMaster 1986), off the New England Coast (Bothner et al. 1981b), off northern Australia (Wolanski and Ridd 1990) and the Great Barrier Reef (Sahl and Marsden 1987), along the French side of the Channel (Brylinski et al. 1984), along the Korean west and south coast (Park et al. 1986; Wells 1988a), and off southern Argentina (Perillo and Cuadrado 1990).

The dispersal of suspended matter from the coast towards offshore, by eddy diffusion and mixing of water masses, is opposed by a complex interaction of

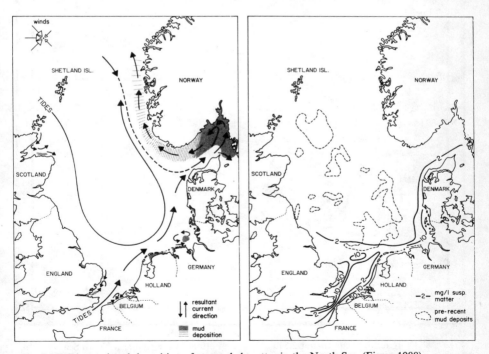

Fig. 7.37. Dispersal and deposition of suspended matter in the North Sea. (Eisma 1988)

forces. For the Dutch North Sea coast, Dronkers et al. (1990) found that the turbidity of the coastal water and the average residual longshore transport are strongly influenced by density- and tide-induced processes and wind effects and only partly related to the average residual water flux. Differences in density between the nearshore water and the water further offshore are present where there is freshwater supply from the land, if only by lateral dispersal from an estuary. Because of the greater buoyancy of the nearshore water, a density stratification will develop with the nearshore water flowing outward along the surface and the offshore water flowing inward along the bottom (Fig. 7.38). Suspended matter, flowing outward with the nearshore water, will tend to settle, particularly when wave action diminishes, and return with the bottom water. Such estuarine-like circulations are known from the southern North Sea and the eastern shelf of the USA (Drake 1976; Eisma and Kalf 1987a; Dronkers et al. 1990). This circulation is constrained (1) by the development of rather stable density fronts that restrict the dispersal of surface water but induce downwelling of the heavier water (Simpson et al. 1978), and (2) by vertical mixing of surface

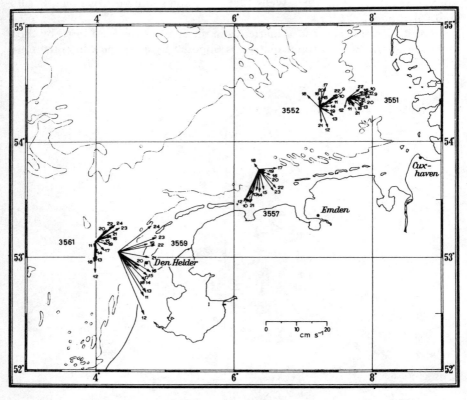

Fig. 7.38. Daily residual currents at 2.5 m above the bottom in the southern North Sea. They all point landward in compensation for seaward displacement of surface water. (Dietrich 1955)

and bottom water by the tides and by wind-generated waves. Forcing by onshore winds will drive water shoreward over the shelf and induce longshore currents and bottom flow where the water is piled up along the coast. As the coast is a barrier for any flow, currents will tend to be aligned parallel to the coast (including tidal currents and those induced by the Coriolis force) with relatively little exchange with the water further offshore. Suspended matter will tend to stay nearshore once it is there and deposition in nearshore or inshore areas (lagoons, intertidal areas, river mouths, the nearshore sea) is favored, particularly where these areas are sheltered against waves. This will be enhanced where the sediment flux over the shelf is normally landward and only seaward during storms. This occurs on the Middle Atlantic Bight of the U.S. (Wright et al. 1991) where low-frequency fluxes over the shelf are just as often seaward as landward. The nearshore concentration of suspended matter is reflected in the distribution of Recent mud deposits: they occur mostly inshore, nearshore and on the inner shelf (Fig. 7.39). The mid and outer shelf are usually covered with relict or reworked sediments, often only containing a Recent admixture of biogenic carbonate and fine-grained organic matter. Recent mid- and outer shelf muds are mostly formed by sediment supplied from a nearby large river (Zaire: Congo-Gabon shelf; Mississippi: NW Gulf of Mexico; Columbia river: Washington-Oregon shelf). A mid-shelf mud deposit off the Ebro delta at present does not receive much sediment from the river: its supply has been reduced by more than 95% because of the construction of dams, reservoirs, and irrigation channels, and most of the present supply is deposited near the river mouth

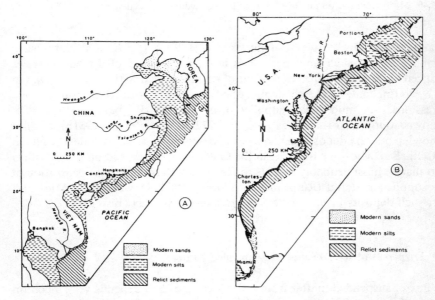

Fig. 7.39A, B. Modern sands and silts and relict sediments on the shelves of eastern Asia (A) and the Atlantic coast of the United States (B). (Emery 1968)

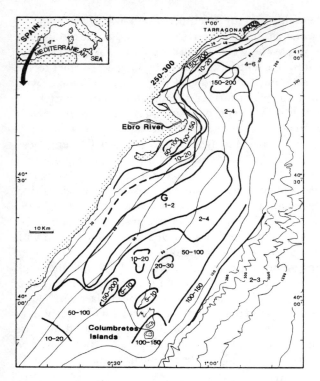

Fig. 7.40. Mean grain size (in μm) of Ebro shelf bottom sediment. (Palanques et al. 1990)

(Palanques and Drake 1990; Palanques et al. 1990; Fig. 7.40). Off Namibia, a nearshore to mid-shelf mud deposit is largely composed of diatom frustules produced in large quantities as a result of strong upwelling (Bremner 1980). On the New England, USA, shelf, mid-shelf muds are formed in topographic depressions, the mud being supplied from the nearby banks, where it is winnowed out by storm waves (Meade 1972; Bothner et al. 1981b). Similar winnowing out and deposition in deeper parts (in this case by the tides) occurs in the Celtic Sea. Mid-shelf deposits in the Gulf de Gascogne and off southeastern Australia are most probably relict (sub-Recent), with some Recent deposition of river-supplied material (Rumeau and Vanney 1968; Davies 1979). Mid- and outer-shelf deposits in the North Sea are almost all relict (Eisma 1981).

7.6.1 Dispersal of River-Supplied Suspended Matter

The flux of suspended matter from rivers over the shelf depends very much on the size of the river (discharge, suspended matter supply) and the circulation in the river mouth. In most river mouths, estuarine mixing is confined to a predominantly inshore area and the export of suspended matter is limited, with much suspended matter deposited inside the estuary and in the freshwater tidal area. Such estuaries are often dominated by an influx of suspended matter from the coastal sea. Even when there is export of suspended material from an estuary, its dispersal is usually limited. Off the Gironde dispersal of suspended matter leaving the estuary is mainly restricted to the inner shelf and deposition to a local area on the inner and central shelf (Castaing et al. 1979; Fig. 7.41). At very large rivers, however, (Amazon, Zaire, Chang Jiang, Hoang He, Mississippi), and at smaller rivers with a saltwedge estuary like the Rhône river, most or all of the mixing and the dispersal of river-supplied suspended sediment takes place in the coastal sea. The dispersal of suspended sediment from most large rivers is reasonably well known, but notable exceptions are the Irrawaddy, Mekong, and Red River, because their mouths and adjacent shelves have been

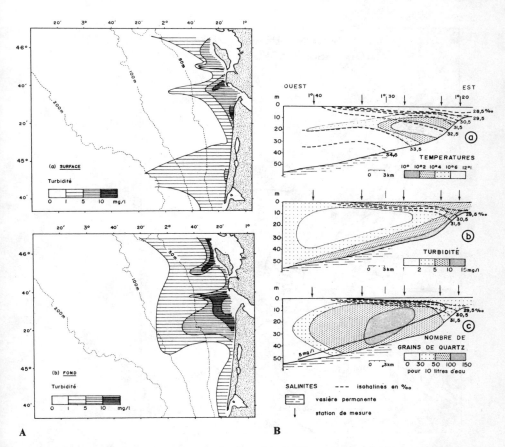

Fig. 7.41. **A** Suspended matter concentration in surface and bottom waters on the inner shelf off the Gironde. **B** Distribution of temperature (**a**), turbidity (**b**), and concentration of quartz grains (**c**) off the Gironde at the end of spring ebb tide. (**A, B** Castaing et al. 1982)

inaccessible for political reasons for decades. For those large rivers whose outflow has been studied, knowledge is largely qualitative and descriptive because of the complexity of the flow off river mouths and of the suspended matter transport processes on the shelf. Thus the dispersal of river-supplied suspended material has been studied mainly by considering the distribution of suspended matter concentrations, or turbidity, and by relating the distribution and composition of fine-grained sediments to a river origin.

At the Amazon, where the outflow is strongly influenced by the tides, onshore waves, and the westward-going Guyana current, a bar is formed off the mouth, where suspended matter and temporarily suspended bed load is dropped as the currents slow down (Nittrouer and DeMaster 1986). The outflow is deflected to the northwest over the shelf and along the north coast of South

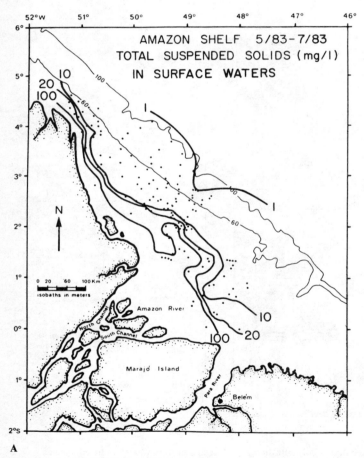

A

Fig. 7.42. A Suspended matter distribution in surface waters on the Amazon shelf. (DeMaster et al. 1986). **B** The Amazon suspended matter dispersal system. (Eisma et al. 1991b)

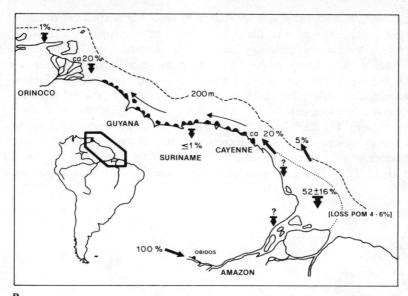

B

Fig. 7.42. (continued)

America by the Guyana current and by waves coming from the NE that are generated by the trade winds (Fig. 7.43). More than half of the suspended matter is deposited on the shelf off the mouth (Fig. 7.43), forming a huge mud deposit, and ca. 20% is moving along the coast of the Guyanas as a turbid zone and as large mud banks. A small part of this coastal transport is deposited along the coast of the Guyanas, and the remainder is deposited in the Orinoco delta, which at its eastern end consists for ca. two-thirds of Amazon mud. A small part is transported further west and deposited in the Gulf of Paria and on the shelf along the north coast of Venezuela. Another small part is transported from the Amazon mouth over the shelf in suspension, and part of this is taken up in the North Equatorial Counter Current and moved eastward through the central Atlantic Ocean. The mud banks that are situated between northwest Brazil and the Orinoco river mouth are peculiar features, as they occur along a high energy coast with a tidal range of ca. 1–2 m and waves that during the period of February to April reach ca. 2 m. On such coasts, normally sandy beaches occur and suspended matter is not deposited (as, e.g., in the southern North Sea). The mud banks are typically 10 to 20 km wide, 50 to 60 km long and ca. 5 m high, with a spacing between them of 30–60 km. They move westward at a speed of 0.5 to 5.5 km . y^{-1} (there were 21 of them in 1961) and they actually protect the coast: waves are damped by the fluid mud that forms the upper layer of the banks, and the incident normal, oscillatory waves change into solitary waves: Wells et al. 1978), while their energy is reduced to only a forward push, as is characteristic for a solitary wave. Between the mud banks, sand beaches are formed, as would have been normal for this coast. The mud banks are probably formed, and remain in existence because of the very large amounts of suspended material that are transported. The mud is concentrated and remains concentrated along the coast by the long shore current system and the onshore waves as well as by the formation of the solitary waves which push the mud landward over the banks. Off Suriname, the surface flow has an offshore component, which is probably topographically induced. This is compensated by a landward component near to the bottom, which also contributes to keeping the suspended material nearshore. The deposition of mud on the Suriname coast is related to the force and direction of the wind: accretion during the past 35 years began when the winds changed to more easterly directions and increased in strength (Eisma et al. 1991). As this implies larger waves from northeasterly directions,

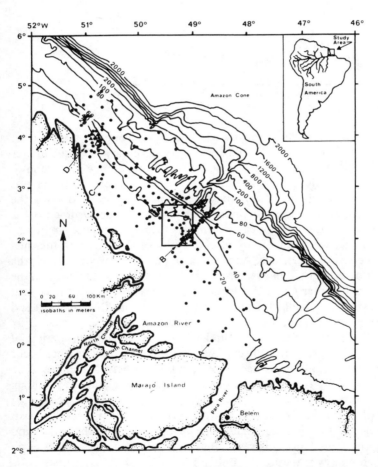

Fig. 7.43. Profiles across the Amazon shelf showing salinity and turbidity distributions and the depth distribution of suspended matter (in mg.l^{-1}). Location of profiles **A** and **B**. (Nittrouer and DeMaster 1986)

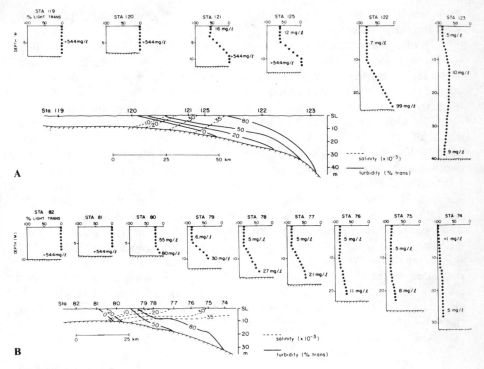

Fig. 7.43. (continued)

more intensive erosion would be expected, but the transformation into solitary waves with more energy means that more mud is transported landward. Erosion of the coast occurs where it is not protected by a mud bank (Augustinus 1978).

The origin of the mud bank system, which has been in existence now for at least ca. 1000 years, is not clear. The coast directly west of the Amazon river mouth is being eroded at present, but there is a strong suspended matter transport along the coast to the west. Deposition begins near to the border with French Guyana (Cayenne). As there is an area of non-deposition on the shelf between the offshore muds deposited off the river mouth and the mud banks along the coast, the mud in the banks is probably all supplied directly from the river mouth in suspension along the coast. Fluid muds with sediment concentrations of more than $400 \, g \cdot l^{-1}$ also occur on the shelf off the mouth and overlie the sea bed. They are present during rising and high river discharge when suspended sediment supply is high (February-June) and reach thicknesses of 1 to 7 m, varying with the tidal phase. During falling river discharge (August) they are absent (AmasSeds Res. Group 1990). The fluid muds occur in spite of strong tidal currents (up to $2 \, m \cdot s^{-1}$) and are probably largely deposited during slack tide. During full tide they are moved over the bed: they have a high density ($> 1.20 \, kg \cdot dm^{-3}$), which together with the thickness of the muds prevents their resuspension at low shear rates (Faas 1986), but induces horizontal displacement with the tidal current.

The dispersal system of the Amazon, the largest river in the world according to discharge and to suspended matter transport, is by far the largest river dispersal system in the world. At the mouth of the Zaire, which is the second largest river according to discharge, but with a relatively low suspended matter load, the river water flows out over a canyon, that begins at the head of the estuary and is already ca. 450 m deep at the mouth (Eisma et al. 1978a; Eisma and Kalf 1984; Fig. 7.44). There no bar can form, and within the river mouth already half of the suspended matter is lost from the outflow, either by deposition in the canyon or by lateral transport into the mangrove swamps that border the estuary. The bed load from the river (large sandbanks migrating seaward) is pushed into the canyon at its head, particularly during high floods. Also much suspended matter is deposited in the canyon and regularly moved down-canyon towards the ocean floor by turbidity currents (Heezen et al. 1964). About 70 to 80% of the yearly supply is deposited on a deep-sea fan that reaches the center of the Angola Basin at ca. 5600 m water depth. Ca. 20 to 30% is deposited on the shelf north of the river. This mud is transported by a coastal current, the Benguela current, which has velocities up to $1 \text{ m} \cdot \text{s}^{-1}$, but is of small volume. As the river outflow forms only a surface layer of a few meters in thickness, the Benguela current dips below the plume without deflecting it very

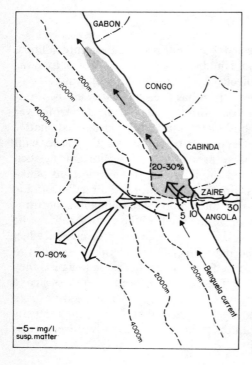

Fig. 7.44 Dispersal and deposition of suspended matter from the Zaire River. *Shaded area* deposit of Zaire supplied mud on the shelf. (Eisma 1988)

much, although in combination with the tides, some strong turbulent mixing can take place at the river mouth. Only a very small amount of suspended matter is moved from the river into the ocean through the surface plume, but after maximum outflow, an admixture of river water and a higher suspended matter concentration can still be detected at ca. 700 km from the mouth, which may be partly due to enhanced primary production.

The third largest suspended sediment supply comes from the Ganges-Brahmaputra (after the Amazon and the Yellow river). Only for a small part is it dispersed over the shelf: most of it is funneled through a canyon and deposited on an enormous submarine fan, which begins near the river mouth, forms the floor of the Bay of Bengal, and extends down to ca. 5000 m water depth in the central Indian Ocean (Curray and Moore 1971). No estimates of suspended matter dispersal have been made, but the amounts dispersed in suspension through the surface waters of the Bay of Bengal are probably very small relative to downflow through the canyon.

The Huang He (Yellow River) not only supplies a very large amount of sediment – it is the second largest river according to sediment discharge – but also supplies it in very high concentrations of ca. $25 \, g.l^{-1}$, increasing to a maximum of ca. $200 \, g.l^{-1}$ during the flood stage of the river. For comparison: the Amazon, the Ganges-Brahmaputra, and the Chang Jiang have average concentrations of 0.19 to $1.72 \, g.l^{-1}$ (Table 2.1). The high concentrations in the Huang He result in the development, directly off the mouth, of high-density bottom flows with suspended matter concentrations of 1 to $10 \, g.l^{-1}$, that descend along the submarine delta front (Wright et al. 1990). Dispersal along the shore is caused by tidal currents (up to $1 \, m.s^{-1}$) and by wind-forced currents which easily develop in the shallow offshore sea (the Bohai Sea). Resuspension of bottom sediment by storm waves can result in high density flows with concentrations up to ca. $100 \, g.l^{-1}$. The high density underflows cover only a relatively short distance, which is attributed to a slowing down because of entrainment of bottom material. In this way, most of the suspended matter supplied by the river remains near to the mouth, resulting in a rapid build-up of the delta. The storm-induced high-concentration bottom flows were not directly observed but inferred from side-scan sonar, sub-bottom data, and bathymetric data, and Wright et al. (1990) point out that the formation of bottom flows is not just down-flow of highly turbid low-salinity water but that the bottom flows that were observed had only slightly reduced salinity. This suggests that settling and resuspension are important intermediate stages.

As well as by turbid bottom flow, suspended matter from the Huang He is dispersed through the water column but at much lower concentrations. Those in the surface water become less than $50 \, mg.l^{-1}$ at short distances from the mouth. Part of the suspended matter is finally deposited on the tidal flats along the delta coast. The sediment dispersal off the Huang He by high-density underflows conforms to the general concept of "hyperpycnal flow" (as opposed to "hypopycnal flow"), which was postulated by Bates (1953) as the way a delta can be built up by a river with high suspended sediment concentrations (Fig. 7.45). The Huang He is the only recent example; all other river outflow, including those of

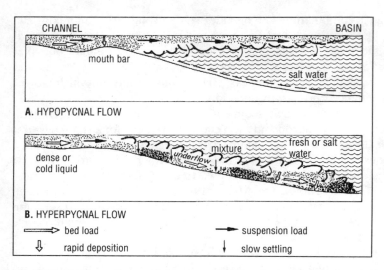

CHANNEL   BASIN

mouth bar

salt water

A. HYPOPYCNAL FLOW

dense or
cold liquid

mixture

fresh or salt
water

underflow

B. HYPERPYCNAL FLOW

⟹ bed load   ⟶ suspension load

⇓   rapid deposition   ↓   slow settling

Fig. 7.45A, B. Types of suspended matter outflow at the mouth of a river. **A** As a surface plume (hypopycnal flow, e.g., Zaire, Mississippi) and **B** as a dense underflow (hyperpycnal flow; e.g., Yellow River). (Chamley 1990; after Bates, in Galloway and Hobday 1983)

the largest rivers, is of the hypopycnal type, but, as noted by Bates (1953), the hyperpycnal type may have been more numerous during the Pleistocene periods of low sea level. At the Rhone river mouth, a bottom nepheloid layer is formed at the fresh-salt water contact, while at the same time part of the suspended matter supplied by the river is transported offshore along the surface with the river water in the plume. The extension of the river plume does not reach further than ca. 60 km offshore and its direction changes with the direction (and the force) of the wind. The bottom nepheloid layer crosses the continental shelf and continues through the submarine canyons (Aloisi et al. 1982; Demarq and Wald 1984).

The dispersal of suspended sediment supplied by the Chang Jiang (Jiang Tse Kiang) is shown in Fig. 7.46. It is very similar to the dispersal from the Amazon: a coastal current (the Jiangsu current) transports the suspended matter to the south, which is already indicated by the orientation of the river mouth. A small part is dispersed eastward, particularly during the summer when the coastal flow tends to change to the north because of the Taiwan current. Off the river mouth, a mud deposit is formed at shallow water depth where the bottom sediment is easily resuspended by the tides or by waves. More than 40% of the supply is deposited in the estuary, resulting in progradation and shoaling (Milliman et al. 1985). Transport is probably by repeated deposition and resuspension, but no mud banks and no fluid muds are formed, except in the river mouth. Concentrations remain low because the suspended matter is more dispersed: it is distributed through numerous bays and estuaries and on the shelf is deposited around the 50 m isobath. The dispersal patterns of the Amazon, the Yellow River, and the Chang Jiang show that even off large rivers the suspended matter hardly

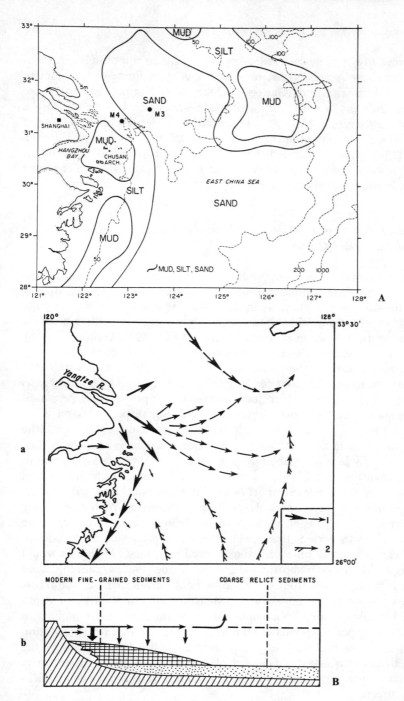

Fig. 7.46. A Distribution of sand and mud on the sea floor off the Chang Jiang river mouth **Ba** Transport direction of sediment supplied from the river (*1*) and from the ocean (*2*) **Bb** Cross-section showing the distribution of modern and relict sediments. (Sternberg et al. 1985, after Chin 1979)

crosses the shelf towards the deep sea, except where the river mouth is near or at the shelf edge (Mississippi, Niger) or where a canyon funnels the sediment directly to the deep sea (Zaire, Ganges-Brahmaputra). Various estimates have indicated that only less than 10% of the river supply reaches the deep sea (Drake 1976; Postma 1980; Eisma 1981). This agrees with a model describing the flux of particulate Al across the shelf of the southeastern USA, developed by Windom and Gross (1989), which indicated that the major transport processes on the shelf are advection and particle settling, and that less than 10% of the river supplied sediment is transported across the shelf by these mechanisms.

7.6.2 Dispersal of Suspended Matter from Coasts (Beaches, Inlets, Eroded Coasts)

Away from the influence of river mouths, surface waves, tidal currents, and wind-forced currents are dominant in transporting and resuspending sediment and in eroding the coast. Turbulence in the surf along the beach is high enough to bring sand in suspension when there are plunging breakers – spilling breakers cause much less turbulence near to the bottom (Miller 1976; Galloway 1988). During storms, waves may also break on offshore shoals and banks, and bring sand in suspension (e.g., on the Norfolk Banks in the North Sea; Stride 1988). From the surf the suspended sand can be transported seaward by tidal currents and rip-currents. The latter, which have a direction approximately perpendicular to the shore, are the most effective in dispersing suspended sand in an offshore direction. Talbot and Bate (1987), however, have shown that the presence of a rip current does not necessarily mean that there is an offshore flux: this may be virtually zero when oblique waves come in pushing the rip current to the side, or when the mouth of a rip between the breakers is closed. It is an unresolved problem whether transport of sand in suspension results in important seawater fluxes from beaches. Intermittent suspension of bottom sand, whereby periods of suspension (related to incident wave periods and wave groups) alternate with periods that are sufficiently long for suspended sand to settle, has been repeatedly observed (summarized in Hanes 1988). This would allow for dispersal only over short distances, but Stride (1988) argues that sand being moved in suspension from one shallow bank to another may be transported by storms up to 100 km seaward. Measurements in 10 m water depth outside the surf along Long Island indicated that mud and silt remained in suspension for long periods relative to the wave period, but that suspended sand remained close to the bed (with very small amounts already at 1 m above the bed) and settled out quickly (Vincent et al. 1983). These measurements also indicated that transport of sand in suspension will not be over large distances in an offshore direction.

In inlets, sediment in suspension is transported by flood and ebb tidal jets. These result in parallel shoals formed by sediment settling out at the side of the jets (Özsoy 1986), but the ebb tidal jet seldom reaches further offshore than ca. 10 km, as it dissipates by friction, and flow reversal occurs at every turn of the tide. Exchange in the absence of tides is usually small and does not easily result in the formation of a strong outgoing flow. Erosion along the coast is mainly by waves, in particularly by storm waves. Eroded material is carried seawards by the undertow and is brought in the near-shore zone where it can be dispersed further by currents. As both the surf zone and the residual transport in the nearby sea respond rapidly to changes in waves and thus to changes in the wind conditions, transport of erosion products in suspension usually is highly variable and with hardly any time lag between the fluctuations in the surf and those on the adjacent shelf (Snedden and Nummedal 1990). Lagoons are usually supplied with suspended sediment from streams or rivers entering them, from in-situ primary production and from surface runoff from the surrounding land. The in- and outflow through the inlet is usually weak because of the weakness (or absence) of the tides. It does not result in significant supply of suspended matter through the inlet from the coastal sea, or to outflow in the reverse direction. As most lagoons are shallow, resuspension of bottom sediment by waves is a common feature. A typical example is described from the French Mediterranean coast by Cataliotti-Valdina (1982).

7.6.3 Suspended Matter and Mud Deposits on the Inner and Middle Shelf

As was seen above, suspended matter tends to remain nearshore and to be deposited in areas that are sheltered against waves, i.e., in-shore and nearshore in areas where it cannot easily be resuspended. Exceptions are the mud banks along the north coast of South America, where the mud is being supplied from the Amazon river (see Sect. 7.6.1). Other anomalous mud deposits – with high suspended matter concentrations in the overlying water – are known from the North Sea, from the Louisiana coast and the U.S. East Coast, and from the coast of Kerala (S.W. India). The mud patch in the German Bight of the North Sea is related to a suspended matter concentration process consisting of a combination of strong tidal asymmetry (with flood dominance) and settling of suspended matter during periods of calm weather (Eisma and Kalf 1987a). The tidal asymmetry develops where the sea floor shallows rapidly and the tidal wave is forced into the Bight. The deposit lies between 15 and 40 m water depth. During storms, the top sediment is eroded particularly on its upper part and the resuspended material is transported into deeper water offshore, from where it is returned along the bottom by the tides. Along the Louisiana coast, mud deposits and high suspended matter concentrations occur along the open coast west of the Mississippi river mouth from where the mud is supplied. As along the coast along the Guyanas, waves are attenuated by the fluid mud that forms the upper part of the mud deposits and, although some resuspension occurs, the re-suspended material is rapidly redeposited (Morgan et al. 1953; Kemp and Wells 1987). Another (small) anomalous mud area is located in Cape Lookout Bight, where fluid muds are present at ca. 5 m waterdepth. Little or no resuspension occurs in response to tidal currents or stormwaves. The area is rather sheltered and acts as a sink for large amounts of suspended material, probably through the formation of a large eddy (Wells 1988b). Off the Kerala coast, high suspended matter concentrations occur as a temporary feature in an area where, under normal conditions, waves and residual currents cause concentrations to reach ca. 0.5 g.l^{-1} (Mallik et al. 1988). This results in muddy deposits that can be resuspended en masse by large monsoon waves. Then a thick suspension is formed that remains more or less in place: lateral dispersal is prevented by the residual current pattern and a fluid mud is formed that absorbs the energy of the incident waves. This prevents the mud from settling, and keeps it fluid. For this fluid mud, as well as for the formation of anomalous muds in other areas where a sandy coast would be normal, suspended matter concentrations have to be initially high, either because of a large supply (Amazon, Mississippi) or because of some concentration mechanism (German Bight, Kerala coast).

Off the Belgian-Dutch coast in the North Sea, the presence of relatively high suspended matter concentrations in combination with a patchy mud deposit is related to residual currents converging to the same area. This patch, as well as similar patches of East Anglia and in the German Bight, are related to the presence of a residual gyre. Such gyres have also been described by Ferentinos and Collins (1979, 1980) from the Western Channel coast and the Bristol Channel. In the southern North Sea, the muddy deposits and high suspended matter concentrations are located at the flank of the residual gyres in relatively shallow water: the shoreward residual component probably causes advection of suspended matter to an area where it is easily deposited. Measurements by Gossé (summarized in Eisma et al. 1982 and Eisma and Kalf 1987a) have indicated that the mud area off the Belgian-Dutch coast is a sink during calm weather but a source of suspended matter during storms, with an over-all net erosion. Small mud patches on a sandy sea floor off the Long Island coast have been attributed by Clark and Swift (1984) to random behavior of resuspended particles. Once the mud patch is there, bottom boundary conditions at the mud patch are different from those at the surrounding sandy sediment because the muddy sediment surface is less rough and the mud has a higher threshold velocity for erosion.

Resuspension is probably a very common feature on shelves. On the New England shelf and the Grand Banks, resuspension of bottom sediment occurs during storms and results in near-bottom suspended matter concentrations up to ca. 15 mg.l^{-1} (Bothner et al. 1981b; Fig. 7.47). Resuspension on the Great Barrier Reef off Australia is caused by the tides and by waves during the dry season. Resuspension by waves, particularly storm waves, is also an important factor in shelf seas like the North Sea, the middle Atlantic Bight off the U.S. east coast, the Gulf of Alaska, the California shelf, and the Bering Sea (Feeley et al. 1979; Drake et al. 1980; Drake and Cacchione 1986, 1989; Eisma 1981; Meade et al. 1975; Young et al. 1981). The degree of resuspension depends very much on the cohesion of the sediment. Measurements by Drake and Cacchione (1986, 1989) on the California and Alaska shelves indicate agreement of the measured threshold shear stress values with the shields threshold criterion, which implies that there is no significant cohesion. Larsen et al. (1981; see also Sect. 5.3) had already found that the Shields (1936) entrainment function for unidirectional currents, when plotted against grain Reynolds number, adequately predicts the threshold of (unconsolidated) grain motion on the shelf. However, also higher threshold shear stresses have been found, which indicates cohesion of the sediment or "armoring" of the fine sediment by coarser material, left as a

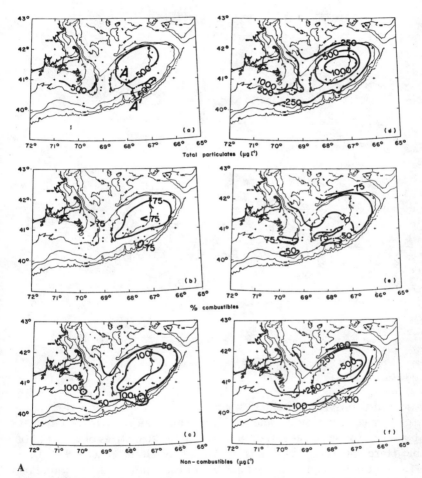

Fig. 7.47. A Suspended matter distribution in early-mid May 1977 on the shelf and the slope off the NE United States. **a–c** Surface water; **d–f** near-bottom water; **a, d** total suspended matter concentration (in $\mu g . l^{-1}$); **b, c** percentage of combustible material; **c, f** concentration of ash (in $\mu g . l^{-1}$). **B** Suspended matter distributions in mid-February 1978 on the shelf and the slope off the NE United States. **a–c** Surface waters; **d–f** near-bottom water; **a, d** total suspended matter concentration (in $\mu g . l^{-1}$); **b, c** percentage of combustible material; **e, f** concentration of ash (in $\mu g . l^{-1}$). **A, B** Bothner et al. 1981b)

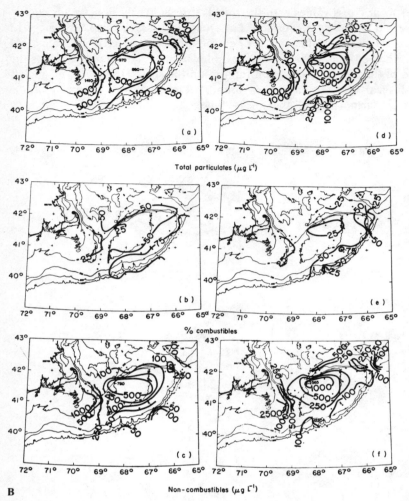

B

Fig. 7.47. (continued)

residual deposit from previous entrainment of finer material. Organisms also can give greater cohesion by producing organic films on the particles and by making burrows. The seasonal variation in threshold shear stress of sediment of Barnstable Harbor (Mass., USA) was attributed by Grant et al. (1982) to a seasonal formation of organic surface films. Sub-Recent muds, such as found off the Ebro river mouth, are consolidated and resistant against resuspension. The influence of organisms and organic material on particle entrainment is shown in Fig. 5.25 (after Nowell et al. 1981), which indicates that Inman's extension of the Shields curve probably gives a good fit.

The data of Palanques and Drake (1990) show a seaward dispersal at mid-depth of suspended matter resuspended from unconsolidated deposits at the

Ebro delta front. Off the Mississippi delta, resuspension is relatively unimportant and overshadowed by the formation of high near-bottom concentrations (> 100 mg.1^{-1} where ca. 5 mg.1^{-1} is normal) by settling from the surface plume during short periods of high turbidity. These are probably related to local shifts in the position of the plume (Adams et al. 1987). Much sediment can be resuspended by bottom trawling: in heavily fished areas like the northeastern shelf of the U.S. and the North Sea, it is estimated that bottom trawls rework the surface sediment in the entire area several times a year down to ca. 5 to 10 cm depth (e.g., Churchill 1989; Rauck 1988, in Puls and Sündermann 1990; Floderus and Pihl 1990). Icebergs touching the bottom also disturb the sediment. Off Alaska, it is estimated that in this way ca. 4×10^3 tons.y^{-1}.km^{-2} is resuspended and transported away by bottom currents (Rearic et al. 1990). In the North Sea and in Hangzhou Bay, some mud deposition areas and high suspended matter concentrations are associated with frontal zones separating water masses of different temperature and salinity (Creutzberg et al. 1984; Su and Wang 1986; Su et al. 1990). Particularly in Hangzhou Bay, it is evident that suspended matter is concentrated at the frontal zone. Flow in and near to this zone is very complex, and it is not clear how suspended matter is concentrated there.

Away from the coast, often a three-layered system develops with relatively high suspended matter concentrations in the surface and bottom waters and less turbid water in between. High concentrations in the surface waters, where not caused by advection from a river, are usually caused by plankton growth, which is strongly related to the availability of sunlight and nutrients (phosphates and nitrogen compounds, for diatoms and other silicious plankton also dissolved silica; calcium and carbonate are always present in excess). Most of the nutrients are supplied by rivers: usually highly productive zones occur off river mouths where the turbidity has become low because the suspended matter supplied by the river has settled out or has been dispersed (e.g., off the Amazon, Chang Jiang). Also in upwelling areas, where water from deeper levels is being moved upwards, plankton growth is high: in the surface water nutrients are usually rapidly used up when light conditions are favorable, so that supply of nutrients from below makes continuing plankton growth possible. The same occurs in temperate zones where during the summer the surface water is heated up and the water becomes stratified. A surface layer develops where nutrients are rapidly depleted by growing plankton, and a deeper layer where nutrients are accumulated because organic particles settle downward and are mineralized there. When during fall the surface waters cool off and storm winds induce stronger mixing of the water column, the stratification disappears and the surface waters become high in nutrients again. When in early spring enough sunlight becomes available, plankton starts to grow again in large numbers. The same can occur during fall when the water column is mixed when enough sunlight is still available. It will be clear that the amounts of plankton in the water can be extremely variable: it can range from virtually nothing to more than 90% of the total mass of material in suspension.

A turbid bottom layer, where present, is caused by resuspension of bottom sediment by waves or currents, or by advection along the seafloor of suspended material from a river. This can clearly be seen on the Oregon shelf, where the development of a bottom nepheloid layer is related to advection of suspended matter from the Columbia river mouth located further north. This southward supply dominates: less than 10% of the suspended matter transport is perpendicular to the coast. Resuspension by waves and currents occurs both on nearshore mud deposits as well as on river-supplied mud at the outer shelf. Bottom nepheloid layers are less frequent when coastal upwelling occurs and ocean water flows in over the shelf floor (Pak and Zaneveld 1977; McCave 1979b). Mid-depth turbid layers have been observed that are related to the presence of a pycnocline over the shelf (Harlett and Kulm 1973). This has also been observed in the North Sea and was found to be related to the presence of plankton at the thermocline.

A turbid bottom layer is often absent. In the North Sea, for instance, it is not found even during the winter, when much bottom sediment is moved by waves down to a water depth of at least 70 m (probably to double that depth). No higher concentrations were found in the bottom water than in the surface water (Eisma and Kalf 1987a, b). Much of the North Sea floor consists of sands and gravels with very small amounts of fine material or of very consolidated old clays. River supply or supply from erosion or resuspension along the coast is not in high amounts or in large concentrations. Vertical mixing by the tides and during storms is strong so that large gradients in suspended matter concentrations do not develop. It should, however, be realized that suspended matter sampling is not done during the winter storms but only before or afterwards so that any suspended sand will have settled already when the sampling is done. As very much work has been done only in this way, and very little has been done with in-situ measuring equipment that records continuously, our knowledge on suspended matter transport on the shelf remains still very incomplete.

7.6.4 Suspended Matter on the Outer Shelf

Near to the shelf edge, current velocities are usually higher than more inward on the shelf, as was summarized by Stanley et al. (1972): (1) internal waves develop in the thermocline which in the ocean is located at 100 to 200 m water depth and interact with the shelf break, (2) the tidal wave that comes in from the ocean suddenly comes in much shallower depth, which results in acceleration, and (3) internal tides can develop: interaction between the tides and the bottom topography near the shelf break, resulting in internal tidal mixing, was described by New and Pingree (1990) in the Bay of Biscay. Therefore on the outer shelf, fine-grained material can be winnowed out and suspended material prevented from settling (Heathershaw et al. 1987). Ocean waters can move in over the shelf by advection or from deeper levels by upwelling: as the ocean water usually has a much lower suspended matter concentration than the shelf waters, the outer shelf is generally an area of relatively clear water, rather coarse and often (reworked) relict bottom deposits, nondeposition and winnowing out of fine material. McCave (1971, 1972, 1985a) has argued that the seaward limit of suspended matter deposition lies where the suspended matter concentrations decrease relative to the rate and frequency of resuspension by waves and currents. This limit is displaced seaward under higher (wave) energy conditions, so that a mid-shelf or outer-shelf mud deposit can be formed. As seen above, a number of mid-shelf mud deposits exist that receive a supply of suspended matter from the coast or the inner shelf. Advection to the outer shelf of

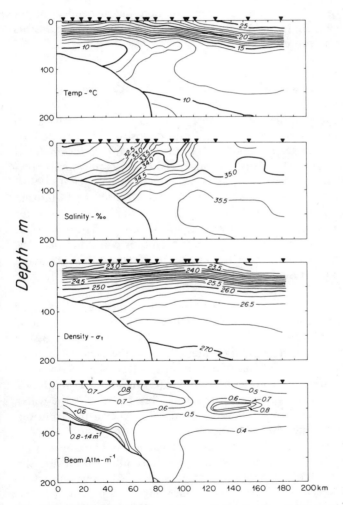

Fig. 7.48. Distribution of temperature, salinity, density, and transmissometer results (beam attenuation) over the shelf edge and upper slope off Cape Cod, Mass., USA (Churchill et al. 1988)

suspended matter resuspended further landward on the shelf off Cape Cod has been observed by Churchill et al. (1988), but on the outer shelf the formation of mud deposits is counteracted by the higher near-bottom velocities and inflow of ocean water.

On the shelf there is a cross-shelf gradient in the bottom stress because of the decrease of surface wave amplitude with depth. On the outer shelf, not the regular waves but the episodic periods of high bottom stress, caused by storms, resuspend the bottom sediment, together with bottom trawling and the more regular interaction of internal waves and tides with the shelf break. The (re)suspended material at the shelf break moves intermittently further offshore over the slope along density surfaces (Fig. 7.48). This also occurs along the shelf edge in the Golfe du Lion (NW Mediterranean), but here the suspended matter moving off the shelf is taken up and diluted by the general circulation towards the SW (Durrieu de Madron et al. 1990).

7.6.5 In-Situ Measurements and Modeling

To understand the suspended matter transport on the continental shelf and to be able to predict suspended matter transport and deposition, the water flow over the shelf has to be understood as well as the behavior of the material suspended in the flow. Surface waves and wind-forced currents play an important role on the shelf as well as density gradients and the tides. Lyne et al. (1990a, b) have shown that neglecting the influence of surface waves off the California coast leads to erroneous results on shelf sediment transport even at a water depth of 80 m. As storm conditions usually cause strong sediment transport but are not favorable for making measurements from a ship or collecting samples, in-situ measurements are required. This also applies to measurements involving suspended matter, usually present as flocs, that easily break up during sampling (see Chap. 6).

Under a combined flow of waves and currents, two distinct boundary layer regions develop: an oscillatory boundary layer with a thickness in the order of cm in the immediate vicinity of the bottom (3–5 cm with small waves, 10–30 cm with large waves), and a much larger boundary layer, called the "planetary boundary layer" with a thickness of tens of meters: at a bottom shear velocity of $1 \, \text{cm} \cdot \text{s}^{-1}$ and at a latitude of 40° its thickness is 40 m (Grant and Madsen 1986). A third layer that can be distinguished is the Ekman layer, which is the water layer where the flow is affected by the deflection caused by the earth's rotation; its thickness can be many times the layer thickness of the bottom boundary. The surface mixed layer over the shelf is typically 20 to 30 m deep, but mixing may go deeper during storms and interaction between the surface mixed layer and the bottom boundary layer will occur. When the water is stratified, or when surface waves break, this will affect the depth of the surface mixed layer as well as the thickness of the bottom boundary layer. Most of the shelf flow is hydrodynamically rough because of the presence of sandy bottom sediments and topography.

As seen above (Chap. 5), sediment in suspension can modify the flow; concentration gradients may result in stratification and change the layer characteristics of the bottom boundary. Here, mainly behavior of suspended matter will be discussed. For a full treatment of the flow of water over the shelf, the reader is referred to the review paper by Grant and Madsen (1986) and to the work cited there.

On the shelf, wave effects are very important for the development of a suspended sediment stratification because of resuspension of bottom material. Grant and Madsen (1986) showed that this stratification is determined by the product of particle concentration and particle fall velocity, not by concentration alone. In practice, both the concentration distribution and the particle fall velocity are difficult to measure in situ because of the difficulty of measuring a reliable mean particle concentration near to the bottom and because of the flocculation of suspended particles. Glenn and Grant (1987) have shown that the stratification induced by suspended sediment influences mainly the lower, near-bottom part of the Ekman layer. The higher part of this layer is mainly influenced by temperature and salinity stratification. The importance of suspended sediment stratification is also reduced by the fact that suspended sand grains remain mostly near to the bottom and suspended finer material is usually distributed more or less uniformly over the water column, forming clouds rather than vertical gradients. In-situ measurements on the Norfolk banks, where wave action dominates (carried out by Vincent and Green 1990), showed that sand is intermittently suspended up to 20–30 cm above the bottom. This is not always associated with wave-induced currents but also with vortices developing at the bottom ripples. The interaction of waves and currents resulted in a strong shoreward transport close to the bottom, offshore transport between 5 and 15 cm above the bottom, and a weak onshore flux above 15 cm. In a similar wave-dominated environment, Hanes and Huntley (1986) had found that suspension of bottom sands was stronger during the onshore phase of wave motion than during the offshore phase, which indicates that flow acceleration may be more important than flow velocity for suspending sand grains. The vertical gradients in the suspended sediment flux resulted in time lags between the moment of resuspension at the bed and the effect of this at a certain height above the bottom. Off Long Island, in-situ measurements indicated that the transport of sand as bed load dominated (Vincent et al. 1983): transport of suspended sand was ca. two orders of magnitude less. The bed load and the coarse suspended load had a strong onshore component, while the fine suspended sediment was transported approximately parallel to the coast. The predominance of sand transport as bed load confirms the distinction, made in practice already for a long time, between bed load and suspended load while disregarding the limited amounts of (re)suspended bottom sands.

The effect of storm waves at greater water depths on the shelf was demonstrated by Drake and Cacchione (1985) by in-situ measurements at 84 m water depth off California. During the summer, wave motion and currents were generally below the threshold level for resuspension of the local bottom sediment (sandy-clayey-silt), but they increased strongly during the winter: resuspension occurred by large waves from distant storms, short waves from local storms, and by currents associated with the outflow of the nearby Russian River. During the winter, suspended matter concentrations reached 150 mg.l^{-1}, where during the summer they were ca. 7 mg.l^{-1}.

A limited validation of descriptive modeling of suspended sediment concentrations and transport has been done at measuring sites by comparing the model results with the measurements and estimating transport under different conditions (e.g., Lyne et al. 1990a, b). Models covering a larger area have been made for Hangzhou Bay, the Irish Sea, and the North Sea (Han and Cheng 1986; Onishi and Thompson 1986; Puls and Sündermann 1990). For Hangzhou Bay, regions of erosion and deposition were identified that agreed with the observations in nature. For the Celtic Sea, the concentration distribution of suspended sand, silt, and clay in the water column and the dispersal patterns were obtained with reasonable success. The North Sea numerical model of Puls and Sündermann (1990) is the most comprehensive one developed so far and is based on an existing 3-D flow model to which wave data have been added. Assuming reasonable figures for settling velocity, bed shear, and suspended matter input, a partial agreement is obtained with measured suspended matter concentration distributions, areas of deposition and deposition rates, dispersal from different sources, and areas of bottom sediment resuspension (erosion). There are still many uncertainties in such modeling: the model of Puls and Sündermann is not very sensitive to processes in the water column but is very sensitive to the assumptions made for bed shear velocities and for deposition, erosion, and bioturbation. It is towards these aspects that also most in-situ measurements are directed. Techniques for in-situ measurement on the shelf of velocity gradients, turbulence, concentration gradients, in-situ particle size, and in-situ settling velocity are available (see Chaps. 5 and 6): by such measurements most of the questions raised in modeling suspended matter transport can be answered in the future. This will eventually lead to predictive modeling, which at present is still very unreliable, if feasible at all.

7.7 Suspended Matter in Canyons and Along the Continental Slope

Suspended matter distributions in submarine canyons typically show high concentrations in the surface water and near to the bottom, with often high concentration layers at mid-depths separated by low concentration layers. The high surface concentrations, when not caused by advection from a coastal source, are related to plankton production in the euphotic zone. The high concentrations in the bottom waters can be explained in different ways: by resuspension (tidal currents, internal waves), by inward concentration of suspended matter, and by preventing suspended matter settling out from above to reach the bottom. Bottom currents in canyons, as far as measured, are strongly oscillatory, related to the tides, and are strong enough at least part of the time to resuspended bottom sediment, or to keep settling particles from being deposited (Keller et al. 1972; Shepard et al. 1979). Inward flow can be stronger than outward flow (as, e.g., in the Zaire River canyon) and in that way concentrate suspended matter in the canyon. Resuspension can occur also by intrusion of cold water moving upwards in the canyon head by internal waves that are focused towards that direction. In Baltimore canyon, this results in resuspension at 200 to 800 m water depth in the canyon axis (Gardner 1989). It is a seasonal phenomenon: the upward surge of the internal waves occurs predominantly in late winter and early spring, when the stratification in the canyon is less pronounced. The resuspended material can settle again in the canyon, but the turbid water, that is formed and remains in existence for some time when the suspended material does not settle out quickly, can flow downward in the canyon because of its higher density (water + suspended particles). When arriving at a density level of the same magnitude, the downward flow can detach itself and flow out more or less horizontally along a surface of equal density (isopycnal surface). In this way, a mid-depth turbid layer can be formed that extends oceanwards and remains in existence until it loses its character by mixing or by settling out of the particles. This is analogous to density-driven flow in the ocean (Drake and Gorsline 1973; Drake et al. 1978; Gardner 1989). Another possibility is that mid-depth turbid layers are formed by settling of particles from the surface water which are retarded at a certain density level. The particulate material itself, however, does not need to be resuspended or settling inorganic material: mid-depth turbid layers described from Wilmington canyon, associated with cold water intrusions, consist of dinoflagellates and silicoflagellates (Gibbs 1982c). This suggests a relationship either with the temperature requirements of the organisms, or with an optimum layer where nutrients and light are both available in amounts that are sufficient for plankton growth.

Besides inward flow there is also outward flow in canyons. Turbidity currents occur episodically, but probably more regular is downward movement because of excess density, which is enhanced by the steep slope. Suspended matter concentrations often are sufficient for such a flow to occur (Drake and Gorsline 1973; Baker 1976). It is limited by the density stratification: the flow should have a higher density than the surrounding water or the flow will stop. At such a point, accumulation occurs and the flow may detach itself but also particulate matter may settle out, thus reducing the density.

Similar processes occur on the continental slope. Turbid layers stretching out from the continental shelf into the adjacent ocean have been described off California at 70 to 90 m water depth (Drake 1971), off Peru (Pak et al. 1980a) at 200 to 400 m water depth, off the Oregon-Washington shelf at the thermocline and at 150 to 200 m water depth (Pak et al. 1980b; Fig. 7.49), and off Cape Cod at 40 to 50 m water depth (Churchill et al. 1988; Fig. 7.48). Off California, the turbid zones fan out from the outer shelf into the ocean. Off Peru, the turbid layers coincide with the oxygen minimum layer. This layer is also high in nitrite

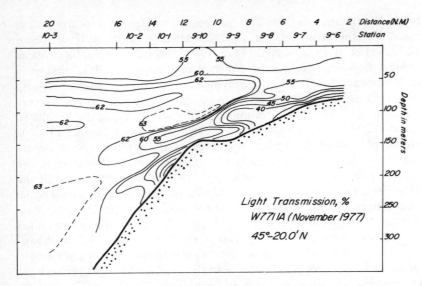

Fig. 7.49. Distribution of light transmission over the Oregon shelf, NW USA, in November 1977. (Pak et al. 1980b)

and ammonia, which indicates mineralization of particulate organic matter, but the particulate matter is assumed to have come from the bottom water over the adjacent outer continental shelf and not from the surface water by settling. Also off Oregon-Washington, the turbid zone lies well below the euphotic zone (where plankton growth is possible) and also here the deeper layer can have been formed by settling out from the surface water or by horizontal advection from the shelf. It was found that the particle size in the deep turbid layer is the same as in the bottom nepheloid layer on the adjacent outer shelf, so that the latter was considered to be the source of the particles (Pak et al. 1980b). Postma (1969) and McCave (1972) have postulated that suspended matter flows can reach deeper levels than those where they are stopped because of reaching waters of approximately the same density. When the turbid layer flows out horizontally, particles will settle out and when these come near to the bed at a deeper level, a new flow can be formed (cascading). The profile off California gives some indications for this (Drake 1971; Fig. 3.16), but it has not been observed elsewhere. On many slopes, no turbid layers at all are found at intermediate depths in the vicinity of the shelf. As pointed out by McCave (1972), such layers are only possible where there are no strong boundary currents (as is the case off Peru, California, and Oregon-Washington). There are indications, however, that also at deeper levels along the slope suspended matter is being dispersed into the ocean, as will be seen below (Sect. 7.8).

7.8 Suspended Matter in the Ocean

The shelf area is ca. 10% of the total area of the sea. The amount of mineral (aluminosilicate) particles that comes from land, crosses the shelf, and escapes into the deep sea is limited to less than 10% of the supply from land ($\sim 1 \times 10^{15}$ g . y^{-1} or 3–8% following the data given in Table 2.6). Primary production on the shelf is 10–14% of the total production in the ocean, planktonic $CaCO_3$ production less than 10%, but opal production on the shelf is almost 50% of the total production of opal in the sea (Table 2.6). Because of rapid remineralization of organic matter and low production of pelagic carbonate, both are only a minor admixture in shelf sediments, but opal deposition rates on the shelf can be rather high, because of slower remineralization of opal, (relative to organic matter) coupled to a rapid settling of opal particles as part of flocs. Ledford-Hoffman et al. (1986) estimate that 10–60% of the total opal production in the sea is deposited on continental margins. This means that $\sim 2 \times 10^{15}$ g opal may be yearly transported from the shelf into the deep sea together with ca. 1×10^{15} g C . y^{-1} of organic carbon which is 10–20% of the organic C production on the shelf (as estimated by Walsh (1989) on the basis of sediment trap data) and a relatively very small amount of planktonic carbonate. The dispersal and sedimentation of particulate matter in the ocean is schematically given in Fig. 7.50 (from Hay 1974). Most of the escape from the shelf is along the bottom through a bottom turbid layer, through mass movements, slumps, and slidings of bottom material along the edge and slope, and by turbidity currents through canyons. Through the surface waters, by advection and mixing, only small amounts of suspended matter are moved off the shelf, even in large river plumes (the Zaire river plume only contains a few percent of the total river supply; Eisma and Kalf 1984). Both cascading of water over the shelf edge and the slope, and low-density suspended matter flows following the density stratification of the water over the slope are not common, and any effects remain within a limited distance from the shelf edge. Only turbidity currents result in substantial suspended matter transport that can extend far into the abyssal plains.

Eolian transport brings fine material (up to 30 μm; Windom 1976) directly into the ocean. Its contribution to the suspended matter on the shelf is usually negligible, except where other sediment sources are small or absent. The eolian supply to the ocean is insignificant compared to the total particle supply, but it is large when only mineral particle supply is considered (Table 2.3). It is of the same order as the supply from ice flows and only little less than the supply from rivers. As most of the eolian material comes from arid and semi-arid regions, this supply is concentrated in the ocean areas directly off these regions (west Africa, east Asia, western Australia, etc.). This is evident from dust clouds observed from satellites (Windom 1975) and from the composition of ocean floor sediments: those in the areas covered by the dust clouds coming out of the dry regions may contain up to 40% of wind-supplied particles (Goldberg 1971; Windom 1975, 1976).

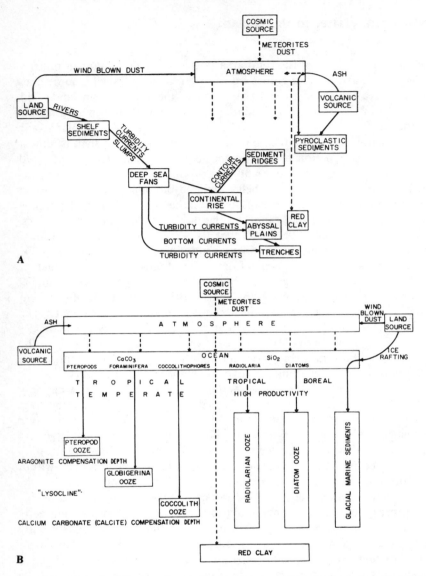

Fig. 7.50. A Scheme of clastic sedimentation in the ocean. **B** Scheme of pelagic sedimentation in the ocean. (**A, B** Hay 1974)

Particle supply in the reverse direction, from the ocean to the shelf, is probably large, although the concentrations of particulate matter in the ocean are low and the net supply is probably small. The concentration mechanisms on the shelf are limited to the area of fresh-water influence. Except off large river mouths, or where river mouths are located on the middle or outer shelf, this area is within the inner shelf. In the North Sea, the freshwater supply is about 1% of the supply of ocean water, the river supply of suspended matter is 10–15% of the total suspended matter supply, whereas the supply from the ocean is 20–25% (Eisma and Irion 1988). The outflow of suspended matter from the North Sea into the Norwegian Sea, however, is about equal to the inflow from the ocean: the effect of the near-shore concentration processes on the inflow from the ocean is within the statistical error of the data. That the net supply of particulate matter from the ocean to the shelf is likely to be small is indicated by the widespread occurrence of relict deposits on the outer shelf and by the very small admixture of silicate material on shelves where carbonate production (reefs, carbonate sand, and muds) prevails.

7.8.1 Particle Transport in the ocean

In the ocean, settling of suspended particles dominates, although there is horizontal transport primarily by surface currents and by the abyssal current system (western boundary currents). Particularly strong suspended sediment transport occurs along the North Atlantic Ocean floor, where bed transport as well as suspended matter transport are generated by contour currents that have their origin in the cold water overflow from the Norwegian Sea over the Scotland-Greenland ridge (Fig. 7.51). A bottom nepheloid layer is formed of up to several hundred meters in thickness, consisting of resuspended particles as well as particles that have settled from above and are kept in suspension. Particle concentrations in the ocean are in the order of $0.01–0.05$ mg.l^{-1}, whereas in the overflow water they are $0.1–0.2$ mg.l^{-1} (Brewer et al. 1976). The bottom nepheloid layer is thicker than the bottom mixed layer, which has a thickness of ca. 12 to 60 m. This has been explained by the existence of bursts of turbid water reaching from the bottom upwards far above the bottom mixed layer (Eittreim et al. 1975), and by advection from the adjacent slope by bottom mixed layers that have become detached from the bottom with only a small supply or loss of particles by vertical diffusion (McCave et al. 1980; McCave 1983; McCave and Tucholke 1986). This may also explain the layered structure of the bottom water with rather uniform layers of 10 to 30 m thickness separated by steep gradients. On the southeastern slope of the Gardar drift, which is a coarse-grained deposit, mud waves are found. The mud is probably deposited out of slowly moving nepheloid layers of ca. 300 m in thickness moving in from the northwest, where the particles were resuspended by the strong contour currents in that area.

Off Nova Scotia on the continental rise at 4000 to 5000 m depth, suspended particle concentrations in the nepheloid layer reach more than 1 mg.l^{-1}. Near-bottom velocities are usually less than 20 cm.s^{-1} but infrequently become higher for several days ("storms": out of 14 storms observed in 20 months, 10 were westward and 4 were eastward, Weatherly and Kelley 1985). Near-bottom currents then may reach 70 cm.s^{-1} (Emery and Ross 1968). General flow is

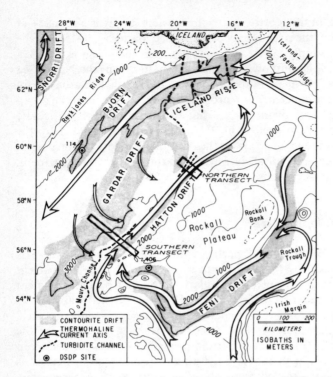

Fig. 7.51. Contour currents in the North Atlantic Ocean. (McCave et al. 1980)

westward, but when a Gulfstream meander or ring passes overhead, current directions become predominantly eastward. Resuspension of bottom sediment begins at 20 cm . s^{-1} (10 cm . s^{-1} at 1 m above the bottom), whereas deposition occurs at lower velocities. In the bottom nepheloid layer the lateral advection of detached bottom mixed layers shows up in changes in the suspended matter size distributions (Fig. 7.52 after McCave 1983). The analyses were done by Coulter counter: the maximum size was ca. 250 μm, but the in-situ size is probably larger, as at least part of the suspended material is flocculated (McCave 1985b).

   Bottom nepheloid layers at 400 to 600 m depth on the slope of Porcupine Bank (off Ireland) are believed to have been formed through resuspension by internal tides and waves (Dickson and McCave 1986). The deep ocean sediment

Fig. 7.52A, B. Vertical profile on the Nova Scotia continental rise (**A**) showing the potential temperature (*left*), the light attenuation coefficient c (*middle*), and particle size spectra at the indicated levels. Particle size was measured with a Coulter counter. *Numbers under each spectrum* indicate total measured particle volume (in ppb = mm^{-3} . m^{-3}). **B** Water flow at the Nova Scotia continental rise: *WBUC-NSOW* western boundary under current of Norwegian Sea Overflow Water; *DWBC-?AABW* deeper western boundary current of probably Antarctic Bottom Water. *Shaded area* region of highest abyssal eddy kinetic energy. (McCave 1983 and 1985b; partly after Schmitz 1984)

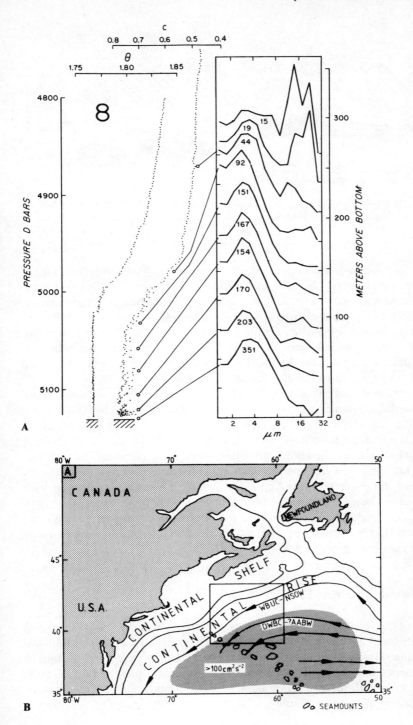

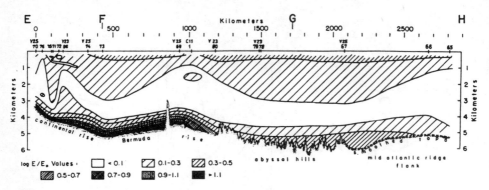

Fig. 7.53. Turbidity (light scattering) profile from the continental rise off New York to the flank of the Mid-Atlantic Ridge. (Spencer 1984)

transport and deposition was theoretically modeled by McLean (1985). Deposition by settling is relatively rapid so that distributions in the surface water are reflected in the deep sea sediments. There are indications, however, for horizontal transport by abyssal currents over considerable distances (> 1000 km; Bishop and Biscaye 1982). This transport is probably by a series of steps, temporary deposition followed by resuspension. The spatial distribution of Antarctic diatom frustules marks the flow of Antarctic Bottom water from the Weddell Sea almost up to the equator (Burckle 1981).

Lateral transport of particulate matter (resuspended bottom sediment) from the continental margin is indicated by the distribution of turbidity (Spencer 1984; Gardner et al. 1984; Fig. 7.53; see also Sect. 3.1.4) and by an increase of the concentration of alumino-silicate particles with water depth (Brewer et al. 1980; Honjo et al. 1982b, c). In addition to this, downward (density) flow (downwelling) and surface water disturbances may reach the ocean floor (storm disturbances, large-scale upward and downward movements associated with large eddies). These phenomena, however, as well as the abyssal western boundary flow and the advection from the continental margins, are regional disturbances. Large parts of the deep ocean are characterized by slow laminar flow in layers of often only a few meters in thickness.

7.8.2 The Vertical Particle Flux

Estimates of the vertical flux of particles have been made on the basis of series of samples collected at different water depths (with sampling bottles, with in-situ large volume filtration, and with sediment traps), and on the basis of bottom sediment deposition rates determined for various fractions: alumino-silicates, organic carbon, carbonate, (foraminifera, coccoliths, pteropods), opal (diatoms, radiolarians), or elements, including radioactive elements. Sampling with bottles or in-situ filtration gives data (concentrations) at a certain point in time as well as in space; sediment traps integrate fluxes over days or months, but sedimentation rates of bottom sediments are integrated over hundreds to thousands of years because of bioturbation and the small amounts yearly deposited in most ocean areas. Rates for terrigenous muds on the continental rise and adjacent abyssal plains are in the order of $7-300$ g.cm^{-2}.1000 y^{-1} ($\approx 50-2000$ mm. 1000 y^{-1}), for calcareous ooze $0.7-9$ g.cm^{-2}.1000 y^{-1} ($\approx 5-60$ mm. 1000 y^{-1}), for silicious ooze $0.3-1.5$ g.cm^{-2}.1000 y^{-1} ($\approx 2-10$ mm. 1000 y^{-1}) and for "red" deep sea clays $0-0.5$ g.cm^{-2}.1000 y^{-1} ($\approx 0-3$ mm. 1000 y^{-1}) (Seibold and Berger 1982). For comparison, deposition rates on the shelf vary from 0 g.cm^{-2}.y^{-1} in areas with relict deposits to high rates off river mouths (more than 800 g.cm^{-2}.y^{-1} off the Huang He). The vertical particle flux is mainly sustained by the sinking down of particles of 10^2-10^4 μm diameter (marine snow), which are abundant and ubiquitous in the ocean (Smayda 1971; McCave 1975; Alldredge 1984), but also other types of particulates are present. They can be subdivided into a biogenic group (fecal pellets, intact organisms and hard parts of plankton, amorphous aggregates), and an inorganic group (clay aggregates, sediment particles, and benthic fluff; Fowler and Knauer 1986). Fecal material occurs as large amorphous aggregates and as fecal pellets that consist of organic material tightly packed together (with some inorganic material that was also ingested by the organism). They range in size from a few micron to several mm and have often a characteristic species-related shape, size, and composition. Most types of pellets are held together by a membrane. This may remain intact for a week or even longer while pellets are deposited in the deep ocean and are incorporated in the bottom sediment. Small silicate particles (clays) that are incorporated make them heavier, increasing their settling rate. In coastal waters, concentrations of fecal pellets range from a few to several hundred per m^3 and downward fluxes of 3×10^5 pellets m^{-2}.day^{-1} have been estimated. In the open ocean, fluxes of 1.5×10^3 pellets m^{-2}.day^{-1} were found, increasing to 5×10^6 m^{-2}.day^{-1} for very small pellets (for references see Fowler and Knauer 1986). The small ($3-50$ μm) pellets, although 10^3 times more in number than the large (> 50 μm) pellets, nevertheless contribute only $11-49\%$ of the carbon flux from the large pellets (Gowing and Silver 1985). Pellets found below the surface waters can come from the surface, but also from organisms producing pellets at deeper levels. In this way particulate material can be "repackaged". Settling rates of pellets are in the order of $31-122$ m.day^{-1} but laboratory measurements indicate that large plankton and fish feces may reach 2700 m.day^{-1}. This is a "potential"settling rate as opposed to real in-situ settling rates, which include the effect of turbulence.

Intact organisms or exoskeletons consisting of $CaCO_3$, opal, or organic matter are subject to degradation (dissolution, remineralization) which will be discussed below. Hard parts (molts, fish scales, macrophytes such as *Sargasso*) are regularly found, but molts and fish scales have a short life (in the order of days). Carcasses are also rapidly consumed. They sink fast to the bottom ($1500-4000$ m.day^{-1}) and together with macrophytes can form a substantial part of the food of deep sea organisms. Pellet production is much higher than molt production (for euphausids 2–5% of the dry body weight against ca. 1% molts). Salp pellet production is 2.5–40 times larger. There is, however, a scarcity of data and the in-situ production rate of hard parts is not adequately known.

Marine snow is most abundant in ocean surface waters, where up to 79 flocs per liter have been observed. Small flakes may be present in concentrations of two to six flakes per liter. Below 200 m water depth, the floc concentration decreases to 0–1 floc . l^{-1} (Alldredge 1984). The downward flux can reach 10^5 flocs . m^{-2} day^{-1}; sinking rates of marine snow are usually in the order of 50–150 m . day^{-1} (Alldredge and Gotschalk 1988b) but large flocs of phytoplankton detritus may reach 300–1020 m . day^{-1}. The downward flux of marine snow is much influenced by the size of the phytoplankton and the size and trophic level of the consumers. Michaels and Silver (1988) have shown that picoplankton, even where it is the dominant producer, contributes little to the downward flux because of the number of trophic steps to the consumers that produce the sinking particles, each step involving a loss of roughly 90% of the available POC. The large plankton, such as sampled with nets, contributes an important amount, even when this is only a small part of the total primary production. In this way, different types of foodwebs may result in a very different downward flux of POC (Frost 1984).

Clay particles of only a few μm diameter, that are present in concentrations of 10 μg . l^{-1} in the ocean (Lal 1980), have settling rates of 0.5 m . day^{-1}, but are thought to become aggregated with large particles that have a sticky surface, to be incorporated in marine snow, or to be packaged into fecal pellets by pelagic organisms. Large-sized inorganic suspended particles are very rare and supposed to be supplied laterally from the continental margin (Brewer et al. 1980; Honjo et al. 1982c; Spencer 1984). Benthic fluff is an unconsolidated floc layer of 2–5 cm in thickness, very fluid, and reddish brown. It is different in composition from the downward particle flux several hundred meters above the bottom and probably represents the transformation of a highly biogenic flux into a bottom sediment that is poor in organic matter or, in the case of deep sea clays, mainly inorganic.

To the downward settling particles coming from the surface water (euphotic zone), a flux of new particles is added that are produced at deeper levels. This was found as an increased flux of organic carbon at different depths below the surface zone (Knauer and Martin 1981; Walsh et al. 1988; Fig. 7.54). It correlates with an increased fecal pellet flux and with the presence of microorganisms associated with large rapidly sinking particles and presumably using NH_4^+ and HS^- as an energy source. This was found in several areas and may be a general feature. It can be explained by chemosynthetic carbon fixation (Karl and Knauer 1984; Karl et al. 1984), by zooplankton feeding and pellet production, or by particle aggregation effects. The over-all effect is an enhanced availability of carbon for other organisms and, where particle size (floc size) increases, a higher settling rate. The role of bacteria, and the transformation of dissolved organic matter into particulate, is not yet clear. Cho and Azam (1988) and Angel (1989)

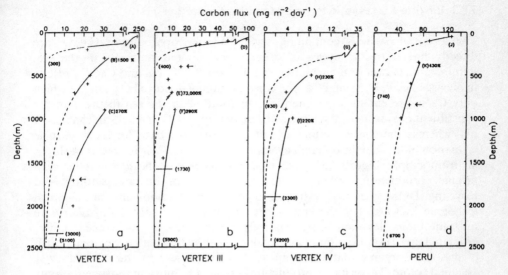

Fig. 7.54a–d. Organic carbon flux at four pacific Ocean trap deployment sites. Capital letters *B, C, E, F, H, I,* and *K* refer to aphotic areas of relative increases in organic carbon flux (indicated by *numbers in percent*). (Knauer et al. 1984)

indicate that 80% to almost 100% of the organic carbon flux is associated with microbes.

The downward particle flux may also be influenced by the vertical migration of small organisms (plankton or nekton). For plankton this migration is limited to the upper 700 m, for nekton to the upper 1000–1500 m. This is particularly of importance when the migration speed of the organisms is higher than the settling rate of the particles they produce. The organic matter that settles on the ocean floor (as fluff when large amounts are involved, otherwise as single flocs) is degraded at the sediment surface or in the upper mm of the sediment mainly within a few months (Reimers 1989). After that, degradation of the residue continues slowly.

Upward fluxes below the turbulent surface water have been thought to occur in the ocean only by diffusion and only for dissolved material, but particles may be moved upwards in vertical shear zones between water masses: in this way material in suspension in the western boundary currents may be moved upward, as was discussed above (Fig. 3.15). There is also some upward flux of buoyant particles with a density of 0.9–1.0 (oily material, fish eggs, eggs and larvae of benthic deep-sea organisms). They may reach the surface from 5000 m depth within a year, probably within a week. This flux was thought to be only 1–4% of the downward flux (Simoneit et al. 1986), but the content of lipids is much higher in the upward flux, so that the ratio between downward and upward transport may be very different for different substances. Recently, Smith et al. (1989) found that the upward flux of organic matter may be as much as 67% of the downward flux.

7.8.3 Particle Fluxes in Relation to Organic Production

The fluxes of particles containing organic matter, opal, and/or $CaCO_3$ are directly linked to the cycles of organic carbon and nutrients in the sea. Remineralization of organic matter and dissolution of opal and $CaCO_3$ results in release of C, P, N, Si, and Ca, and consumption of dissolved O_2, among which only Ca is present in large quantities in solution in the sea (being a major constituent of seawater). Uptake or release of C (as HCO_3^-) results in changes in pH, whereas the concentrations of nutrients form the basis for (new) particle formation. But while the different cycles are interlinked, their interrelationship is far from simple. Figure 7.55 gives a general scheme of the relations between primary production, grazing, carnivores, particle production, settling, and recycling. DOM is added as an unknown. Recent developments in analytical technique indicate that the concentrations of large dissolved organic, more resistant molecules containing relatively much nitrogen, have been underestimated up to now (Sugimura and Suzuki 1988), so that actually twice as much of dissolved organic matter, or more, may be present in the sea than was assumed before. The primary production is related to nutrient (nitrogen) supply and the availability of sunlight, which are determined by hydrodynamical and climatological conditions: upwelling, river outflow, vertical and horizontal mixing, convergence, and divergence. In the temperate zone this results in spring plankton blooms, followed by steady-state conditions during the summer that are characterized by regeneration (recycling). The trade wind zones are characterized by upwelling along (mainly) the eastern side of the continents grading into the steady-state recycling of the oceanic gyres. Both summer conditions in the temperate zone and conditions in the oceanic gyres at lower latitudes are characterized by the presence of a thermocline.

Upwelling, and winter mixing followed by spring light increase, are characterized by plankton bloom populations with a low diversity of species, that are present in large quantities, and incomplete recycling, leaving organic material to be moved out of the area or to be deposited on the sea floor (Peinert et al. 1989).

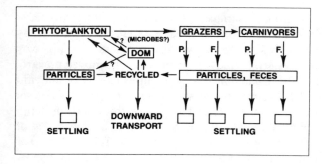

Fig. 7.55. General scheme of the relations between primary production, grazing, carnivores, particle production, settling, and recyling, *P*. particles; *F*. feces. Possible upward transport is not indicated

The plankton populations are geared to an abundance of nutrients, food, and sunlight, and consist primarily of diatoms and coccolithophorids.

Under steady state, conditions prevail that are characterized by recycling in the surface water. The plankton populations are multispecific, with low numbers of specimens and geared to low concentrations of nutrients and food; the species producing skeletons are mainly foraminifera and pteropods. Many plankton organisms can regulate the conditions in their cells in accordance with the environmental conditions, but there is still a large lack of knowledge on such mechanisms. Diatoms are more buoyant when conditions are favorable, and less buoyant under bad conditions. Old diatom cells produce mucus, which results in aggregation of cells and rapid settling. Coccoliths develop a large form, which also leads to rapid settling. Pteropods were found to settle in sediment traps in the Norwegian Current only in autumn, which may have been caused by starvation, by old age, or by predators. Also flagellates can settle in large amounts to the sea floor, as was found for colonial Phaeocystis in the Barents Sea.

Besides the primary production, the type of grazing influences the amount of particulate matter that settles out of the surface water. Swarm-grazers (fish, krill, salps), which have a patchy distribution, produce large feces (rapid settling) and favor export of particulate matter from the surface. Grazing by copepods, which have a widespread and more even distribution than the swarm-grazers, produce small feces (slow settling), and recycling is favored. When grazing starts during a bloom, and when it can more or less keep pace with a phytoplankton production, recycling is favored. When grazing starts late, or even after the bloom, much particulate organic matter has had the chance to settle out. In this way, the relation between the life cycles of the grazers and the phytoplankton organisms may strongly influence the downward particle flux. In general, therefore, the downward flux of material reflects the whole system (as schematically indicated in Fig. 7.55) and not only the primary production, so that often there is no direct relation between the primary production (or the upward flux of nitrate) and the downward flux of biogenic material. Large differences in downward flux may develop, but Knauer et al. (1984) showed on the basis of sediment trap data that at a similar range of primary production rates in the surface water, the downward flux of C and N at the lower limit of the euphotic zone is much larger in coastal waters off central California and Peru than in oceanic areas off Mexico and Hawaii and in the North East Pacific. This was tentatively explained by a possible two-layer structure of the euphotic zone in the ocean, with a very productive surface layer and a much less productive one below it. Only the material left over from the lower layer would leave the euphotic zone.

Six years of sediment trap measurements in several ocean basins (Sargasso Sea, Panama Basin) have shown an annual periodicity coupled to great differences in flux level for different years and areas (Deuser and Ross 1980; Billett et al. 1983; Deuser 1984, 1986). The seasonal fluctuations indicate rapid settling of the particles ($50–100$ m . day^{-1}). The downward flux is inversely related to surface water temperatures; lower temperatures coincide with greater mixed layer depth and greater availability of nutrients.

7.8.4 Compositional Changes

During settling, the composition of the organic matter changes. In marine plankton the mean ratios of C:N:P are 106:16:1 (Redfield et al. 1963). Many N and P compounds degrade more rapidly than the organic carbon compounds so that the ratios of C:N and C:P in particles containing organic matter increase with waterdepth. In Section 2.4, simplified and tentative cycles of organic matter, SiO_2 and $CaCO_3$, and N and P, were discussed. The formation of new organic matter particles was not included as this cannot yet be quantified on a world scale. Living organisms that use or release dissolved organic material are associated with settling organic matter (Fellows et al. 1981; Cho and Azam 1988). The changes in the organic matter composition of particles settling from the surface occur predominantly in the upper 1000 m and become much less in deeper water.

The dissolution of opal and calcium carbonate in the deep sea is also not a uniform process. The solution of opal varies with particle origin: diatoms generally dissolve in the upper few hundred meters (Honjo et al. 1982a), but there are thin-walled diatoms that dissolve more rapidly than thick-walled ones. Radiolarians and silicoflagellates are more resistant than diatoms: in the North Equatorial Pacific, ca. 80% of the radiolarians and 100% of the silicoflagellates is dissolved not in the surface waters but at the bottom (Cobler and Dymond 1980; Takahashi 1987). $CaCO_3$ is present as calcite (coccoliths, foraminifera) and as aragonite (pteropods), which is less stable. In the North Pacific, 90% of the aragonite is rapidly dissolved in the upper 2200 m (Betzer et al. 1984), while calcite is mainly regenerated on the ocean floor (Dymond and Lyle 1985). In general, the fluxes of opal and $CaCO_3$ below the euphotic zone are related to the seasonal and regional production patterns in the ocean surface waters so that the distribution of diatoms, radiolarians, coccoliths, etc. in the surface waters as well as in the oceanic bottom sediments reflect the production areas. The paucity of data, however, on the primary-secondary production system and the influence of recycling, and the uncertainties on the processes that influence the vertical particle flux in the ocean below 1000 m depth, (called "transient flux" by Berger et al. 1989), make it difficult to generalize and to model particle fluxes.

Modeling, based on particle size, Stoke's settling velocities, and dissolution/remineralization rates, was done to understand particle solution and recycling (Krishnaswami and Lal 1973; Lal and Lerman 1973; Lerman et al. 1974, 1977; Brun-Cottan 1976, 1985; Lerman and Lal 1977; Lal 1980). This was mainly done for particles of only a few micron in size up to several hundred micron (plankton skeletons). On the basis of solution rates, residence times of 60 $\pm$ 20 y were found for small particles in the deep ocean, whereas on the basis of Stokes' settling rates the residence time would be $> 10^2$ y (Lerman et al. 1977).

This agrees with a residence time of Pb-210 (which is adsorbed onto particles) of 50–100 y, as found in the deep Pacific. In the surface water the residence times are 0.66 and 4.5 y, respectively. Brun-Cottan (1976),however, made clear that settling following Stokes' law occurs only for particles larger than 4–5 μm and Hawley (1982) found that particles < 100 μm generally settle faster than Stokes' law predicts, the deviation being more pronounced for smaller particles. This can be explained by aggregation of the particles, whereby the smaller particles that settle slower are scavenged by the faster-settling larger ones (Lal 1980; Hawley 1982). According to Lal (1980), the very fine particles (1–10 μm), with settling rates of only a few meters per day (or less), mainly interact with the chemistry of the ocean water, for which the larger particles settle too fast. Brun-Cottan (1985) arrived at a mean settling rate of ca. 200 m . y^{-1} for particles with a median size of 1.7 μm. As was seen above, however, it is not known to what extent the small particles measured by Coulter counter or seen under a microscope (Lambert et al. 1981) exist in nature as single particles: the break-up of flocs during sampling and analysis may have produced them.

Postscript on Sediments

The ultimate fate of suspended material is to be deposited, or mineralized (organic matter), or dissolved (opal, carbonate). Although suspended particles, taken up in turbulent whirls, may remain in suspension (indefinitely) as long as the turbulence remains, the residence time of suspended particles, even in the ocean, is in the order of years or less. This in contrast to "dissolved" material smaller than 0.4 μm, which may be removed quickly, but also may remain in solution much longer, up to 10^8 or 10^9 years, or as long as seawater exists on earth.

Fine-grained bottom sediments are formed out of suspended material and have a composition that reflects the composition of the original suspended material, but there are also differences. Mineral particles, being resistant weathering residues, are rarely changed during or after deposition, but contents of organic matter, opal, biogenic carbonate, and substances adsorbed onto particles can be considerably lower, or higher, in the bottom sediments. Also, the floc structure of the suspended material is changed or lost soon after deposition, and the grain size (the size of the constituent grains) can be different because of the size selection during deposition or (repeated) resuspension. The changes on deposition may be relatively rapid, and may take hours or days (e.g., mineralization of organic material), or a much longer time (e.g., accumulation of manganese in the top sediment, which takes place on a time scale of months or years). Further alteration may take 10^3 to 10^6 years.

As well as being an indication of the composition of the material in suspension, bottom sediments can give an indication of the fluxes of suspended material. This can be seen in the deposition rates, which are $10^1 - 10^2$ mm . y^{-1} in deltas and fjords (Nichols 1989), $1-10$ mm . y^{-1} in lakes, estuaries, tidal marshes, coastal lagoons, and some shelf deposits. On the shelf and in the ocean they decrease to less than 1 mm . y^{-1}, reaching zero on parts of the shelf, particularly the outer shelf. This distribution reflects the dispersal pattern of suspended matter as discussed in Chapters 2, 3, and 7.

Although bottom sediment can give general indications, it will be clear from this discussion that only in-situ studies can give insight into the composition and dispersal processes of suspended matter that forms the basis of fine-grained deposits but at the same time is part of – and reflects – the geochemical and biological as well as geological processes operating in the aquatic environment. Modeling of the behaviour of suspended matter is still too uncertain to provide more than, at most, a reasonable description of the existing situation.

References

Abraham G (1988) Turbulence and mixing in stratified tidal flows. In: Dronkers J, van Leussen W (eds) Physical processes in estuaries. Springer, Berlin Heidelberg New York, pp 149–180

Abrahamse J, Joenje W, van Leeuwen-Seelt N (1976) Waddenzee. Landelijke Vereniging tot Behoud van de Waddenzee, Harlingen. 368 pp

Adams ChE, Jr Swift DJP, Coleman JM (1987) Bottom currents and fluviomarine sedimentation on the Mississippi prodelta shelf: February–May 1984. J Geophys Res 92:C13 14.595–14.609

Adams ChE Jr, Wells JT, Park Y-A (1990) Internal hydraulics of a sediment-stratified channel flow. Mar Geol 95:131–145

Admiraal W, Breugem P, Jacobs DMLHA, de Ruyter van Steveninck ED (1990) Fixation of dissolved silicate and sedimentation of biogenic silicate in the lower Rhine during diatom blooms. Biogeochemistry 9:175–185

Airy W (1834) Min Proc Inst Civ Eng, p 227

Ali W, O'Melia CR, Edzwald JK (1984) Colloidal stability of particles in lakes: measurement and significance. Water Sci Technol 17:701–712

Alldredge AL (1972) Abandoned larvacean houses: a unique food source in the pelagic environment. Science 177:885–887

Alldredge AL (1976) Discarded appendicularian houses as sources of food, surface habitats, and particulate organic matter in plankton environments. Limnol Oceanogr 21:14–23

Alldredge AL (1979) The chemical composition of macroscopic aggregates in two neritic seas. Limnol Oceanogr 24:855–866

Alldredge AL (1984) Macroscopic organic aggregates. In: Worksh Global Ocean Flux Study, Woodshole, pp 167–179

Alldredge AL, Gotschalk CC (1988a) Direct observations of the mass flocculation of diatom blooms: characteristics, settling velocities and formation of diatom aggregates. Deep Sea Res 36:159–171

Alldredge AL, Gotschalk CC (1988b) In situ settling behavior of marine snow. Limnol Oceanogr 33:339–351

Alldredge AL, Silver MW (1988) Characteristics, dynamics and significance of marine snow. Progr Oceanogr 20:41–82

Allen GP, Castaing P, Klingebiel A (1974) Suspended sediment transport and deposition in the Gironde estuary and adjacent shelf. Mem Inst Geol Bassin Aquitaine 7:27–36

Allen GP, Sauzay G, Castaing P, Jouanneau JM (1976) Transport and deposition of suspended sediment in the Gironde estuary, France. In: Wiley M (ed) Estuarine processes II. Academic Press, New York, London, pp 63–81

Allen GP, Castaing P, Jouanneau J-M (1977) Mécanismes de remise en suspension et de dispersion des sédiments fins dans l'estuaire de la Gironde. Bull Soc Geol Fr XIX (2):167–176

Allen GP, Salomon JC, Bassoulet P, du Penhoat Y, de Grandpré C (1980) Effects of tides on mixing and suspended sediment transport in macrotidal estuaries. Sediment Geol 26:69–90

Allen T (1966) A critical evaluation of the Coulter counter. Soc Anal Chem Lond 84:110–127

Allersma E (1968) Mud on the oceanic shelf off Guiana In: Symp Invest and Res Caribbean Sea and Adjacent Regions, UNESCO, Paris, pp 193–203

Aloisi JC, Cambon JP, Carbonnel J, Cauwet G, Millot C, Monaco A, Pauc H (1982) Origine et rôle du néphéloide profond dans le transfer des particules au milieu marin. Ocean Acta 5:481–491

AmasSeds Research Group (ed) (1990) A multidisciplinary Amazon Shelf Sediment Study. EOS 71(45):1771–1777

Amos CL (1987) Fine-grained sediment transport in Chignecto Bay, Bay of Fundy, Canada. Cont Shelf Res 7:1295–1300

Amos CL, Mosher DC (1985) Erosion and deposition of fine-grained sediments from the Bay of Fundy. Sedimentology 32:815–832

Angel MV (1989) Does mesopelagic biology affect the vertical flux? In: Berger WH, Smetacek VS, Wefer G (eds) Productivity of the ocean: present and past. John Wiley & Sons, New York Chichester, pp 155–173

Angino EE, Billings GK (1967) Atomic absorption spectrometry in geology. Elsevier, Amsterdam, p 144

Anwar HO (1975) Turbulent dispersion and meandering of a surface plume. In: Proc 16th Congr IAHR, Sao Paulo, vol 1, pp 367–376

Anwar HO, Atkins E (1980) Turbulence measurements in simulated tidal flow. J Hydraul Div 106:1273–1289

Argaman Y, Kaufman WJ (1970) Turbulence and flocculation. J Sanit Eng Div Proc Am Soc Civ Eng 96:223–241

Ariathurai R, Krone RB (1976a) Mathematical modeling of sediment transport in estuaries. In: Wiley M (ed) Estuarine processes II. Academic Press, New York London, pp 98–106

Ariathurai R, Krone RB (1976b) Finite element model for cohesive sediment transport. J Hydraul Div, Proc Am Soc Civil Eng 102 HY3:323–338

Armstrong FAJ, Atkins WRG (1951) The suspended matter of sea water. J Mar Biol Assoc UK 29:139–143

Atkins WGR (1926) Seasonal changes in the silica content of natural waters in relation to the phytoplankton. J Mar Biol Assoc UK:89–99

Augustinus PGEF (1978) The changing shoreline of Surinam (South America). Thesis, State Univ Utrecht

Avnimelech Y, Troeger BW, Reed LW (1982) Mutual flocculation of algae and clay: evidence and implications. Science 216:63–65

Avoine J, Allen GP, Nichols M, Salomon JC, Larsonneur C (1981) Suspended-sediment transport in the Seine estuary, France: effect of man-made modifications on estuary-shelf sedimentology. Mar Geol 40:119–137

Bacon MP, Anderson FR (1982) Distribution of thorium isotopes between dissolved and particulate forms in the deep sea. J Geophys Res 87:2045–2056

Bagnold RA (1962) Auto-suspension of transported sediment; turbidity currents. Proc R Soc London Ser A 265:315–319

Bagnold RA (1963) Mechanics of marine sedimentation. In: Hill MN (ed) The sea, vol 3, Wiley-Interscience, New York, pp 507–528

Bagnold RA (1966) An approach to the sediment transport problem from general physics. Geol Surv Prof Pap 422-I, Washington

Baker ET (1976) Distribution, composition and transport of suspended particulate matter in the vicinity of Willapa submarine canyon, Washington. Bull Geol Soc Am 87:625–632

Bale AJ, Morris AW (1987) In situ measurement of particle size in estuarine waters. Est Coast Shelf Sci 24:253–263

Barbaroux L, Gallenne B, Ottman F, Margarel JP (1974) Evolution de l'estuaire de la Loire au Quaternaire. Mem Inst Geol Bassin Aquitaine 7:267–275

Barber R (1966) Interaction of bubbles and bacteria in the formation of organic aggregates in sea-water. Nature (Lond) 211 (5046):257–258

Barham E (1979) Giant larvacean houses: observations from deep submersibles. Science 205:1129–1131

Barishnikov NB (1967) Sediment transportation in river channels with flood plains. Publ Int Assoc Sci Hydrol 75:404–413

Bartholdy J (1984) Transport of suspended matter in a bar-built Danish estuary. Est Coast Shelf Sci 18:527–541

Bartholdy J, Pfeiffer Madsen P (1985) Accumulation of fine-grained material in a Danish tidal area. Mar Geol 67:121-137

Bates CC (1953) Rational theory of delta formation. Bull Am Assoc Petrol Geol 37:2119-2161

Bayazit M (1972) Random walk model for motion of a solid particle in turbulent open-channel flow. J Hydraul Res 10:1-14

Baylor ER, Hirschfeld DS (1962) Adsorption of phosphates onto bubbles. Deep Sea Res 9: 120-124

Baylor ER, Sutcliffe WH Jr (1963) Dissolved organic matter in seawater as a source of particulate food. Limnol Oceanogr 8:369-371

Bechteler W (1980) Stochastische Modelle zur Simulation des Transportes suspendierter Feststoffe. Wasserwirtschaft 6:70

Bechteler W (1986) Models for suspended sediment concentration distribution. In: Proc 3rd Int Symp River Sed, Univ Mississippi, pp 787-795

Bechteler W, Färber K (1982) Stochastic model for particle movement in turbulent open channel flow. In: Euromech 156, Istanbul, pp 165-171

Bechteler W, Färber, K (1985) Stochastic model of suspended solid dispersion. Proc ASCE, 111 J Hydraul Eng 1:64-78

Bechteler W, Vetter M (1983) Comparison of suspended sediment transport models with measurements. In: Proc 2nd Int Symp River Sediment, Nanjing, China, pp 402-420

Beebe W (1934) Half mile down. Harcourt, Brace, New York, p 344

Been K, Sills GC (1981) Self weight consolidation of soft soils: an experimental and theoretical study. Geotechnique 31:519-535

Berger H (1903) Geschichte der wissenschaftlichen Erdkunde der Griechen, 2nd edn. Von Veit, Leipzig, 662 pp

Berger WH, Smetacek VS, Wefer G (eds) (1989) Productivity of the oceans: present and past. John Wiley & Sons, New York, p 471

Berthois L (1961) Observations directes des articules sédimentaires fins dans l'eau. Rev Geogr Phys Geol Dyn IV, 1:39-42

Betzer PR, Byrne RH, Acker JG, Lewis CS, Jolly RR, Feely RA (1984) The oceanic carbonate system: a reassessment of biogenic controls. Science 226:1074-1077

Beverage JP, Culbertson JK (1964) Hyperconcentrations of suspended sediment. J Hydraul Div Proc ASCE 90 HY6:117-128

Biddanda BA (1988) Microbial aggregation and degradation of phytoplankton-derived detritus in seawater. I: Microbial succession. II: Microbial metabolism. Mar Ecol Prog Ser 42: 79-95

Biddle P, Miles JH (1972) The nature of contemporary silts in British estuaries. Sediment Geol 7:23-33

Biggs RB (1970) Sources and distribution of suspended sediment in northern Chesapeake Bay. Mar Geol 9:187-201

Billett DSM, Lampitt RS, Rice AL, Mantoura RFC (1983) Seasonal sedimentation of phytoplankton to the deep-sea benthos. Nature (Lond) 302:520-522

Biscaye PE, Eittreim SL (1977) Suspended particulate loads and transports in the nepheloid layer of the abyssal Atlantic Ocean. Mar Geol 23:155-172

Biscaye PE, Olsen CR (1976) Suspended particulate concentrations and compositions in the New York Bight. In: Grant Gross M (ed) Proc Symp Middle Atlantic Continental Shelf and the New York Bight. Am Soc Limnol Oceanogr Spec Symp 2:124-137

Bishop JKB, Biscaye PE (1982) Chemical characterization of individual particles from the nepheloid layer in the Atlantic Ocean. Earth Planet Sci Lett 58:265-275

Bishop JKB, Edmond JM (1976) A new, large volume in situ filtration system for sampling oceanic particulate matter. J mar Res 34:181-198

Bogdanov YA (1968) Quantitative distribution and granulometric content of suspended matter in the Pacific Ocean. Okeanol Issl 18:42-52 (in Russian)

Bohlen WF (1976) Shear stress and sediment transport in unsteady turbulent flows. In: Wiley M (ed) Estuarine processes, vol 2. Academic Press, New York London, pp 109-123

Bokuniewicz HJ, Gordon RB (1980) Sediment transport and deposition in Long Island Sound. In: Saltzman B (ed) Estaurine physics and chemistry: studies in Long Island Sound. Academic Press, New York London, pp 69–106 (Advances in geophysics 22)

Bokuniewicz HJ, Gebert J, Gordon RB (1976) Sediment mass balance of a large estuary, Long Island Sound. Est Coast Mar Sci 4:523–536

Bolin B, Degens ET, Duvigneaud P, Kempe S (1979) The global biogeochemical carbon cycle. In: Bolin B, Degens ET, Kempe S, Ketner P (eds) The global carbon cycle. John Wiley & Sons, New York, pp 1–56 (Scope 13)

Boon JD III (1978) Suspended-solids transport in a saltmarsh creek – an analysis of errors. In: Kjerve BJ (ed) Estuarine transport processes. Univ SC Press, Columbia, pp 147–160

Borah DK, Balloffet A (1986) Sediment routing model for estaurine harbors. In: 3rd Int Symp on River Sediment, Univ Mississippi, pp 443–452

Botazzi EM, Schreiber B, Bowen VT (1971) Acantharia in the Atlantic Ocean, their abundance and preservation. Limnol Oceanogr 16(4):677–684

Bothner MH, Spiker EC, Johnson PP, Rendigs RR, Aruscavage PJ (1981a) Geochemical evidence for modern sediment accumulation on the continental shelf off southern New England. J Sediment Petrol 51:281–292

Bothner MH, Parmenter CM, Milliman JD (1981b) Temporal and spatial variations in suspended matter in continental shelf and slope waters off the north-eastern United States. Est Coast Shelf Sci 13:213–234

Bowden KF (1980) Physical factors: salinity, temperature, circulation and mixing processes. In: Olausson E, Cato I (eds) Chemistry and biogeochemistry of estuaries. John Wiley & Sons, New York, pp 37–70

Bowen R (1988) Isotopes in the earth sciences. Elsevier, Amsterdam London, 647 pp

Boyle EA, Edmond JM, Sholkovitz ER (1977) The mechanism of iron removal in estuaries. Geochim Cosmochim Acta 41:1313–1324

Brahms A (1757) Anfangsgründe der Deich- und Wasserbaukunst. Aurich, Hannover

Bremner JM (1980) Physical parameters of the diatomaceous mud belt off South West Africa. Mar Geol 34:M67–M76

Brewer PG, Spencer DW, Biscaye PE, Hanley A, Sachs PL, Smith CL, Kadar S, Fredericks J (1976) The distribution of particulate matter in the Atlantic Ocean. Earth Planet Sci Lett 32:393–40

Brewer PG, Nozaki Y, Spencer DW, Fleer AP (1980) Sediment trap experiments in the deep North Atlantic: isotopic and elemental fluxes. J Mar Res 38:703–728

Brodie JW, Irwin J (1970) Morphology and sedimentation in Lake Wakatipu, New Zealand. NZ J Mar Freshwater Res 4:479–496

Broecker WS, Peng T-H (1982) Tracers in the sea. Columbia Univ Press, New York, 690 pp

Brooks NH (1965) Calculation of suspended load discharge from velocity and concentration parameters. In: Proc Fed Interagency Sediment Conf Miscell Publ 970, Washington, pp 229–237

Brun-Cottan JC (1976) Stokes settling and dissolution rate model for marine particles as a function of size distribution. J Geophys Res 81:1601–1606

Brun-Cottan JC (1985) Vertical transport of particles within the ocean. In: NATO-ASI Meet, Carcans

Brydsten H, Jansson M (1989) Studies of estuarine sediment dynamics using ^{137}Cs from the Tjernobyl accident as a tracer. Est coast Shelf Sci 28:249–259

Brylinski JM, Dupont J, Bentley D (1984) Conditions hydrobiologiques au large du Cap Gris Nez (France):premiers résultats. Oceanol Acta 7:315–322

Buat LG du (1786) Principes d'hydraulique, vérifiés par un grand nombre d'expériences faites par ordre du gouvernement, 2 vols. Paris.

Burban P-Y, Lick W, Lick J (1989) The flocculation of fine-grained sediments in estuarine waters. J Geophys Res 94:C6 8323–8330

Burckle LH (1981) Displaced Antarctic diatoms in the Almirante Passage. Mar Geol 39:M39–M43

Burt TN (1986) Field settling velocities of estuarine muds. In: Mehta AJ (ed) Estuarine cohesive sediment dynamics. Springer, Berlin Heidelberg New York, pp 126–150 (Lecture notes on coastal and estuarine studies, vol 14)

Byrne JV, Kulm LD (1967) Natural indicators of estuarine sediment movement. J Waterways Harbors Div Proc Am Soc Civ Eng 93 WW2:181–194

Cadée GC (1985) Macroaggregates of *Emiliana huxleyi* in sediment traps. Mar Ecol Prog Ser 24:193–196

Calvert SE (1983) Sedimentary geochemistry of silicon. In: Aston SR (ed) Silicon geochemistry and biogeochemistry. Academic Press, New York London pp 143–186

Carder KL, Schlemmer EC II (1973) Distribution of particles in the surface waters of the eastern Gulf of Mexico: an indicator of circulation. J Geophys Res 78:6286–6299

Carson B, Carney KC, Meglis AJ (1988) Sediment aggregation in a salt-marsh complex, Great Sound, New Jersey. Mar Geol 82:83–96

Carson RL (1951) The sea around us. Staples, London, p 230

Carter L (1976) Seston transport and deposition in Pelorus Sound, South Island, New Zealand. NZ J Mar Freshwater Res 10:263–282

Castaing P (1989) Co-oscillating tide controls long-term sedimentation in the Gironde estuary, France. Mar Geol 89:1–9

Castaing P, Allen GP (1981) Mechanisms controlling seaward escape of suspended sediment from the Gironde: a macrotidal estuary in France. Mar Geol 40:101–118

Castaing P, Allen GP, Houdart M, Moign Y (1979) Etude par télédétection de la dispersion en mer des eaux estuariennes issues de la Gironde et du Pertuis de Maumusson. Oceanol Acta 2:459–468

Castaing P, Philippo I, Weber O (1982) Répartition et dispersion des suspensions dans les eaux du plateau continental aquitain. Oceanol Acta 5:85–96

Cataliotti-Valdina D (1982) Evolution de la turbidité des eaux du complexe lagunaire de Bages-Sigean-Port-la-Nouvelle (Aude, France). Oceanol Acta 5:411–420

Chamley H (1990) Sedimentology. Springer, Berlin Heidelberg New York, p 285

Chandler JA (1977) X-ray microanalysis in the electron microscope. Practical methods in electron microscopy, vol 5, pt 2. Elsevier, Amsterdam, pp 317–547

Chang FH, Simons DB, Richardson EV (1967) Total bed material discharge in alluvial channels. In: Proc 12th Congr IAHR, Fort Collins, Colorado, pp 132–140

Chase RRP (1979) Settling behavior of natural aquatic particulates. Limnol Oceanogr 24:417–426

Chesselet R, Jedwab J, Darcourt C, Dehairs F (1976) Barite as discrete suspended particles in the Atlantic Ocean. Abstr EOS 57:255

Chester R (1990) Marine geochemistry. Unwin Hyman, London, p 698

Chézy M (1921) Mémoire sur la vitesse de l'eau conduite dans une rigole donnée. Ann Ponts Chaussées Mem 1921-II, pp 241–251

Chin Y-S (1979) A study on sediment and mineral compositions of the sea floor of the East China Sea. Ocean Selections 2:130–142.

Chisholm SW, Olson RJ, Yentsch CM (1988) Flow cytometry in oceanography: status and prospects. Eos 3:562–572

Chiu Ch-L (1967) Stochastic model of motion of solid particles. Proc ASCE HY 5:203–218

Cho BC, Azam F (1988) Major role of bacteria in biogeochemical fluxes in the oceans's interior. Nature (Lond) 332:441–443

Churchill JH (1989) The effect of commercial trawling on sediment resuspension and transport over the Middle Atlantic Bight continental shelf. Cont Shelf Res 9:841–864

Churchill JH, Biscaye PE, Aikman F III (1988) The character and motion of suspended particulate matter over the shelf edge and upper slope off Cape Cod. Cont Shelf Res 8:789–809

Clarke ThL, Swift DJP (1984) The formation of mud patches by nonlinear diffusion. Cont Shelf Res 3:1–7

Cloern JE (1987) Turbidity as a control on phytoplankton biomass and productivity in estuaries. Cont Shelf Res 7(11/12):1367–1381

Coale KH, Bruland KW (1985)^{234}TH:^{238}U disequilibria within the California current. Limnol Oceanogr 30:22–33

Coale KH, Bruland KW (1987) Oceanic stratified euphotic zone as elucidated by ^{234}Th:^{238}U disequilibria. Limnol Oceanogr 32:189–200

Cobler R, Dymond J (1980) Sediment trap experiment on the Galapagos spreading center, Equatorial Pacific. Science 209:801–803

Cochran JK (1984) The fates of uranium and thorium decay series nuclides in the estuarine environment. In: Kennedy V (ed) The estuary as a filter. Academic Press, New York London, pp 179–220

Colby BR (1963) Fluvial sediments – a summary of source, transportation and measurement of sediment discharge. Geol Surv Bull 1181-A, Washington, p 47

Coleman NL (1969) A new examination of sediment suspension in open channels. J Hydraul Res 7:67–82

Coleman NL (1970) Flume studies of the sediment transfer coefficient. Water Resource Res 6:801–809

Coleman NL (1986) Effect of suspended sediment on channel velocity distribution. Euromech 192:1985

Coles SL, Stratham R (1973) Observations on coral mucus "flocs" and their potential trophic significance. Limnol Oceanogr 18:673–678

Conomos TJ, Peterson DH (1976) Suspended-particle transport and circulation in San Francisco Bay: an overview. In: Wiley M (ed) Estuarine processes II. Academic Press, New York London pp 82–97

Cornaglia P (1891) Sul regime della spiagge e sulla regolazione dei porti. Stampere Taravia, Torino, p 220

Coulter WH (1956) High speed automatic blood cell counter and cell size analyzer. Proc Natl Electr Conf 12:1034–1042

Court Stevenson J, Ward LG, Kearney MS (1988) Sediment transport and trapping in marsh systems: implications of tidal flux studies. Mar Geol 80:37–59

Creutzberg F, Postma H (1979) An experimental approach to the distribution of mud in the southern North Sea. Neth J Sea Res 13:211–261

Creutzberg F, Wapenaar P, Duineveld G, Lopez Lopez N (1984) Distribution and density of the benthic fauna in the southern North Sea in relation to bottom characteristics and hydrographic conditions. Rapp PV Réun Cong Int Explor Mer 183:101–110

Curray JR, Moore DG (1971) Growth of the Bengal deep-sea fan and denudation in the Himalayas. Bull Geol Soc Am 82:563–572

Dahl HJ, Heckenroth H (1978) Landespflegerisches Gutachten zur Emsumleitung durch den Dollart. Naturschutz Landschaftspflege Niedersachsen 6:214

d'Anglejean B, Ingram RG (1976) Time-depth variations in tidal flux of suspended matter in the St. Lawrence estuary. Est Coast Mar Sci 4:401–406

Davies PJ (1979) Marine geology of the continental shelf off southeast Australia. Bull Bur Mineral Res Aust 195:1–51

Degens ET (1989) Perspectives on biogeochemistry. Springer, Berlin Heidelberg New York, 423 pp

Degens ET, Ittekkot V (1985) Particulate organic carbon: an overview. Mitt Geol Paläontol Inst Univ Hamburg, SCOPE/UNEP Sonderbd 58:7–27

Degens ET, Mopper K (1976) Factors controlling the distribution and early diagenesis of organic material in marine sediments. In: Riley JD, Chester R (eds) Chem Oceanogr 6:59–113

Dehairs F, Chesselet R, Jedwab J (1980) Discrete suspended particles of barite and the barium cycle in the open ocean. Earth Planet Sci Lett 49:528–550

Delichatsios MA, Probstein RF (1975) Coagulation in turbulent flow: theory and experiment. J Colloid Interface Sci 51:394–405

Demarq H, Wald L (1984) La dynamique superficielle du panache du Rhône d'après l'imagerie infrarouge satellitaire. Oceanol Acta 7:159–162

DeMaster DJ, Kuehl SA, Nittrouer ChA (1986) Effects of suspended sediments on geochemical processes near the mouth of the Amazon River: examination of biological silica uptake and the fate of particle-reactive elements. Cont Shelf Res 6 (1/2): 107–125

Deuser WG (1984) Seasonality of particulate fluxes in the ocean's interior. In: Worksh Global Ocean Flux Study, Woods Hole, pp 221–236

Deuser WG (1986) Seasonal and interannual variations in deep-water particle fluxes in the Sargasso Sea, and their relation to surface hydrography. Deep Sea Res 33: 225–246

Deuser WG, Ross EH (1980) Seasonal change in the flux of organic carbon to the deep Sargasso Sea. Nature (Lond) 283: 364–365

De Vooys CGN (1979) Primary production in aquatic environments. In: Bolin B, Degens ET, Kempe S, Ketner P (eds) The global carbon cycle. SCOPE Vol 13, John Wiley & Sons, New York, pp 259–292

Dickson RR, McCave IN (1986) Nepheloid layers on the continental slope west of Porcupine Bank. Deep Sea Res 33: 791–818

Dietrich G (1955) Ergebnisse synoptischer ozeanographischen Arbeiten in der Nordsee. Dtsch Geogr Tag Hamburg, Aug 1955, pp 376–383

Dietrich WE (1982) Settling velocity of natural particles. Water Resour Res 18(6): 1615–1626

Dominik J, Burrus D, Vernet J-P (1983) A preliminary investigation of the Rhône river plume in eastern Lake Geneva. J Sediment Petrol 53: 159–163

Douglas I (1967) Man, vegetation and the sediment yields of rivers. Nature (London) 215: 925–928

Drake DE (1971) Suspended sediment and thermal stratification in Santa Barbara Channel, California. Deep Sea Res 18: 763–769

Drake DE (1976) Suspended sediment transport and mud deposition on continental shelves. In; Stanley DJ, Swift DJP (eds) Marine sediment transport and environmental management. John Wiley & Sons, New York, pp 127–158

Drake DE, Cacchione DA (1985) Seasonal variation in sediment transport on the Russian River shelf, California. Cont Shelf Res 4: 495–514

Drake DE, Cacchione DA (1986) Field observations of bed shear stress and sediment resuspension on continental shelves, Alaska and California. Cont Shelf Res 6: 415–429

Drake DE, Cacchione DA (1989) Estimates of the suspended sediment reference concentration (Ca) and resuspension coefficient (gq) from near-bottom observations on the California shelf. Cont Shelf Res 9: 51–64

Drake DE, Gorsline DS (1973) Distribution and transport of suspended particulate matter in Hueneme, Redondo, Newport and La Jolla submarine canyons, California. Bull Geol Soc Am 84: 3949–3968

Drake DE, Hatcher PK, Keller GH (1978) Suspended particulate matter and mud deposition in the upper Hudson submarine canyon. In: Stanley DJ, Kelling G (eds) Sedimentation in submarine canyons, fans and trenches. Dowden, Hutchinson & Ross, Stroudsburg Pa, pp 33–41

Drake DE, Cacchione DA, Muench RD, Nelson CH (1980) Sediment transport in Norton Sound, Alaska. Mar Geol 36: 97–126

Dronkers J (1986a) Tide-induced residual transport of fine sediment In: v.d. Kreeke J (ed) Physics of shallow estuaries and bays. Springer, Berlin Heidelberg New York, pp 228–244

Dronkers J (1986b) Tidal asymmetry and estuarine morphology. Neth J Sea Res 20: 117–131

Dronkers J, Zimmerman JTF (1982) Some principles of mixing in tidal Lagoons. In: Oceanol Acta Proc Int Symp Coastal Lagoons, Bordeaux, pp 107–117

Dronkers J, van Alphen JSLJ, Borst JC (1990) Suspended sediment transport processes in the southern North Sea. In: Cheng RT (ed) Coastal and estuarine studies, vol 38 Springer, Berlin Heidelberg, New York, pp 302–320

Duck RW (1987) Aspects of physical processes of sedimentation in Loch Earn, Scotland. In: Gardiner V (ed) International geomorphology 1986, pt 1, pp 801–821

Duck RW, McManus J (1984) A marker horizon in reservoir sediment cores resulting from wartime agricultural changes in central Scotland. Scot Geogr Mag 100: 184–189

Duck RW, McManus J (1989) Variations in reservoir sedimentation in Scotland in response to land use changes. Arch Hydrobiol Beih 33: 19–26

Durrieu de Madron X, Nyffeler F Godet CH (1990) Hydrographic structure and nepheloid spatial distribution in the Gulf of Lions continental margin. Cont Shelf Res 10:915–929

Dussart B (1966) Limnologie. Gauthier-Villars, Paris

Dyer KR (1986) Coastal and estuarine sediment dynamics. John Wiley & Sons, Chichester, 342 pp

Dyer KR (1978) The balance of suspended sediment in the Gironde and Thames estuaries. In: Kjerve BJ (ed) Estuarine transport processes. Univ S C Press, Columbia, pp 135–145

Dyer KR ed (1979) Estuarine hydrography and sedimentation. Univ Press, Cambridge, p 230

Dyer KR (1989) Sediment processes in estuaries: future research requirements. J Geophys Res 94:C10 14.237–14.339

Dyer KR, New AL (1986) Intermittency in estuarine mixing. In: DA Wolfe (ed) Estuarine variability. Academic Press, New York London, pp 321–339

Dymond J, Lyle K (1985) Flux comparisons between sediments and sediment traps in the eastern tropical Pacific. Implications of atmospheric CO_2 variations during the Pleistocene. Limnol Oceanogr 30:699–712

Eaton A, Grant V, Grant Gross M (1980) Chemical tracers for particle transport in the Chesapeake Bay. Est Coast Mar Sci 10:75–83

Edzwald JK (1972) Coagulation in estuaries. Univ N C Sea Grant Program Publ UNC–SG–72–06, Chapel Hill

Edzwald JK, O'Melia Ch R (1975) Clay distributions in recent estuarine sediments. Clays Clay Minerals 23:39–44

Edzwald JK, Upchurch JB, O'Melia CR (1974) Coagulation in estuaries. Environ Sci Technol 8:58–63

Ehlers J (1988) The morphodynamics of the Waddensea. Balkema, Rotterdam, p 397

Eidsvik KJ, Brørs B (1989) Self-accelerated turbidity current prediction based upon (k − ε) turbulence. Cont Shelf Res 9 (7):617–627

Einstein HA (1950) The bed load function for sediment transportation in open channels. US Dep Agric Soil Conserv Serv Tech Bull 1026

Einstein HA, Chien N (1952) Second approximation to the solution of the suspended load theory. Inst Eng Res Univ Calif Ser No 47, No 2, Berkeley, CA 31/1/52

Einstein HA, Krone RB (1962) Experiments to determine modes of cohesive sediment transport in salt water. J Geophys Res 67:1451–1464

Einstein HA, Anderson AG, Johnson JW (1940) A distinction between bed load and suspended load in natural systems. Trans Am Geophys Union Annu Meet 21 (pt 2), pp 628–633

Eisma D (1968) Composition, origin and distribution of Dutch coastal sands between Hoek van Holland and the island of Vlieland. Neth J Sea Res 4:123–267

Eisma D (1981) Supply and deposition of suspended matter in the North Sea. Spec Publs Int Assoc Sediment 5:415–428

Eisma D (1986) Flocculation and deflocculation of suspended matter in estuaries. Neth J Sea Res 20:183–199

Eisma D (1987) An introduction to the geology of continental shelves. In: Zijlstra JJ, Postma H (eds) Continental shelves. Elsevier, Amsterdam, pp 39–91

Eisma D (1988) Transport and deposition of suspended matter in estuaries and the nearshore sea. In: Lerman A, Meybeck M (eds) Physical and chemical weathering in geochemical cycles. Kluwer, Amsterdam, pp 271–298

Eisma D (1990) Dispersal of Mahakam river suspended sediment in Makassar Strait, Indonesia. In: Ittekkot V, Kempe G, Michaelis W, Spitzy A (eds) Facets of modern biogeochemistry. Springer, Berlin Heidelberg New York, pp 127–146

Eisma D, Irion G (1988) Suspended matter and sediment transport. In: Salomons W, Bayne BL, Duursma EK, Furstner U (eds) Pollution of the North Sea. Springer, Berlin Heidelberg New York, pp 20–35

Eisma D, Kalf J (1984) Dispersal of Zaire river suspended matter in the estuary and the Angola Basin. Neth J Sea Res 17:385–411

Eisma D, Kalf J (1987a) Dispersal, concentration and deposition of suspended matter in the North Sea. J Geol Soc Lond 144:161–178

Eisma D, Kalf J (1987b) Distribution, organic content and particle size of suspended matter in the North Sea. Neth J Sea Res 21:265–285

Eisma D, van der Marel HW (1971) Marine muds along the Guiana coast and their origin from the Amazon. Contrib Mineral Petrol 31:321–334

Eisma D, Kalf J, van der Gaast SJ (1978a) Suspended matter in the Zaire estuary and the adjacent Atlantic Ocean. Neth J Sea Res 12:382–406

Eisma D, van der Gaast SJ, Martin JM, Thomas AJ (1978b) Suspended matter and bottom deposits of the Orinoco delta: turbidity, mineralogy and elementary composition. Neth J Sea Res 12:224–251

Eisma D, Kalf J, Veenhuis M (1980) The formation of small particles and aggregates in the Rhine estuary. Neth J Sea Res 14:172–191

Eisma D, Cadée GC, Laane R (1982) Supply of suspended matter and particulate and dissolved organic carbon from the Rhine to the coastal North Sea. Mitt Geol paläontol Inst Univ Hamburg 52:483–505

Eisma D, Boon J, Groenewegen R, Ittekkot V, Kalf J, Mook WG (1983) Observations on macroaggregates, particle size and organic composition of suspended matter in the Ems estuary. Mitt Geol Paläontol Inst Univ Hamburg 55:295–314

Eisma D, Berger GW, Chen W-Y, Shen J (1989) Pb-210 as a tracer for sediment transport and deposition in the Dutch-German Waddensea. In: Proc Coastal Lowlands Symp Den Haag, May 1987, pp 237–253

Eisma D, Schuhmacher T, Boekel H, van Heerwaarden J, Franken H, Laan M, Vaars A, Eijgenraam F, Kalf J (1990) A camera and image analysis system for in-situ observation of flocs in natural waters. Neth J Sea Res 27:43–56

Eisma D, Bernard P, Cadée GC, Ittekkot V, Kalf J, Laane R, Martin JM, Mook WG, Van Put A, Schuhmacher T (1991a) Suspended-matter particle size in some west-European estuaries. Part I. Particle-size distribution. Part II. A review of floc formation and break-up. Neth J Sea Res 28 (3) 193–214, 215–220

Eisma D, Augustinus PGEF, Alexander C (1991b) Recent and sub-recent changes in the dispersal of Amazon mud. Neth J Sea Res 28 181–192

Eittreim S, Biscaye PE, Amos AF (1975) Benthic nepheloid layers and the Ekman thermal pump. J Geophys Res 80:5061–5067

Ellenberger F (1988) Histoire de la géologie. I. Des anciens à la première moitié du XVIIe siècle. Lavoisier, Paris, 352 pp

Elliott AJ (1976) Response of the Patuxent estuary to a winter storm. Chesapeake Sci 17:212–216

Elliot AJ (1978) Observations of the meteorologically induced circulation in the Potomac estuary. Est Coast Mar Sci 6:285–290

Elmore D, Phillips FM (1987) Accelerator mass spectrometry for measurement of long lived radioisotopes. Science 236:543–550

Emery KO (1968) Relict sediments on continental shelves of the world. Bull Am Assoc Petrol Geol 52:445–464

Emery KO, Milliman JD (1978) Suspended matter in surface waters: influence of river discharge and of upwelling. Sedimentology 25:125–140

Emery KO, Ross DA (1968) Topography and sediments of a small area of the continental slope south of Martha's Vineyard. Deep Sea Res 15:415–422

Emery KO, Milliman JD, Uchupi E (1973) Physical properties and suspended matter of surface waters in the southeastern Atlantic Ocean. J Sediment Petrol 43:822–837

Emery KO, Lepple F, Toner L, Uchupi E, Rioux RH, Pople W, Hulburt EM (1974) Suspended matter and other properties of surface waters of the northeastern Atlantic ocean. J Sediment Petrol Vol 44:1087–1110

Eppler B, Neis U, Hahn HH (1975) Engineering aspects of the coagulation of colloidal particles in natural waters. Prog Water Technol 7:207–216

Etter RJ, Hoyer RP, Partheniades E, Kennedy JF (1968) Depositional behavior of kaolinite in turbulent flow. J Hydraul Div Proc ASCE, pp 1439–1452

Euler L (1755) Principes généraux du mouvement de l'état déquilibre des fluides. Hist Acad Berlin

Ewing M, Thorndike E (1965) Suspended matter in deep ocean water. Science 147:1291–1294

Faas RW (1986) Mass-physical and geotechnical properties of surficial sediments and dense nearbed sediment suspensions on the Amazon continental shelf. Cont Shelf Res 6:189–208

Faas RW, Wells JT (1990) Rheological control of fine-sediment suspension. Cape Lookout Bight N C J Coast Res 6:503–515

Fairbridge RW (1980) The estuary: its definition and geodynamic cycle. In: Olausson E, Cato I (eds) Chemistry and biogeochemistry of estuaries. John Wiley & Sons, New York pp 1–35

Fairchild JC (1977) Suspended sediment in the littoral zone at Ventnor, New Jersey, and Nags Head, North Carolina. CERC Tech Rep 77–5

Faktorovitch ME (1967) Transformation de lits à l'aval des usines hydroélectriques de grande puissance de l'URSS. Publ Int Assoc Sci Hydrol 75:401–403

Falconer RA, Owens PH (1989) Mathematical modelling of tidal currents and sediment transport rates in the Humber estuary. BHRA, Fluid Eng Centre, Crossfield, Bedford, pp 55–73

Farmer DM, Osborn TR (1976) The influence of wind on the surface layer of a stratified inlet. Part I. Observations. J Phys Oceanogr 6:931–940

Feely RA, Baker ET, Schumacher JD, Massoth GJ, Landing WM (1979) Processes affecting the distribution and transport of suspended matter in the northeast Gulf of Alaska. Deep Sea Res 26A:445–464

Fellows DA, Karl DM, Knauer GA (1981) Large particle fluxes and the vertical transport of living carbon in the upper 1500 m of the northeast Pacific Ocean. Deep Sea Res 28A:921–936

Ferentinos G, Collins M (1979) Tidally induced secondary circulations and their associated sedimentation processes. J Oceanogr Soc Jpn 35:65–74

Ferentinos G, Collins M (1980) Effects of shoreline irregularities on a rectilinear tidal current and their significance in sedimentation processes. J Sediment Petrol 50:1081–1094

Festa JF, Hansen DV (1976) A two-dimensional numerical model of estuarine circulation: the effects of altering depth and river discharge. Est Coast Mar Sci 4:309–323

Festa J, Hansen DV (1978) Turbidity maxima in partially mixed estuaries: a two-dimensional numerical model. Est Coast Mar Sci 7:347–359

Feuillet JP, Fleischer P (1980) Estuarine circulation: controlling factor of clay mineral distribution in James river estuary, Virginia. J Sediment Petrol 50:267–280

Finlayson BL (1978) Suspended solids transport in a small experimental catchment. Z Geomorphol NF 22:192–210

Firth BA, Hunter RJ (1976) Flow properties of coagulated colloidal suspensions. J Colloid Interface Sci 57:248–275

Fischer J (1983) Remote sensing of suspended matter, phytoplankton and yellow substances over coastal waters. Part 1. Aircraft measurements. Mitt Geol Paläontol Inst Univ Hamburg 55:85–95

Fischer J, Doerffer, R, Grassl H (1988) Remote sensing of suspended matter, phytoplankton and yellow substance over coastal waters. Part 2: Satellite measurements. Mitt Geol,Paläontol Inst Univ Hamburg 66:31–41

Floderus S, Pihl L (1990) Resuspension in the Kattegat: impact of variation in wind climate and fishery. Est Coast Shelf Sci 31:487–498

Fontugne MR, Jouanneau JM (1981) La composition isotopique du carbone organique des matières en suspension dans l'estuaire de la Gironde. CR Acad Sci Paris 293 II, 5:389–392

Forbes RJ (1963) Studies in ancient technology, VII. Brill, Leiden, 253 pp

Forel FA (1892–1901) Le Léman. Monographie limnologique, 3 parts. Lausanne

Förstner U (1986) Solid/solution relations of contaminants in surface waters. In: Bechteler W (ed) Transport of suspended solids in open channels. Proc Euromach 192. Balkema, Rotterdam, pp 209–220

Förstner U, Wittmann GTW (1979) Metal pollution in the aquatic environment. Springer, Berlin Heidelberg New York, 486 pp

Fowler SW, Knauer GA (1986) Role of large particles in the transport of elements and organic compounds through the oceanic water column. Progr Oceanogr 16:147–194

Francois RJ, van Haute AA (1985) Structure of hydroxide flocs. Water Res 19:1249–1254

Frey DG (1963) Limnology in North America. Univ Wisc Press, Madison
Friedlander SK (1977) Smoke, dust and haze. Fundamentals of aerosol behavior. John Wiley & Sons, New York, p 317
Friedman GM, Sanders JF (1978) Principles of sedimentology. John Wiley & Sons, New York
Frost BW (1984) Utilization of phytoplankton production in the surface layer. In: Global Ocean Flux Study worksh, Woods Hole, pp 125–135
Gaarder T, Gran HH (1927) Investigations of the production of plankton in the Oslo Fjord. Rapp Proc Verb Reun Cons Int Explor Mer 42:1–48
Galappatti G, Vreugdenhil CB (1985) A depth integrated model for suspended sediment transport. J Hydr Res IAHR 23:359–377
Gallenne B (1974) Study of fine material in suspension in the estuary of the Loire and its dynamic grading. Est Coast Mar Sci 2:261–272
Galloway JS (1988) Field investigation of suspended sediment clouds under plunging breakers. Est Coast Shelf Sci 27:119–130
Galloway WE, Hobday DK (1983) Terrigenous clastic depositional systems. Springer, Berlin Heidelberg New York 423 pp
Gao S, Xie Q-Ch, Feng Y-J (1990) Fine-grained sediment transport and sorting by tidal exchange in Xiangshan Bay. Zhejiang, China. Est Coast Shelf Sci 31:397–409
Gardner WD (1980a) Sediment trap dynamics and calibration: a laboratory evaluation. J Mar Res 38:17–39
Gardner WD (1980b) Field assessment of sediment traps. J Mar Res 38:41–52
Gardner WD (1989) Periodic resuspension in Baltimore Canyon by focusing of internal waves. J Geophys Res 94:C12 18.185–18.194
Gardner WD, Hinga KR, Marra J (1983) Observations on the degradation of biogenic material in the deep ocean with implications on accuracy of sediment trap fluxes. J Mar Res 41:195–214
Gardner WD, Bishop JKB and Biscaye PE (1984) Nephelometer and current observations at the STIE site, Panama Basin. J Mar Res 42:207–219
Garnier JM, Lipiatou E, Martin JM, Mouchel JM, Thomas AJ (1990) Surface properties of particulates and distribution of selected pollutants in the Rhine delta and the Gulf of Lion. In: Martin JM, Barth H (eds) Proc EROS-2000 Worksh, Blanes (Spain), Febr 6–9, 1990. Water Poll Res Rep. 20, Environ Res· Progr EEC Direct. Sci, Res and Dev. Brussels, pp 501–552
Garrels RM, Mackenzie FT (1971) Evolution of sedimentary rocks. Norton, New York, 397 pp
Ghosh SN (1986) An investigation on the nature of variation of sediment and momentum transfer coefficients in natural river. 3rd Int Symp River Sed Univ Mississippi, March–April 1986, pp 187–194
Gibbs RJ (1973) Mechanisms of trace metal transport in rivers. Science 180:71–73
Gibbs RJ (1977a) Transport phases of transition metals in the Amazon and Yukon rivers. Bull Geol Soc Am 88:829–843
Gibbs RJ (1977b) Clay mineral segregation in the marine environment. J Sediment Petrol 47:237–243
Gibbs RJ (1981) Floc breakage by pumps. J Sediment Petrol 51:670–672
Gibbs RJ (1982a) Floc breakage during HlAC light-blocking analysis. Environ Sci Technol 16:298–299
Gibbs RJ (1982b) Floc stability during Coulter-counter size analysis. J Sediment Petrol 52:657–660
Gibbs RJ (1982c) Turbid horizons at mid-depth in the Wilmington Canyon area. Est Coast Mar Sci 14:313–324
Gibbs RJ (1983) Effect of natural organic coatings on the coagulation of particles. Environ Sci Technol 17:237–240
Gibbs RJ (1985) Estuarine flocs: their size, settling velocity and density. J Geophys Res 90:3249–3251
Gibbs RJ, Konwar LN (1982) Effect of pipetting on mineral flocs. Environ Sci Technol 16:119–121
Gibbs RJ, Tsudy DM, Konwar L, Martin JM (1989) Coagulation and transport of sediments in the Gironde Estuary. Sedimentology 36:987–999

Gieskes WWC, Veth C, Woehrmann A, Graefe M (1987) Secchi disc visibility world record shattered. EOS Trans Am Geophys Union 68 (9):123

Gilbert GK (1917) Hydraulic-mining debris in the Sierra Nevada. US Geol Surv Prof Pap 105:154

Gilbert R (1983) Sedimentary processes of Canadian arctic fjords. Sediment Geol 36:147–175

Gilmer RW (1972) Free-floating mucus webs: a novel feeding adaptation for the open ocean. Science 176:1239–1240

Gilmer RW (1974) Some aspects of feeding in thecosomatous pteropod molluscs. J Exp Mar Biol 15:127–144

Glangeaud L (1938) Transport et sédimentation dans l'estuaire et à l'embouchure de la Gironde. Bull Soc Geol Fr 8:599–630

Glenn SM, Grant WD (1987) A suspended sediment stratification correction for combined wave and current flows. J Geophys Res 92:8244–8264

Gobeil Ch, Sundby Bj, Silverberg N (1981) Factors influencing particulate matter geochemistry in the St Lawrence estuary turbidity maximum. Mar Chem 10:123–140

Goetz A, Tsuneishi N (1951) The application of molecular filter membranes to the bacteriological analysis of water. J Am Water Works Assoc 43:943–969

Goldberg ED (1971) Atmospheric dust, the sedimentary cycle and man: comments on earth sciences. Geophysics 1:117–132

Goldberg ED, Baker M, Fox DL (1952) Microfiltration in oceanographic research. I. Marine sampling with the molecular filter. J Mar Res 11:194–204

Goldstein S (1929) The steady flow of viscous fluid past a fixed spherical obstacle at small Reynolds numbers. Proc R Soc Lond A 123.

Gons HJ (1982) Structural and functional characteristics of epiphyton and epipelon in relation to their distribution in Lake Vechten. Hydrobiologia 95:79–114

Gons HJ, Kromkamp JC (1984) Dynamics and structural characteristics of epipelon in Lake Vechten and Lake Maarsseveen (The Netherlands). Verh Int Verein Limnol 22:897–903

Gons HJ, Otten J, Rijkeboer M (1991) The significance of wind resuspension for the predominance of filamentous cyanobacteria in a shallow, eutrophic lake. Mem Ist Ital Idrobiol, 33

Gordeev VV (1963) Quantitative distribution of suspended matter in the surface water of the East Atlantic. Dokl Akad Nauk USSR 149:181–184 (in Russian)

Gordeev VV (1964) Quantitative distribution of suspended matter in the surface water of the North Indian Ocean. Trudy Inst Okeanol 64:202–213 (in Russian)

Gordeev VV (1983) River inputs to the oceans and their chemical characteristics. Nauka, Moscow, p 160 (in Russian)

Gordon CM (1975) Sediment entrainment and suspension in a tidal flow. Mar Geol 18:M57–M64

Gordon DC (1970) A microscopic study of organic particles in the North Atlantic Ocean. Deep Sea Res 17:175–185

Gowing MM, Silver MW (1985) Minipellets: a new and abundant size class of marine fecal pellets. J Mar Res 43:395–418

Graf WH (1971) Hydraulics of sediment transport. McGraw-Hill Book Cy

Grant WD, Madsen OS (1986) The continental-shelf bottom boundary layer. Annu Rev Fluid Mech 18:265–305

Grant WD, Boyer L, Sanford LP (1982) The effect of bioturbation on the initiation of motion of intertidal sand. J Mar Res 40:659–677

Green Th (1986) The double-diffusive aspects of sedimentation. In: 3rd Int Symp River Sediment, Univ Mississippi, pp 371–377

Greensmith JT, Tucker EV (1973) Holocene transgressions and regressions on the Essex coast, Outer Thames estuary. Geol Mijnbouw 52:193–202

Gregory J (1978) Effects of polymers on colloid stability. In: Ives KJ (ed) The scientific basis of flocculation. Sijthoff and Noordhoff, Alphen aan de Rijn, The Netherlands, pp 89–99

Griffin JJ, Windom H, Goldberg ED (1968) The distribution of clay minerals in the World Ocean. Deep Sea Res 15:433–459

Gry H (1942) Das Wattenmeer bei Skallingen. Quantitative Untersuchungen über den Sinkstofftransport durch Gezeitenströmungen. Folia Geogr Dan II.1:13

Gunnerson CG, Emery KO (1962) Suspended sediment and plankton over San Pedro Basin, California. Limnol Oceanogr 7:14–20

Gust G (1976) Observations on turbulent-drag reduction in a dilute suspension of clay in seawater. J Fluid Mech 75:29–47

Guy HP (1964) An analysis of some storm-period variables affecting stream sediment transportation. US Geol Surv Prof Pap 462-E, pp E1–E46

Hamner WM, Madin LB, Alldredge AL, Gilmer RW, Hamner PP (1975) Underwater observations of gelatinous zooplankton: sampling problems, feeding biology, and behavior. Limnol Oceanogr 20:907–917

Han Z, Cheng H (1986) Two-dimensional sediment mathematical model of Hangzhou Bay. In: 3rd Int Symp River Sediment, Univ Mississippi, pp 463–471

Hanes DM (1988) Intermittent sediment suspension and its implications to sand tracer dispersal in wave-dominated environments. Mar Geol 81:175–183

Hanes DM, Huntley DA (1986) Continuous measurements of suspended sand concentration in a wave-dominated nearshore environment. Cont Shelf Res 6:585–596

Hannah SA, Cohen JM, Robeck GG (1967) Measurement of floc strength by particle counting. J AWWA 59, July 1967:843–858

Hansen DV, Rattray M Jr (1966) New dimensions in estuarine classification. Limnol Oceanogr 11:319–326

Harlett JC, Kulm LD (1973) Suspended sediment transport on the northern Oregon continental shelf. Bull Geol Soc Am 84:3815–3826

Harris JE (1977) Characterization of suspended matter in the Gulf of Mexico-II. Particle size analysis of suspended matter from deep water. Deep Sea Res 24:1055–1061

Hathaway JC (1972) Regional clay mineral facies in estuaries and continental margin of the United States East Coast. In: BW Nelson (ed) Environmental framework of coastal plain estuaries. Mem Geol Soc Am 133:203–316

Hawley N (1982) Settling velocity distribution of natural aggregates. J Geophys Res 87 C12:9489–9498

Hay W (1974) Introduction. Studies in paleo-oceanography. SEPM Spec Publ 20:1–5

Hayes MO (1975) Morphology of sand accumulation in estuaries. In: Cronin LE (ed) Estuarine research II. Academic Press, New York London, pp 3–22

Hayes MO (1978) Impact of hurricanes of sedimentation in estuaries, bays and lagoons. In: Wiley ML (ed) Estuarine interactions. Academic Press, New York London, pp 323–346

Hayes MO, Michel J (1982) Shoreline sedimentation within a forearc embayment, Lower Cook Inlet, Alaska. J Sediment Petrol 52:251–263

Healy RG, Pye K, Stoddart DR, Bayliss-Smith TP (1981) Velocity variations in salt marsh creeks, Norfolk, England. Est Coast Shelf Sci 13:535–545

Healy TW, La Mer VK (1962) The adsorption-flocculation reactions of a polymer with an aqueous colloidal dispersion. J Phys Chem 66:1835–1838

Hearne V (1705) Memorabilia nonnulla lacus Vetteri. Philos Trans 24:1938–1946

Heath RA (1973) Flushing of coastal embayments by changes in atmospheric conditions. Limnol Oceanogr 18:849–862

Heathershaw AD, New AL, Edwards PD (1987) Internal tides and sediment transport at the shelf break in the Celtic Sea. Cont Shelf Res 7:485–517

Heezen BC, Menzies RJ, Schneider ED, Ewing WM, Granelli NCG (1964) Congo submarine canyon. Bull Am Assoc Petrol Geol 48:1126–1149

Heim A (1990) Der Schlammabsatz am Grunde des Vierwaldstättersees. Vierteljahresschr Naturforsch Ges Zürich 45:164–182

Herring JR (1980) Waste water particle dispersion in the southern California offshore region. In: Kavanaugh MC, Leckie JC (eds) Particulates in water. Am Chem Soc Washington Adv Chem Sea 189:283–304

Hjulström F (1935) Studies in the morphological activity of rivers as illustrated by the River Fyris. Bull Geol Inst Univ Ups 25:221–527

Holeman JN (1968) Sediment yield of major rivers of the world. Water Resourc Res 4:737–747

Holeman JN (1980) Erosion rates in the US estimated by the Soil Conservation Service's inventory (Abstr). EOS 61:954

Holmes ChW (1982) Geochemical indices of fine sediment transport, northwest Gulf of Mexico. J Sediment Petrol 52:307–321

Holmes PW (1968) Sedimentary studies of late Quaternary material in Windermere Lake (Great Britain). Sediment Geol 2:191–200

Honeyman BD, Balistrieri LS, Murray JW (1988) Oceanic trace metal scavenging: the importance of particle concentration. Deep Sea Res 35:227–246

Honjo S (1978) Sedimentation of materials in the Sargasso Sea at a 5367-m-deep station. J Mar Res 36:469–492

Honjo S, Roman MR (1978) Marine copepod fecal pellets: production, preservation and sedimentation. J Mar Res 36:45–57

Honjo S, Manganini SJ, Cole JJ (1982a) Sedimentation of biogenic matter in the deep ocean. Deep Sea Res 29:609–625

Honjo S, Manganini SJ, Poppe LJ (1982b) Sedimentation of lithogenic particles in the deep ocean. Mar Geol 50:199–220

Honjo S, Spencer DW, Farrington JW (1982c) Deep advective transport of lithogenic particles in the Panama basin. Science 216:516–518

Honjo S, Doherty KW, Agrawal YC, Asper VL (1984) Direct optical assessment of large amorphbous aggregates (marine snow) in the deep ocean. Deep Sea Res 31:67–76

Hübbe F (1860) Über die Eigenschaften und das Verhalten des Schlicks. Z Bauwesen 10:491–520

Hughes DA (1980) Floodplain inundation: processes and relationships with channel discharge. Earth Surface Processes 5:297–304

Hunt JN (1954) The turbulent transport of suspended sediment in open channels. Proc R Soc Lond A224:322–335

Hunt JR (1980) Prediction of oceanic particle size distributions from coagulation and sedimentation mechanisms In: Kavanaugh MC, Leckie JC (eds) Particulates in water. Am Chem Soc Washington Adv Chem Ser 189:243–257

Hunt JR (1986) Particle aggregate break-up by fluid shear. In: Mehta AJ (ed) Estuarine cohesive sediment dynamics. Springer, Berlin Heidelberg New York, pp 85–109

Hunter KA, Liss PS (1979) The surface charge of suspended particles in estuarine and coastal waters. Nature (Lond) 282:823–825

Hunter KA, Liss PS (1982) Organic matter and the surface charge of suspended particles in estuarine waters. Limnol Oceanogr 27:322–335

Hurd DC, Spencer DW (eds) (1991) Marine particles: Analysis and Characterization. Amer Geophys Union Monograph Series 63, Washington DC, pp 472

Hutchinson GE (1957) A treatise on limnology, vol 1. John Wiley & Sons, New York, p 1015

Huzzey LM, Brubaker JM (1988) The formation of longitudinal fronts in a coastal plain estuary. J Geophys Res 93:C2 1329–1334

Inglis CC, Allen FH (1957) The regimen of the Thames estuary as affected by currents, salinities and river flow. Proc Inst Civil Eng 7:827–868

Inman DL (1949) Sorting of sediment in light of fluid mechanics. J Sediment Petrol 19:51–70

Ittekkot V (1988) Global trends in the nature of organic matter in river suspensions. Nature (Lond) 332:436–438

Ittekkot V, Deuser WG, Degens ET (1984a) Seasonality in the fluxes of sugars, amino acids, and amino sugars to the deep ocean: Sargasso Sea. Deep Sea Res 31:1057–1069

Ittekkot V, Degens ET, Honjo S (1984b) Seasonality in the fluxes of sugars, amino acids, and amino sugars to the deep ocean: Panama Basin. Deep Sea Res 31:1071–1083

Jackson GA (1989) Simulation of bacterial attraction and adhesion to falling particles in an aquatic environment. Limnol Oceanogr 34:514–530

Jackson GA (1990) A model of the formation of marine algal flocs by physical coagulation processes. Deep Sea Res 37:1197–1211

Jackson RG (1976) Sedimentological and fluid-dynamic implications of the turbulent bursting phenomenon in geophysical flows. J Fluid Mech 77:531–560

Jacobs MB, Ewing M (1969) Suspended particulate matter: concentration in the major oceans. Science 163:380–383

Jannasch HW (1973) Bacterial content of particulate matter in offshore surface waters. Limnol Oceanogr 18:340–341

Jarvis J, Riley C (1987) Sediment transport in the mouth of the Eden Estuary Est Coast Shelf Sci 24:463–481

Jeandel C, Martin J-M, Thomas AJ (1981) Plutonium and other artificial radionuclides in the Seine estuary and adjacent areas. In: IAEA (ed) Techniques for identifying transuranic speciation in aquatic environments. IAEA, Vienna, pp 89–102

Jedwab J (1980) Rare anthropogenic and natural particles suspended in deep ocean waters. Earth Planet Sci Lett 49:551–564

Jeffrey DJ (1982) Aggregation and break-up of clay flocs in turbulent flow. Adv Colloid Interface Sci 17:213–218

Jenkins R (1976) An introduction to X-ray spectrometry. Heyden, London New York, p 163

Johannes RE (1967) Ecology of organic aggregates in the vicinity of a coral reef. Limnol Oceanogr 12:189–195

Johnson BD, Cooke RC (1979) Bubble populations and spectra in coastal waters: a photographic approach. J Geophys Res 84:3761–3766

Johnson BD, Cooke RC (1980) Organic particle and aggregate formation resulting from the dissolution of bubbles in seawater. Limnol Oceanogr 25:653–661

Johnson BD, Wangersky PJ (1985) A recording backward scattering meter and camera system for examination of the distribution and morphology of macroaggregates. Deep Sea Res 32:1143–1150

Johnson BD, Wangersky PJ (1987) Microbubbles: stabilization by monolayers of adsorbed particles. J Geophys Res 92:C13 14.641–14.647

Johnson BD, Zhou X, Wangersky PJ (1986) Surface coagulation in the sea. Neth J Sea Res 20:201–210

Johnson BH, Trawle MJ, Kee PG (1986) Discussion of a laterally averaged numerical model for computing salinity and shoaling with an application to the Savannah estuary. In: 3rd Int Symp River Sediment, Univ Mississippi, pp 1443–1459

Jones BF, Bowser CJ (1978) The mineralogy and related chemistry of lake sediments. In: Lerman A (ed) Lakes; chemistry, geology, geophysics. Springer, Berlin Heidelberg New York, pp 179–235

Jordan PR (1965) Fluvial sediment of the Mississippi River at St Louis, Missouri. Geol Surv Water Supply Pap 1802:89

Jouanneau JM (1982) Matières en suspension et oligo-éléments métalliques dans le système estuarien Girondin: comportement et flux. Thesis, Univ Bordeaux

Judson SN (1968) Erosion of the land. Am Sci 56:356–374

Jumars PA, Nowell ARM, Self RFL (1981) A simple model of flow-sediment-organism interaction. Mar Geol 42:155–172

Kajihara M (1971) Settling velocity and porosity of large suspended particles. J Oceanogr Soc Jpn 27:158–162

Kalle K (1935) Meereskundliche chemische Untersuchungen mit Hilfe des Zeisschen Pulfrich-Photometers. 5. Mitt. Die Bestimmung des Gesamt-Phosphorgehaltes, des Plankton-Phosphorgehaltes (lebende Substanz) und Trübungsmessungen. Ann Hydr Mar Meteorol 63, V:195–204

Kalle K (1937) Nährstoff-Untersuchungen als hydrographisches Hilfsmittel zur Unterscheidung von Wasserkörpern. Ann Hydr Mar Meteorol 65, I:1–18

Kamps LF (1962) Mud distribution and land reclamation in the Eastern Wadden Shallows. Rijkswaterstaat Comm 4, Den Haag, 73 pp

Karl DM, Knauer GA (1984) Vertical distribution, transport and exchange of carbon in the northeast Pacific Ocean: evidence for multiple zones of biological activity. Deep Sea Res 31:221–243

Karl DM, Knauer GA, Martin JH, Ward BB (1984) Chemolithotrophic bacterial production in association with rapidly sinking particles: implications to oceanic carbon cycles and mesopelagic food webs. Nature (Lond) 309:54–56

Keller GH, Lambert D, Rowe G, Staresinic N (1972) Bottom currents in the Hudson Canyon. Science 180:181–183

Kelts K, Hsü KJ (1978) Freshwater carbonate sedimentation. In: Lerman A (ed) Lakes; chemistry, geology, geophysics. Springer, Berlin Heidelberg New York, pp 295–323

Kelvin Lord, (Sir William Thomson) (1869) On vortex motion. Trans R Soc Edinburgh 25:217–260.

Kemp GP, Wells JT (1987) Observations of shallow-water waves over a fluid mud bottom: implications to sediment transport. In Coastal sediments 87. WW Div ASCE, New Orleans, pp 363–378

Kempe S (1985) Compilation of carbon and nutrients discharge from major rivers. Mitt Geol Paläontol Inst Univ Hamburg 58:29–32

Kendrick MP, Derbyshire DV (1983) Factors affecting the supply and distribution of sediment in some tropical ports. Can J Fish Aquat Sci 40 (Suppl):35–43

Kennedy RG (1895) The prevention of silting in irrigation canals. Proc Inst Civil Eng 119:281–290

Kennett JP (1982) Marine geology. Prentice Hall, Englewood Cliffs, 813 pp

Kennish MJ (1986) Ecology of estuaries, vol 1. CRC, Boca Raton, 254 pp

Kineke GC, Sternberg RW (1989) The effect of particle settling velocity on computer suspended sediment concentration profiles. Mar Geol 90:159–174

Kineke GC, Sternberg RW, Johnson R (1989) A new instrument for measuring settling velocities in situ. Mar Geol 90:149–158

King IP, Granat MA, Ariathurai CR (1986) An inundation algorithm for finite element hydrodynamic and sediment transport modeling. In: 3rd Int Symp River Sediment Univ Mississippi, pp 1583–1593

Kirby R (1988) High concentration suspension (fluid mud) layers in estuaries. In: Dronkers J, van Leussen W (eds) Physical processes in estuaries. Springer, Berlin Heidelberg New York, pp 463–487

Kirby R, Parker WR (1977) The physical characteristics and environmental significance of fine sediment suspensions in estuaries. In: Studies in geophysics. Estuaries geophysics and the environment. Natl Res Council NAS Washington DC, pp 110–120

Kirby R, Parker WR (1982) A suspended sediment front in the Severn estuary. Nature (London) 295:396–399

Kirby R, Parker WR (1983) The distribution and behaviour of fine sediment in the Severn estuary and Inner Bristol Channel. Can J Fish Aquat Sci 40 (Suppl):83–95

Kitchen J, Zaneveld JRV, Pak H (1978) The vertical structure and size distributions of suspended particles off Oregon during the upwelling season. Deep Sea Res 25:453–468

Kitchener JA (1972) Principles of action of polymeric flocculants. Br Polymer 4:217–229

Kjerve BJ (1975) Tide and fair weather wind effects in a bar-built Louisiana estuary. In: Cronin LE (ed) Estuarine Res 2:47–62

Klenova MV (1959) Suspended matter in the northern part of the Atlantic Ocean between Scotland and Iceland. Dolk Akad Nauk USSR 127:435–439 (in Russian)

Klenova MV (1964) Some results of investigations of suspended matter in the Atlantic and in the southern part of the Indian Ocean (in Russian). Trudy Mon Gidrofis Inst 30:62–76

Klug HP, Alexander LE (1974) X-ray diffraction procedures. Wiley-Interscience, New York, p 966

Knauer GA, Martin JH (1981) Primary production and carbon-nitrogen fluxes in the upper 1500 m of the northeast Pacific. Limnol Oceanogr 26:181–182

Knauer GA, Hebel D, Cipriano F (1982) Marine snow: major site of primary production in coastal waters. Nature (London) 300:630–631

Knauer GA, Martin JH, Karl DM (1984) The flux of particulate organic matter out of the euphotic zone. Worksh Global Ocean Flux Study, Woodshole 1984, pp 136–150

Knox GA (1986) Estuarine ecosystems: a system approach, vols 1,2. CRC, Boca Raton, p 230, 289

Knudsen M (1922) The penetration of light into the Sea. Pub de Circonstance 76, Cons Perm Int Explor Mer, Copenhagen, 16 pp

Koblentz-Mishke OJ, Volkovinski VV, Kabanova JG (1970) Plankton primary production of the world ocean. In: Wooster WS (ed) Scientific exploration of the South Pacific. Natl Acad Sci USA, New York, pp 183–193

Kolkwitz R (1912) Plankton und seston. Ber Dtsch Bot Ges 30(42):334–346; (68):574–577

Komar PD (1970) The competence of turbidity current flow. Bull Geol Soc Am 81:1555–1562

Komar PD, Miller MC (1973) The threshold of sediment movement under oscillatory water waves. J Sediment Petrol 43:1101–1110

Komar PD, Miller MC (1974) Sediment threshold under oscillatory waves. Proc 14th Coastal Eng Conf Copenhagen, Denmark, vol 3, pp 756–775

Komar PD, Reimers CE (1978) Grain shape effects on settling rates. J Geol 86:193–209

Komar PD, Morse AP, Small LF, Fowler SW (1981) An analysis of sinking rates of natural copepod and euphausid fecal pellets. Limnol Oceanogr 26:172–180

Koppelman LE, Weyl PK, Grant Gross M, Davies DWS (1976) The urban sea: Long Island Sound. Praeger, New York p 223

Kostaschuk RA, Luternaner JL, Chruch MC (1989) Suspended sediment hysteresis in a salt-wedge estuary: Fraser River, Canada. Mar Geol 87:273–285

Kranck K (1973) Flocculation of suspended sediment in the sea. Nature (London) 246:348–350

Kranck K (1975) Sediment deposition from flocculated suspensions. Sedimentology 22:111–123

Kranck K (1980) Experiments on the significance of flocculation in the settling of fine-grained sediment in still water. Can J Earth Sci 17:1517–1526

Kranck K (1981) Particulate matter grain-size characteristics and flocculation in a partially mixed estuary. Sedimentology 28:107–114

Kranck K (1984a) The role of flocculation in the filtering of particulate matter in estuaries. In: Kennedy V (ed) The estuary as a filter. Academic Press, New York London, pp 159–175

Kranck K (1984b) Grain-size characteristics of turbidites. In: Stow DAV, Piper DJW (eds) Fine-grained sediments: deep water processes and facies. Geol Soc; and Blackwell, London, pp 83–92

Kranck K (1986) Settling behaviour of cohesive sediment. In: Mehta AJ (ed) Estuarine cohesive sediment dynamics. Springer, Berlin Heidelberg New York, pp 151–169

Kranck K, Milligan TG (1980) Macroflocs: production of marine snow in the laboratory. Mar Ecol Prog Ser 3:19–24

Kranck K, Milligan TG (1983) Grain size distributions of inorganic suspended river sediment. Mitt Geol Paläontol Inst Univ Hamburg; SCOPE/UNEP Sonderbd 55:525–534

Kranck K, Milligan TG (1985) Origin of grainsize spectra of suspended deposited sediment. Geo Mar Lett 5:61–66

Kranck K, Milligan TG (1988) Macroflocs from diatoms: in situ photography of particles in Bedford Basin, Nova Scotia. Mar Ecol Prog Ser 4:183–189

Krey J (1949) Über Art und Menge des Seston im Meere. Verh Dtsch Zool, Mainz; Akademischer Verlag, Leipzig, pp 295–301

Krey J (1950) Eine neue Methode zur quantitativen Bestimmung des Planktons. Kieler Meeresforsch 7:1

Krey J (1952) Untersuchungen zum Sestongehalt des Meerwassers. I. Der Sestongehlat in der westlichen Ostsee und unter Helgoland. Ber Dtsch Kommiss Meeresforsch NF 12:431–456

Krey J (1953) Plankton- und Sestonuntersuchungen in der südwestlichen Nordsee auf der Fahrt der "Gauss" Februar/März 1952. Ber Dtsch Kommiss Meeresforsch 13:136–153

Krey J (1961) Der Detritus im Meere. J Cons Int Explor Mer 26:263–280

Krishnaswami S, Lal D (1973) Oceanic particulate matter: size distribution in water and in sediment. Spec Publ Mar Biol Assoc India: 139–148

Krishnaswami S, Lal D (1977) Particulate organic carbon in Atlantic surface waters. Nature (Lond) 266:713–716

Krishnaswami S, Sarin MM (1976) Atlantic surface particulates: composition, settling rates and dissolution in the deep sea. Earth Planet Sci Lett 32:430–440

Krishnaswami S, Lal D, Somayajulu BLK (1976) Investigations of gram quantities of Atlantic and Pacific surface particulates. Earth Planet Sci Lett 32:403–419

Krone RB (1962) Flume Studies on the transport of sediment in estuarial shoaling processes. Univ Calif Hyd Eng Lab and Sanit Eng Res Lab Berkeley, 110 pp

Krone RB (1963) A study of rheologic properties of estuarial sediments. Rep 63-8 Hydraul Eng Lab; and San Eng Res Lab Univ Cal, Berkeley, p 191

Krone RB (1978) Aggregation of suspended particles in estuaries. In: Kjerve B (ed) Estuarine transport processes. Univ SC Press, Columbia, pp 171–190

Krone RB (1979) Sedimentation in the San Francisco Bay system. In: Conomos TJ (ed) San Francisco Bay: the urbanized estuary. 58th Ann Mett Pac Div AAAS, San Francisco 1977, pp 85–96

Kuo A, Nichols MM, Lewis J (1978) Modeling sediment movement in the turbidity maximum of an estuary. Virginia Water Resource Center Bull 111, p 76

Lacey G (1930) Stable channels in alluvium. Proc Inst Civil Eng 229:259–292

Lal D (1977) The oceanic microcosm of particles. Science 198 (4321): 997–1009

Lal D (1980) Comments on some aspects of particulate transport in the oceans. Earth Planet Sci Lett 49:520–527

Lal D, Lerman A (1973) Dissolution and behavior of particulate biogenic matter in the ocean: some theoretical considerations. J Geophys Res 78:7100–7111

Lamb H (1932) Hydrodynamics, 6th edn (1st edn 1879). Univ Press, Cambridge, 738 pp

Lambert AM, Kelts KR, Marshall NF (1976) Measurements of density underflows from Walensee, Switzerland. Sedimentology 23:87–105

Lambert ClE, Jehanno C, Silverberg N, Brun-Cottan JC, Chesselet R (1981) Log-normal distributions of suspended particles in the open ocean. J Mar Res 39:77–98

Lane EN, Kalinske AA (1939) The relation of suspended to bed material in rivers. Trans Am Geophys Union Washington 4:637–641

Lane EW, Carlson EJ, Hanson OS (1949) Low temperature increases sediment transportation in Colorado River. Civil Eng ASCE 19 (9):45–46

Langmuir I (1938) Surface motion of water induced by wind. Science 87:119–123

Larsen LH, Sternburg RW, Shi NC, Marsden MAH, Thomas L (1981) Field investigations of the threshold of grain motion by ocean waves and currents. Mar Geol 42:105–132

Lavelle JW, Cudaback CN, Paulson AJ, Murray JW (1991) A rate for the scavenging of fine particles by macroaggregates in a deep estuary. J Geophys Res 96:C1 783–790

Ledford-Hoffman PA, DeMaster DJ, Nittrouer CA (1986) Biogenic-silica accumulation in the Ross Sea and the importance of Antarctic continental-shelf deposits in the marine silica budget. Geochim Cosmochim Acta 50:2099–2110

Lee C, Cronin C (1984) Particulate amino acids in the sea: effects of primary productivity and biological decompositions. J Mar Res 42:1075–1097

Le Hir P, Bassoulet P, L'Yavanc J (1989) Modelling mud transport in a macrotidal estuary. In: Palmer MH (ed) Advances in water modelling and measurement. BHRA, Fluid Eng Centre, Crossfield, Bedford, pp 43–54

Leighly JB (1934) Turbulence and the transportation of rock debris by streams. Geogr Rev 24:453–464

Leliavsky S (1955) An introduction to fluvial hydraulics. Constable, London, p 257

Leonardo da Vinci (1948) The notebooks. MacCurdy E (ed), 2.vols, 5th edn. Cape, London, pp 639, 655

Leopold LB (1962) Rivers. Am Sci 50(4):511–537

Leopold LB, Wolman MG, Miller JP (1964) Fluvial processes in geomorphology. Freeman, San Francisco

Lerman A, Lal D (1977) Regeneration rates in the ocean. Am J Sci 277:238–258

Lerman A, Lal D, Dacey MF (1974) Stokes' settling and chemical reactivity of suspended particles in natural waters. In: Gibbs RJ (ed) Suspended solids in water. Plenum, New York, pp 17–47

Lerman A, Carder KL, Betzer PR (1977) Elimination of fine suspensions in the oceanic water column. Earth Planet Sci Lett 37:61–70

Lesieur M (1987) Turbulence in fluids. Nijhoff, Den Haag, p 286

Lewis RE, Lewis JO (1987) Shear stress variations in an estuary. Est Coast Shelf Sci 25:621–635

Li R-H, Shen H-W (1975) Solid particle settlement in open-channel flow. J Hydraul Div Proc ASCE 101, HY7:917–931

Li Y-H (1981) Ultimate removal mechanisms of elements from the ocean. Geochim Cosmochim Acta 45:1659–1664

Lick W, Lick J (1988) Aggregation and disaggregation of fine-grained lake sediments. J Great Lakes Res 14:514–523

Lieth H (1975) Historical survey of primary productivity research. In: Lieth H, Whittaker RH (eds) Primary productivity of the biosphere. Springer, Berlin Heidelberg New York, pp 7–16

Likens GE (1975) Primary productivity of inland aquatic ecosystems. In: Lieth H, Whittaker RH (eds) Primary productivity of the biosphere. Springer, Berlin Heidelberg New York, pp 185–202

Lincoln JM, Fitzgerald DM (1988) Tidal distortions and flood dominance at five small tidal inlets in southern Maine. Mar Geol 82:133–148

Linley EAS, Field JG (1982) The nature and ecological significance of bacterial aggregation in a nearshore upwelling ecosystem. Est Coast Shelf Sci 14:1–11

Lisitzin AP (1959) Suspended matter in seawater. In: Int Oceanogr Cong Preprints AAAS Washington, pp 470–471

Lisitzin AP (1961) Distribution and composition of suspended material in seas and oceans. Sovremennie Osadki Morei i Okeanov, Akad Nauk USSR, pp 175–231 (in Russian)

Lisitzin AP (1972) Sedimentation in the World Ocean. SEPM Spec Publ 17:218

Liu A-K, Davies SH (1977) Viscous attenuation of mean drift in water waves. J Fluid Mech 81 63–84

Lochte K, Turley CM (1988) Bacteria and cyanobacteria associated with phytodetritus in the deep sea. Nature (London) 333:67–69

Loder TC, Liss PS (1985) Control by organic coatings of the surface charge of estuarine suspended particles. Limnol Oceanogr 30:418–421

Loeb GI, Neihof RA (1975) Marine conditioning films. Adv Chem Ser 145:319–335

Loeb GI, Neihof RA (1977) Adsorption of an organic film at the platinum-seawater interface. J Mar Res 35:283–291

Long Y, Guishu X (1981) Sediment measurement in the Yellow River. Proc IAHS Symp Erosion and Sediment Transport Measurement. IAHS Publ 133:275–285

Long Y, Xiong G (1981) Sediment measurement in the Yellow River. In: Erosion and sediment transport measurement. Proc Florence Symp June 1981, IAHS Publ No 133, pp 275–285

Lugt HJ (1983) Vortex flow in nature and technology. John Wiley & Sons, New York p 297

Lyell Ch (1830) Principles of geology, being an attempt to explain the former changes of the Earth's surface, by reference to causes now in operation. Murray, London, p 511

Lyne VD, Butman B, Grant WD (1990a) Sediment movement along the US east coast continental shelf I: Estimates of bottom stress using the Grant-Madsen model and near-bottom wave and current measurements. Cont Shelf Res 10:397–428

Lyne VD, Butman B, Grant WD (1990b) Sediment movement along the US east coast continental shelf II. Modelling suspended sediment concentration and transport rate during storms. Cont Shelf Res 10:429–460

Maa P-Y, Mehta AJ (1987) Mud erosion by waves: a laboratory study. Cont Shelf Res 7:1269–1284

Magara Y, Nambu S, Utosawa K (1976) Biochemical and physical properties of an activated sludge on settling characteristics. Water Res 10:71–77

Mallik TK, Mukherji KK, Ramachandran KK (1988) Sedimentology of the Kerala mud banks (fluid muds?). Mar Geol 80:99–118

Manheim FT, Meade RH, Bond GC (1970) Suspended matter in surface waters of the Atlantic continental margin from Cape Cod to the Florida Keys. Science 167:371–376

Manheim FT, Hathaway JC, Uchupi E (1972) Suspended matter in surface water of the northern Gulf of Mexico. Limnol Oceanogr 17:17–27

Manohar M (1955) Mechanics of bottom sediment movement due to wave action. Beach Erosion Board Tech Memo 75

Mantoura RFC, Woodward EMS (1983) Conservative behaviour of riverine dissolved organic carbon in the Severn Estuary: chemical and geochemical implications, Geochim Cosmochim Acta 47:1293–1309

Mantz PA (1977) Incipient transport of fine grains and flakes by fluids—an extended Shields diagram. ASCE J Hydraul Div 103:601–615

Martin JH, Knauer GA, Karl DM, Broenkow WW (1987) VERTEX: carbon cycling in the northeast Pacific. Deep Sea Res 34: 267–285

Martin J-M (1971) Contribution à l'étude des apports terrigènes d'oligo-éléments stables et radioactifs à l'océan. Thesis, Univ Paris, p 155

Martin J-M, Meybeck M (1979) Elemental mass balance of material carried by major world rivers. Mar Chem 7: 173–206

Martin J-M, Whitfield M (1983) The significance of the river input of chemical elements to the ocean. In: Wong CS, Boyle E, Bzuland KW, Burton JD, Goldberg ED (eds) Trace metals in sea water. Plenum, New York, pp 265–296

Martin JM, Meybeck M, Heuzel M (1970) A study of the dynamics of suspended matter by means of natural radioactive tracers, an application to the Gironde estuary. Sedimentology 14: 27–37

Martin JM, Mouchel JM, Nirel P (1986) Some recent developments in the characterization of estuarine particulates. Water Sci Technol 18: 83–92

Martin TR (1962) Discussion on "experiments on the scour resistance of cohesive sediments". J Geophys Res 67: 1447–1449

Mayer LM (1982) Aggregation of colloidal iron during estuarine mixing: kinetics, mechanism and seasonality. Geochim Cosmochim Acta 46: 2527–2535

McAnally WH, Hayter EJ (1990) Estuarine boundary layers and sediment transport. In: Cheng R'ı (ed) Estuarine studies 38. Springer, Berlin Heidelberg New York, pp 260–275

McAnally WH, Thomas WA, Letter JV Jr (1980) Physical and numerical modeling of estuarine sedimentation. Int Symp River Sediment, Beijing, pp 1071–1099

McAnally WH, Letter JV Jr, Thomas WA (1986) Two- and three-dimensional modeling systems for sedimentation. In: Proc 3rd Int Symp River Sediment, Univ Mississippi, pp 399–411

McCave IN (1970) Deposition of fine-grained suspended sediment from tidal currents. J Geophys Res 75: 4151–4159

McCave IN (1971) Wave effectiveness at the seabed and its relationship to bed-forms and deposition of mud. J Sediment Petrol 41: 89–96

McCave IN (1972) Transport and escape of fine-grained sediment from shelf areas. In: Swift DJP, Duane DB, Pilkey OH (eds) Shelf sediment transport. Dowden, Hutchinson & Ross, Stroudsburg, 10: 225–248

McCave IN (1975) Vertical flux of particles in the ocean. Deep sea Res 22: 491–502

McCave IN (1979a) Suspended sediment. In: Dyer KR (ed) Estuarine hydrography and sedimentation. Univ Press, Cambridge, pp 131–185

McCave IN (1979b) Suspended material over the central Oregon continental shelf in May 1974. I. Concentration of organic and inorganic components. J Sediment Petrol 49: 1181–1194

McCave IN (1983) Particulate size spectra, behavior, and origin of nepheloid layers over the Nova Scotian Continental Rise. J Geophys Res 88: C12, 7647–7666

McCave IN (1984a) Erosion, transport and deposition of fine-grained marine sediments. In: Stow DAV, Piper DJW (eds) Fine-grained sediments: deep water processes and facies. Geol Soc London Spec Publ 15: 35–69

McCave IN (1984b) Size spectra and aggregation of suspended particles in the deep ocean. Deep sea Res 31: 329–352

McCave IN (1985a) Recent shelf clastic sediments. In: Brenchley PJ, Williams BPJ (eds) Sedimentology. Recent developments and applied aspects. Geol Soc London Spec Publ 18: 49–65

McCave IN (1985b) Properties of suspended sediment over the HEBBLE area on the Nova Scotian Rise. Mar Geol 66: 169–188

McCave IN, Jarvis J (1973) Use of the model T coulter counter in size analysis of fine to coarse sand. Sedimentology 20: 305–315

McCave IN, Swift SA (1976) A physical model for the rate of deposition of fine-grained sediments in the deep sea. Bull Geol Soc Am 87: 541–546

McCave IN, Tucholke BE (1986) Deep current controlled sedimentation in the western North Atlantic. In: The geology of North America, vol M chap 27: The western North Atlantic region. Geol Soc Am, Boulden, Colorado pp 451–468

McCave IN, Lonsdale PF, Hollister CD, Gardner WD (1980) Sediment transport over the Hatton and Gardar Contourite drifts. J Sediment Petrol 50:1049–1062

McClimans TA (1988) Estuarine fronts and river plumes. In: Dronkers J, van Leussen W (eds) Physical processes in estuaries. Springer, Berlin Heidelberg New York, pp 55–69

McDowell DN, O'Connor BA (1977) Hydraulic behaviour of estuaries. MacMillan, London, p 292

McHenry JR, Coleman NL, Willis JC, Gill AC, Sansom OW, Carroll BR (1970) Effect of concentration gradients on the performance of a nuclear sediment concentration gage. Water Resourc Res 6:538–548

McLean SR (1985) Theoretical modelling of deep ocean sediment transport. Mar Geol 66:243–265

McNown JS, Lin PN (1952) Sediment concentration and fall velocity. Proc 2nd Midwest Conf Fluid Mech, Ohio State Univ, pp 401–411

Meade RH (1969) Landward transport of bottom sediments in estuaries of the Atlantic coastal plain. J Sediment Petrol 39:222–234

Meade RH (1972) Transport and deposition of sediments in estuaries. Geol Soc Am Mem 133:91–120

Meade RH (1979) Suspended-sediment and velocity data, Amazon river and its tributaries, June–July 1976 and May–June 1977. US Geol Surv Open-File Rep 79–515:42

Meade RH (1982) Sources, sinks and storage of river sediments in the Atlantic drainage of the United States. J Geol 90:235–252

Meade RH (1985) Suspended sediment in the Amazon River and its tributaries in Brazil during 1982–1984. US Geol Surv Open file Rep 85–492 Denver, p 41

Meade RH (1988) Movement and storage of sediment in river systems. In: Lerman A, Meybeck M (eds) Physical and chemical weathering in geochemical cycles. Kluwer, Dordrecht, pp 165–179

Meade RH (in press) River-sediment inputs to major deltas In: Milliman JD, Sabhasri S (eds) Sea-level rise and coastal subsidence: problems and strategies. John Wiley & Sons, New York

Meade RH, Parker RS (1985) Sediment in rivers of the United States. Natl Water Summary 1984. US Geol Surv Water Supply Pap 2275:49–60

Meade RH, Trimble SW (1974) Changes in sediment loads in rivers of the Atlantic drainage of the United States since 1900. In: Effects of Man on the interface of the hydrological cycle with the physical environment. Proc Paris Symp, Sept 1974, IAHS 113:99–104

Meade RH, Sachs PL, Manheim FT, Hathaway JC, Spencer DW (1975) Sources of suspended matter in waters of the Middle Atlantic Bight. J Sediment Petrol 45:171–188

Meade RH, Nordin CF Jr, Curtis WF, Costa Rodrigues FM, do Vale CM, Edmond JM (1979) Sediment loads in the Amazon river. Nature (Lond) 278:161–163

Meade RH, Yuzyk TR, Day TJ (1990) Movement and storage of sediment in rivers of the United States and Canada. In: Wolman, MG Riggs HC (eds) The geology of North America, vols 0–1. Surface water hydrology. Geol Soc Am, Boulder, Colorado pp 255–380

Medwin H (1977) In situ acoustic measurements of micro-bubbles at sea. J Geophys Res 75:971–975

Mehta AJ (1973) Depositional behavior of cohesive sediments. Thesis, Univ Florida, Gainesville, March 1973

Mehta AJ (1988) Laboratory studies on cohesive sediment deposition and erosion. In: Dronkers J, van Leussen W (eds) Physical processes in estuaries. Springer, Berlin Heidelberg New York, pp 427–445

Mehta AJ (1989) On estuarine cohesive sediment suspension behavior. J Geophys Res 94:C10 14.303–14.314

Mehta AJ, Maa PY (1986) Waves over mud: modeling erosion. 3rd Int Symp River Sed Univ Mississippi, March–April 1986, pp 588–601

Mehta AJ, Partheniades E (1975) An investigation of the depositional properties of flocculated fine sediment. J Hydrol Res 13:361–381

Mei ChC, Liu K-F (1987) A Bingham-plastic model for a muddy seabed under long waves. J Geophys Res 92:C13 14.581–14.594

Meybeck M (1977) Dissolved and suspended matter carried by rivers: composition, time and space variations, and work balance. In: Golterman HL (ed) Interactions between sediments and fresh water. Proc Int Symp Amsterdam 1976, pp 25–32

Meybeck M (1988) How to establish and use world budgets of riverine materials. In: Lerman A, Meybeck M (eds) Physical and chemical weathering in geochemical cycles. Kluwer, Dordrecht, pp 247–272

Michaels AS, Bolger JC (1962) Settling rates and sediment volumes of flocculated kaoline suspensions. Indian Eng Chem Fund 1:24–33

Michaels EF, Silver MW (1988) Primary production, sinking fluxes and the microbial food web. Deep Sea Res 35:473–490

Migniot C (1968) Etude des propriétés physiques de différents sédiments très fins et de leur comportement sous des actions hydrodynamiques. Houille Blanche (in French with English Abstr) 7:591–620

Miller RL (1976) Role of vortices in surf zone prediction: sedimentation and wave forces. In: Davis RA, Ethington RL (eds) Beach and nearshore sedimentation. SEPM Spec Publ 24, Tulsa, Oklahoma, pp 92–114

Milliman JD (1980) Sedimentation in the Fraser River and its estuary, southwestern British Columbia (Canada). Est Coast Mar Sci 10(6):609–633

Milliman JD, Meade RH (1983) World-wide delivery of river sediment to the oceans. J Geol 91:1–21

Milliman JD, Syvitski JPM (in press) Geomorphic and tectonic control of sediment discharge to the ocean: the importance of small mountainous rivers. J Geol

Milliman JD, Hsueh Y, Hu D-X, Pashinski DJ, Shen H-T, Yang Z-S, Hacker P (1984) Tidal phase control of sediment discharge from the Yangtze river. Est Coast Shelf Sci 19:119–128

Milliman JD, Beardsley RC, Yang Z-S, Limeburner R (1985) Modern Huanghe-derived muds on the outer shelf of the east China Sea: identification and potential transport mechanisms. Cont Shelf Res 4:175–188

Milliman JD, Qin Y-Sh, Ren M-E, Saito Y (1987) Man's influence on the erosion and transport of sediment by Asian rivers: the Yellow River (Huang He) example. J Geol 95:751–762

Milner HB (1962) Sedimentary petrography, I, II. 4th revised edn. Allen & Unwin, London, pp 643, 715

Miquel A (1975, 1980) La géographie humaine du monde musselman jusqu'au milieu du 11e siècle. Mouton, Den Haag, vol 2 (1975): p 705; vol 3 (1980): p 543

Mook WG, Tan FC (1991) Stable carbon isotopes in rivers and estuaries. In: Degens ET, Kempe S, Richey JE (eds) Biogeochemistry of major world rivers. John Wiley & Sons, New York, pp 245–264

Morgan JP, van Lopik JR, Nichols LG (1953) Occurrence and development of mudflats along the western Louisiana coast. Coast Stud Inst Baton Rouge Tech Rep 2:1–34

Morisawa M (1968) Streams, their dynamics and morphology. McGraw-Hill, New York

Morris AW, Loring DH, Bale AJ, Howland RJM, Mantoura RFC, Woodward EMS (1982) Particle dynamics, particulate carbon and the oxygen minimum in an estuary. Oceanol Acta 5:349–353

Mouret G (1921) Antoine Chézy, histoire d'une formule hydraulique. Ann Ponts Chaussées 1921–II

Murray J (1891) Deep sea deposits. Rep Sci Res Voyage HMs Challenger, 1873–1876, p 525

Murray J, Hjort J (1912) The depths of the ocean. Macmillan, London, p 820

Nakagawa H, Nezu I (1975) Turbulence of open channel flow over smooth and rough beds. Proc Jpn Soc Civ Eng 241:155–168

Nakajima K (1969) Suspended particulate matter in the waters on both sides of the Aleutian Ridge. J Oceanogr Soc Jpn 25(5):239–248

Narkis N, Rebhun M (1975) The mechanisms of flocculation processes in the presence of humic substances. J AWWA 67:101–108

NEDECO (ed) (1965) A study on the siltation of the Bangkok Port Channel. II: The field investigation. NEDECO, Den Haag, p 474

NEDECO (ed) (1968) Surinam transport study. Report on hydraulic investigation. NEDECO Den Haag, p 293

Needham J (1959, 1971) Science and civilization in China, vol 3 (1959): p 877, vol 4, pt III (1971): p 378, University Press, Cambridge

Neihof RA, Loeb GI (1972) The surface charge of particulate matter in seawater. Limnol Oceanogr 17:7–16

Neihof RA, Loeb GI (1974) Dissolved organic matter in seawater and the electric charge of immersed surfaces. J Mar Res 32:5–12

Nelsen TA (1981) The application of Q-mode factor analysis to suspended particulate matter studies: examples from the New York Bight apex. Mar Geol 39:15–31

Nelson BW (1960) Clay mineralogy of the bottom sediments, Rappahannock river, Virginia. In: Swineford A (ed) Clays and clay minerals. Proc 7th Natl Conf, Pergamon, New York, pp 135–147

Nelson DM, Goering JJ (1977) Near surface silica dissolution in the upwelling region off northwest Africa. Deep Sea Res 24:65–74

Nelson DM, Gordon LJ (1982) Production and pelagic dissolution of biogenic silica in the Southern Ocean. Geochim Cosmochim Acta 46:491–501

New AL, Pingree RD (1990) Evidence for internal tidal mixing near the shelf break in the Bay of Biscay. Deep Sea Res 37:1783–1803

Newton I (1687) Principia mathematica philosophiae naturalis. London

Nichols MM (1972) Sediments of the James River Estuary, Virginia In: Nelson BW (ed) Environmental framework of coastal plain estuaries. Geol Soc Am Mem 133:169–212

Nichols MM (1989) Sediment accumulation rates and relative sea-level rise in lagoons. Mar Geol 88:201–209

Nichols MM (1985) Fluid mud accumulation processes in an estuary. Geol Mar Lett 4:171–176

Nichols MM (1986) Storage efficiency of estuaries. In: 3rd Int Symp River Sediment, Univ Mississippi, pp 273–289

Nielsen P (1984) Field measurements of time-averaged suspended sediment concentration under waves. Coastal Eng 8:51–72

Nishizawa S, Fukuda M, Inoue N (1954) Photographic study of suspended matter and plankton in the sea. Bull Fak Fish Hokkaido Univ 5:36–40

Nittrouer CA, DeMaster DJ (1986) Sedimentary processes on the Amazon continental shelf: past, present and future research. Cont Shelf Res 6:5–30

Nittrouer CA, Curtin TB, DeMaster DJ (1986) Concentration and flux of suspended sediment on the Amazon Continental Shelf. Cont Shelf Res 6(1/2):151–174

Nolting RF, Eisma D (1988) Elementary composition of suspended particulate matter in the North Sea. Neth J Sea Res 22:219–236

Nordin CF (1985) The sediment load of rivers. In: Rodda JC (ed) Facets of hydrology, vol 2. John Wiley & Sons, New York, pp 183–204

Nowell ARM, Jumars PA, Eckman JE (1981) Effects of biological activity on the entrainment of marine sediments. Mar Geol 42:133–153

Nydegger P (1967) Untersuchungen über Feinstofftransport in Flüssen und Seen, über Entstehung von Trübungshorizonten und zufluß-bedingten Strömungen in Brienzersee und einigen Vergleichsseen. Schweiz Geot Kommiss Hydrol Kommiss Schweiz Naturforsch Ges. Beitr Geol Schweiz Hydrol 16:92

Nyffeler UP, Li Y-H, Santschi PS (1984) A kinetic approach to describe trace element distribution between particles and solution in natural aquatic systems. Geochim Cosmochim Acta 48:1513–1522

O'Brien N (1971) Fabric of kaolinite and illite floccules. Clays Clay Minerals 19:353–359

Odd NVH (1988) Mathematical modelling of mud transport in estuaries. In: Dronkers J, van Leussen W (eds) Physical processes in estuaries. Springer, Berlin Heidelberg New York, pp 503–531

Odd NVH, Owen MW (1972) A two-layer model of mud transport in the Thames estuary. Proc Inst Civil Eng London Pap 7517 S:175–205

Officer ChB (1976) Physical oceanography of estuaries. John Wiley & Sons, New York, p 465

Officer ChB (1980) Box models revisited. In: Hamilton P, MacDonald KB (eds) Estuarine and wetland processes. Plenum, New York, pp 65–114

Officer ChB, Nichols MM (1980) Box model application to a study of suspended sediment distributions and fluxes in partially mixed estuaries. In: Kennedy VS (ed) Estuarine perspectives. Academic Press, New York London, pp 329–340

O'Melia ChR (1985) The influence of coagulation and sedimentation on the fate of particles, associated pollutants and nutrients in lakes In: Stumm W (ed) Chemical processes in lakes. John Wiley & Sons, New York, pp 207–224

Onishi Y, Thompson FL (1986) Sediment and contaminant transport in a marine environment. In: 3rd Int Symp River Sediment Univ Mississippi, pp 569–577

Oseen CW (1927) Neuere Methoden und Ergebnisse in der Hydrodynamik, Akad Verlagsges, Leipzig

Owen MW (1970) A detailed study of the settling velocities of an estuary mud. Hydraul Res St Wallingford Rep INT 78:25

Owen MW (1971) The effect of turbulence on the settling velocities of silt flocs. In: Proc 14th Congr IAHR, Paris, 4, D4-1–D4-5

Özsoy E (1986) Ebb-tidal jets: a model of suspended sediment and mass transport at tidal inlets. Est Coast Shelf Sci 22:45–62

Paerl HW (1973) Detritus in Lake Tahoe: structural modification by attached microflora. Science 180:496–498

Paerl HW (1974) Bacterial uptake of dissolved organic matter in relation to detrital aggregation in marine and freshwater systems. Limnol Oceanogr 19:966–972

Pak H, Zaneveld JRV (1977) Bottom nepheloid layers and bottom mixed layers observed on the continental shelf off Oregon. J Geophys Res 82:3921–3931

Pak H, Codispoti LA, Zaneveld JRV (1980a) On the intermediate particle maxima associated with oxygen poor water off western South America. Deep Sea Res 27A:783–797

Pak H, Zaneveld JRV, Kitchen J (1980b) Intermediate nepheloid layers observed off Oregon and Washington. J Geophys Res 85:C11 6697–6708

Pak H, Zaneveld JRV, Spinrad RW (1984) Vertical distribution of suspended particulate matter in the Zaire river, estuary and plume. Neth J Sea Res 17:412–425

Palanques A, Drake DE (1990) Distribution and dispersal of suspended particulate matter on the Ebro continental shelf, northwest-Mediterranean Sea. Mar Geol 95:193–206

Palanques A, Plana F, Maldonado A (1990) Recent influence of man on the Ebro margin sedimentation system, northwestern Mediterranean Sea. Mar Geol 95:247–263

Pandya JD, Spielman LA (1982) Floc breakage in agitated suspensions: Theory and data processing strategy. J Colloid Interface Sci 90:517–531

Pantin HH (1979) Interaction between velocity and effective density in turbidity flow: phase-plane analyses, with criteria for autosuspension. Mar Geol 31:59–99

Park YA, Kim SC, Choi JH (1986) The distribution and transportation of fine-grained sediments on the inner continental shelf off the Keum River estuary, Korea. Cont Shelf Res 5:499–519

Parker DS, Kaufman WJ, Jenkins D (1972) Floc breakup in turbulent flocculation processes. J San Eng Div Proc Am Soc Civ Eng 98:79–99

Parker WR (1987) Observation on fine sediment transport phenomena in turbid coastal environments. Cont Shelf Res 7:1285–1293

Partheniades E (1965) Erosion and deposition of cohesive soils. J Hydraul Div ASCE 91 (HY1) Proc Pap 4204:105–139

Partheniades E (1971) Erosion and deposition of cohesive materials. In: Hsieh Wen Shen (ed) River mechanics, II. Colorado State Univ, Fort Collins, pp 1–91

Partheniades E, Kennedy JF (1966) Depositional behavior of fine sediment in a turbulent fluid motion. In: Proc 10th Conf Coastal Eng, Tokyo, pp 707–729

Paterson DM (1989) Short-term changes in the erodibility of intertidal cohesive sediments related to the migratory behavior of epipelic diatoms. Limnol Oceanogr 34(1):223–234

Pavoni JL, Tenney MW, Echelberger WF (1972) Bacterial exocellular polymers and biological flocculation. J Water Pollut Control Fed 44:414–431

Peinert R, von Bodungen B, Smetacek VS (1989) Food web structure and loss rate In: Berger WH, Smetacek VS, Wefer G (eds) Productivity of the ocean: present and past John Wiley & Sons, New York, pp 35–48

Pejrup M (1986) Parameters affecting fine-grained suspended sediment concentrations in a shallow micro-tidal estuary. Ho Bugt, Denmark. Est Coast Shelf Sci 22:241–254

Pejrup M (1988) Suspended sediment transport across a tidal flat. Mar Geol 82:187–198

Pellenbarg R (1979) Silicones as tracers for anthropogenic additions to sediments. Mar Pollut Bull 10:267–269

Penck A (1984) Morphologie der Erdoberfläche. 1. Teil. Van Engelhoorn, Stuttgart, p 471

Perillo GME, Cuadrado DG (1990) Nearsurface suspended sediments at Monte Hermoso Beach, Argentina, I. Descriptive characteristics. J Coast Res 6:981–990

Pethick JS (1980) Velocity surges and asymmetry in tidal channels. Est Coast Mar Sci 11:331–345

Petterson H (1934) A transparency meter for sea water. Göteborg Kung Vetensk Vitterh Sämh Hand 5, Ser B 5:1–17

Phleger CF, Soutar A (1971) Free vehicles and deep-sea biology. Am Zool 11:409–418

Picard GL, Felbeck GT Jr (1976) The complexation of iron by marine humic acid. Geochim Cosmochim Acta 40:1347–1350

Piccard J, Dietz RS (1957) Oceanographic observations by the bathyscaphe Trieste (1953–1956). Deep Sea Res 4:221–229

Pickard GL, Rodgers K (1959) Current measurements in Knight Inlet, British Columbia. J Fish Res Board Can 16:635–678

Pickrill RA, Irwin J, Shakespeare BS (1981) Circulation and sedimentation in a tidal-influenced fjord lake: Lake McKerrow, New Zealand. Est Coast Shelf Sci 12:23–37

Pierce JW, Nichols MM (1986) Change of particle composition from fluvial into an estuarine environment. Rappahannock River, Virginia. J Coast Res 2:419–425

Piper DJW, Letson JRJ, Delure AM, Barrie CQ (1983) Sediment accumulation in low-sedimentation, wave-dominated, glacial inlets. Sed Geol 36:195–215

Plank WS, Zaneveld JRV, Pak H (1973) Distribution of suspended matter in the Panama Basin. J Geophys Res 78:7113–7120

Platt T, Subba Rao DV (1975) Primary production of marine microphytes In: Cooper JP (ed) Photosynthesis and productivity of different environments. Univ Press, Cambridge, pp 249–280

Pocklington R, Leonard JD (1979) Terrigenous organic matter in sediments of the St Larence estuary and the Saguenay Fjord. J Fish Res Board Can 36:1250–1255

Pomeroy LR, Deibel D (1980) Aggregation of organic matter by pelagic tunicates. Limnol Oceanogr 25:643–652

Postma H (1954) Hydrography of the Dutch Wadden Sea. Arch Neerl Zool 10:405–511

Postma H (1961) Transport and accumulation of suspended matter in the Dutch Wadden Sea. Neth J Sea Res 1:148–190

Postma H (1967) Sediment transport and sedimentation in the marine environment. In: Lauff GH (ed) Estauries AAAS Publ 83:158–179

Postma H (1969) Suspended matter in the marine environment. In: UNESCO (ed) Morning review lectures, 2nd Int Oceanogr Congr, Moscow 1966, pp 213–219

Postma H (1980) Sediment transport and sedimentation. In: Olausson E, Cato I (eds) Chemistry and biogeochemistry of estuaries. John Wiley & Sons, New York, pp 153–186

Postma H, Kalle K (1955) Die Entstehung von Trübungszonen im Unterlauf der Flüsse, speziell im Hinblick auf die Verhältnisse in der Unterelbe. Dtsch Hydr Z 8:137–144

Powers MC (1954) Clay diagenesis in the Chesapeake Bay area. In: Swinford A, Plummer NV (eds) Clays and clay minerals. Proc 2nd Natl Conf Natl Acad Sci, Natl Res Counc Publ 327:68–80

Prahl FG, Muehlhausen LA (1989) Lipid biomarkers as geochemical tools for paleoceanographic study. In: Berger WH, Smetacek VS, Wefer G (eds) Productivity of the ocean: present and past. John Wiley & Sons, New York, pp 271–289

Prandtl L (1905) Über Flüssigkeitsbewegung bei sehr kleiner Reibung. Verh III Int Math Kongr Heidelberg 1904, Leipzig

Pravdić V (1970) Surface charge characterization of sea sediments. Limnol Oceanogr 15:230–233

Pritchard DW (1967) Observations of circulation in coastal plain estuaries. In: Lauff GH (ed) Estuaries. Am Assoc Adv Sci Publ 83:37–44

Puls W, Kuehl H (1986) Field measurements of the settling velocities of estuarine flocs. In: Proc 3rd Int Symp River Sediment, Univ Mississippi, pp 525–536

Puls W, Sündermann J (1990) Simulation of suspended sediment dispersion in the North Sea. In: Cheng RT (ed) Coastal and estuarine studies 38. Springer, Berlin Heidelberg New York, pp 356–372

Puls W, Kuehl H, Heymann K (1988) Settling velocity of mud flocs: results of field measurements in the Elbe and the Weser estuary. In: Dronkers J, van Leussen W (eds) Physical processes in estuaries. Springer, Berlin Heidelberg New York, pp 404–424

Rabinowitch E (1971) An unfolding discovery. Proc Natl Acad Sci USA 63:2875–2876

Ramirez-Muñoz J (1968) Atomic-absorption spectroscopy. Elsevier, Amsterdam, p 493

Rashid MA (1985) Geochemistry of marine humic compounds. Springer, Berlin Heidelberg New York, p 300

Rauck G (1988) Welchen Einfluß haben Grundschleppnetze auf den Meeresboden und Bodentiere? Inf Fischwesen 35:104–106

Ray J (1673) Observations made in a journey through part of the Low-Countries, Germany, Italy and France. London

Rayleigh Lord (1893) On the flow of viscous fluids especially in two dimensions. Philos Mag 5, 36:354

Rearic DM, Barnes PW, Reimnitz E (1990) Bulldozing and resuspension of shallow-shelf sediment by ice keels: implications for arctic sediment transport trajectories. Mar Geol 91:133–147

Redfield AC, Ketchum BH, Richards FA (1963) The influence of organisms on the composition of seawater. In: Hill MN (ed) The sea, vol 2, Wiley Interscience, New York, pp 26–77

Reimers CE (1989) Control of benthic fluxes by particulate supply. In: Berger WH, Smetacek VS, Wefer G (eds) Productivity of the ocean: present and past. John Wiley & Sons, New York, pp 217–233

Relling O, Nordseth K (1979) Sedimentation of a river suspension into a fjord basin, Gaupnefjord in western Norway. Nor Geogr Tidsskr 33:187–203

Ren M-E (ed) (1986) Modern sedimentation in the coastal and nearshore zones of China. China Ocean Press, Beijing; and Springer, Berlin Heidelberg New York, p 466

Ren M-E, Shi Y-L (1986) Sediment discharge of the Yellow River (China) and its effect on the sedimentation of the Bohai and the Yellow Sea. Cont Shelf Res 6:785–810

Rendon-Herrero O (1974) Estimation of washload produced on certain small watersheds. J Hydraul Div Proc ASCE 100:HY7, 835–848

Reuther JH (1981) Chemical interactions involving the biosphere and fluxes of organic material in estuaries. In: Martin J-M, Burton JD, Eisma D (eds) River inputs to ocean systems. UNESCO, Paris, pp 239–242

Reynolds CS, Wiseman SW, Gardner WD (1980) Aquatic sediment traps and trapping methods. Freshwater Biol Assoc Occ Publ 11:54

Reynolds O (1883) An experimental investigation of the circumstances which determine whether the motion of water shall be direct or sinous and of the law of resistance in parallel channels. Philos Trans R Soc Lond 174:935–982

Reynolds O (1895) On the dynamical theory of incompressible viscous fluids and the determination of the criterion. Philos Trans R Soc Lond 186A:123–164

Ribelin BW, Collier AW (1979) In: Livingston RJ (ed) Ecological processes in coastal and marine systems. Plenum, New York, pp 47–68

Richards K (1982) Rivers. Methuen, London New York, Chap 3

Richardson MJ, Gardner WD (1985) Analysis of suspended-particle-size distributions over the Nova Scotian Continental rise. Mar Geol 66:189–203

Ridge MJH, Carson B (1987) Sediment transport on the Washington continental shelf: estimates of dispersal rates from Mount St Helens ash. Cont Shelf Res 7:759–772

Riebesell U (1991a) Particle aggregation during a diatom bloom. I. Physical aspects. Mar Ecol Prog Ser 69:273–280

Riebesell U (1991b) Particle aggregation during a diatom bloom. II. Biological aspects. Mar Ecol Prog Ser 69:281–291

Riley GA (1963) Organic aggregates in seawater and dynamics of their formation and utilization. Limnol Oceanogr 8:372–381

Riley GA (1970) Particulate organic matter in sea water. Adv Mar Biol 82:1–110

Riley GA, Wangersky PJ, van Hemert D (1964) Organic aggregates in tropical and subtropical surface waters of the North Atlantic Ocean. Limnol Oceanogr 9:546–550

Riley GA, van Hemert D, Wangersky RJ (1965) Organic aggregates in surface and deep waters of the Sargasso Sea. Limnol Oceanogr 10:354–363

Rodi W (1980) Turbulence models and their application in hydraulics. Int Assoc Hydr Res, Delft

Rodi W (1986) Use of advanced turbulence models for calculating the flow and pollutant spreading in rivers. In: Proc 3rd Int Conf River Sediment, Univ Mississippi, pp 1369–1382

Roehl JW (1962) Sediment source areas, delivery ratios and influencing morphological factors. Publ Int Assoc Sci Hydrol 59:202–213

Ross MA (1988) Vertical structure of estuarine fine sediment suspensions. Thesis, Univ Florida, Gainesville (in Mehta 1989)

Rouse H (1937) Modern conceptions of the mechanics of fluid turbulence. Trans ASCE 102 (1965):463–543

Rouse H (1938) Fluid mechanics for hydraulic engineers. Constable, London, p 422

Rouse H, Ince S (1957) History of hydraulics. Iowa Inst Hydraul Res, Iowa State Univ, p 269

Rubey WW (1938) The forces required to move particles on a stream bed. US Geol Surv Prof Pap 189E, pp 121–142

Rumeau JL, Vanney JR (1968) Eléments-traces de vases marines du plateau continental atlantique au sud-ouest du massif Armoricain (Grande Vasière). Bull Centre Rech Pau–SNPA 2: 69–81

Sahl LE, Marsden MAH (1987) Shelf sediment dispersal during the dry season, Princess Charlotte Bay, Great Barrier Reef, Australia. Cont Shelf Res 7:1139–1159

Saliot A, Ulloa-Guevara A, Viets TV, de Leeuw JW, Schenck PA, Boon JJ (1984) The application of pyrolysis-mass spectrometry and pyrolysis-gas chromatography-mass spectrometry to the chemical characterization of suspended matter in the ocean. Org Geochem 6:295–304

Salomons W (1975) Chemical and isotopic composition of carbonates in recent sediments and soils from western Europe. J Sediment Petrol 45:440–449

Salomons W, Eysink WD (1981) Pathways of mud and particulate trace metals from rivers to the southern North Sea. Spec Publ Int Assoc Sedimentol 5:429–450

Salomons W, Förstner U (1984) Metals in the hydrocycle. Springer, Berlin Heidelberg New York, p 349

Salomons W, Mook WG (1977) Trace metal concentrations in estuarine sediments: mobilization, mixing and precipitation. Neth J Sea Res 11(2):119–129

Salomons W, Mook WG (1981) Field observations of the isotopic composition of particulate organic carbon in the Southern North Sea and adjacent estuaries. Mar Geol 41:M11–M20

Salomons W, Hofman P, Boelens R, Mook WG (1975) The oxygen isotopic composition of the fraction less than 2 microns (clay fraction) in recent sediments from Western Europe. Mar Geol 18:M23–M28

Salomons W, de Bruin M, Duin RPW, Mook WG (1978) Mixing of marine and fluvial sediments in estuaries. In: Abstr 16th Coastal Eng Conf, Hamburg, 4 pp

Samaga BR, Ranga Raju KG, Garde RJ (1985) Concentration distribution of sediment mixtures in open-channel flow. J Hydraul Res 23:467–483

Santema P (1953) Enkele beschouwingen over het slibtransport van de Rijn (in Dutch). Bouw Waterbouwkd 3:B37–B41

Sauberer F, Ruttner F (1941) Die Strahlungsverhältnisse der Binnengewässer. Becker, Leipzig

Scheidegger AE (1961) Theoretical geomorphology. Springer, Berlin Heidelberg New York, p 333

Schlesinger WH, Melack JM (1981) Transport of organic carbon in the world's rivers. Tellus 33:172–187

Schmitz WJ Jr (1984) Abyssal eddy kinetic energy in the North Atlantic. J Mar Res 42:509–536

292   References

Schoelhamer DH (1986) Comparison of mean longitudinal suspended-sediment and stream velocities. In: Proc 3rd Int Symp River Sediment, Univ Mississippi, pp 796–803
Schofield RK, Samson HR (1954) Flocculation of kaolinite due to the attraction of oppositely changed crystal faces. Disc Faraday Soc 18:135–145
Schrimpf W (1986) The influence of the mixing length assumption on calculating suspended sediment transport. In: Proc 3rd Int Symp River Sediment, Univ Mississippi, pp 757–765
Schubel JR (1969) Distribution and transport of suspended sediment in Upper Chesapeake Bay. Tech Rep 60 Ref 69-13 Chesapeake Bay Inst, John Hopkins Univ
Schubel JR (1974) Effects of tropical storm Agnes on the suspended solids in water. Plenum, New York, pp 113–132
Schubel JR, Carter HH (1976) Suspended sediment budget for Chesapeake Bay. In: Wiley M (ed) Estuarine processes II. Academic Press, New York London, pp 48–62
Schubel JR, Kana TW (1972) Agglomeration of fine-grained suspended sediment in northern Chesapeake Bay. Powder Technol 6:9–16
Schumm SA, Hadley RF (1961) Progress in the application of landform analysis in studies of semi-arid erosion. US Geol Surv Circ 437, p 14
Secchi P (1865) Relazione delle esperienze fatte a bordo della pontificia Pirocorvetta *Immacolate Concezione* per determinare la transparenza del mare. In: Cialdi A (ed) Sul mota ondoso del mare, e su le correnti, 2nd edn. Roma (1866), p 693
Seibold E, Berger WH (1982) The sea floor. Springer Verlag, p 288
Seki H (1972) The role of microorganisms in the marine food chain with reference to organic aggregate. Mem Inst Ital Idrobiol Suppl 29:247–259
Shanks AL, Edmondson EW (1989) Laboratory-made artificial marine snow – a biological model of the real thing. Mar Biol 101:463–470
Shanks AL, Trent JD (1980) Marine snow: sinking rates and potential role in vertical flux. Deep Sea Res A 27:137–143
Sharma GD (1979) The Alaskan Shelf: hydrographic, sedimentary and geochemical environment. Springer, Berlin Heidelberg New York, 498 pp
Sheldon RW (1968) Sedimentation in the estuary of the river Crouch, Essex, England. Limnol Oceanogr 13:72–83
Sheldon RW, Prakash A, Sutcliffe WH Jr (1972) The size distribution of particles in the ocean. Limnol Oceanogr 17:327–340
Shelford VE, Gail FW (1922) A study of light penetration into sea water made with the Kunz photo-electric cell with particular reference to the distribution of plants. Publ Puget Sound Biol Stn 3(65):141–176
Shepard FP (1973) Submarine geology. Harper and Row, New York, p 517
Shepard FP, Marshall NF, McLoughlin PA, Sullivan CG (1979) Currents in submarine canyons and other sea valleys. Am Assoc Petrol Geol (Tulsa, Oklahoma) Stud Geol 8:1–173
Sherman I (1953) Flocculent structure of sediment suspended in Lake Mead. Trans Am Geophys Union 34:394–406
Shi YL, Yang W, Ren M-E (1985) Hydrological characteristics in the Chang Jiang and its relation to sediment transport to the sea. Cont Shelf Res 4(1/2):5–15
Shideler GL (1984) Suspended sediment responses in a wind-dominated estuary of the Texas Gulf Coast. J Sediment Petrol 54:731–745
Shields A (1936) Anwendung der Ähnlichkeitsmechanik und der Turbulenzforschung auf die Geschiebebewegung. Mitt Preuss Versuchamt Wasserbau Schiffbau (Berlin)
Shimp NF, Witters J, Potter PE, Schleicher JA (1969) Distinguishing marine and freshwater muds. J Geol 77:566–580
Sholkovitz ER (1976) Flocculation of dissolved organic and inorganic matter during the mixing of river water and seawater. Geochim Cosmochim Acta 40:831–845
Sholkovitz ER (1978) The flocculation of dissolved Fe, Mn, Al, Cu, Ni, Co and Cd during estuarine mixing. Earth Planet Sci Lett 41:77–86
Sholkovitz ER, Boyle EA, Price NB (1978) The removal of dissolved humic acids and iron during estuarine mixing. Earth Planet Sci Lett 40:130–136

Shultz DJ, Calder JA (1976) Organic carbon $^{13}C/^{12}C$ variations in estuarine sediments. Geochim Cosmochim Acta 40:381–385

Sieburth JMcN (1965) Organic aggregation in seawater by alkaline precipitation of inorganic nuclei during the formation of ammonia by bacteria. J Gen Microbiol 41:20

Siegbahn K (1966) Alpha-, beta- and gamma-ray spectroscopy, 2 vols. North Holland, Amsterdam, p 1742

Silver MW, Alldredge AL (1981) Bathypelagic marine snow: deep-sea algal and detrital community. J Mar Res 39:501–530

Silver MW, Shanks AL, Trent JD (1978) Marine snow: microplankton habitat and source of small-scale patchiness in pelagic populations. Science 201:371–373

Simoneit BRT, Grimalt JO, Fischer K, Dymond J (1986) Upward and downward flux of particulate organic material in abyssal waters of the Pacific Ocean. Naturwissenschaften 73:322–325

Simpson JH, Allen CM, Morris NCG (1978) Fronts on the Continental Shelf. J Geophys Res 83 (C9):4607–4614

Skoog DA (1985) Principles of instrumental analysis. 3rd edn. Saunders, Philadelphia, p 879

Sleath JFA (1984) Sea bed mechanics. John Wiley & Sons, New York, p 335

Sly PG (1978) Sedimentary processes in lakes. In: Lerman A (ed) Lakes; chemistry, geology, physics. Springer, Berlin Heidelberg New York, pp 78–89

Smayda ThJ (1971) Normal and accelerated sinking of phytoplankton in the sea. Mar Geol 11:105–122

Smerdon ET, Beasley RP (1961) Critical tractive forces in cohesive soils. Agric Eng:26–29

Smetacek VS (1985) Role of sinking in diatom life-history cycles: ecological, evolutionary and geological significance. Mar Biol 84:239–251

Smith JD (1977) Modeling of sediment transport on continental shelves. In: Goldberg ED, McCave IN, O'Brian JJ, Steele JH (eds) The sea, vol 6. Wiley-Interscience, New York, pp 539–577

Smith JD, Hopkins TS (1972) Sediment transport on the continental shelf off Washington and Oregon in light of recent current measurements. In: Swift DJP, Duane DB, Pilkey OH (eds) Shelf sediment transport. Dowden Hutchinson and Ross, Stroudsburg, pp 61–82

Smith KL Jr, Williams PM, Druffel ERM (1989) Upward fluxes of particulate organic matter in the deep North Pacific. Nature (Lond) 337:724–726

Smith ND, Syvitski JPM (1982) Sedimentation in a glacier-fed lake: the role of pelletization on deposition of fine-grained suspensates. J Sediment Petrol 52:503–513

Snedden JW, Nummedal D (1990) Coherence of surfzone and shelf current flow on the Texas (USA) coastal margin: implications for interpretation of paleo-current measurement in ancient coastal sequences. Sediment Geol 67:221–236

Song W, Yoo D, Dyer KR (1983) Sediment distribution, circulation and provenance in a macrotidal Bay: Garolim Bay, Korea. Mar Geol 52:121–140

Soo SL (1967) Fluid dynamics of multiphase systems. Blaisdell Waltham, MA, 517 pp

Spencer CP (1983) Marine biogeochemistry of silicon. In: Aston SR (ed) Silicon geochemistry and biogeochemistry. Academic Press, New York London, pp 101–141

Spencer DW (1984) Aluminium concentrations and fluxes in the ocean. In: Worksh Global Ocean Flux Study, Woods Hole, pp 206–220

Spinrad RW, Bartz R, Kitchen JC, (1989) In situ measurements of marine particle settling velocity and size distributions using the remote optical settling tube. J. Geophys Res 94: C1 931–938

Stacey MW, Bowen AJ (1988) The vertical structure of turbidity currents and a necessary condition for self-maintenance. J Geophys Res 93:C4 3543–3553

Stanley DJ, Fenner P, Kelling G (1972) Currents and sediment transport at the Wilmington Canyon shelf break, as observed by underwater television. In: Swift DJP, Duane DB, Pikey OH (eds) Shelf sediment transport. Dowden, Hutchinson & Ross, Stroudsbury, pp 621–644

Staunton G (1797) Embassy to China, II (4to), London (in Lyell 1830)

Steemann Nielsen E (1954) On organic production in the oceans. J Conserv Perm Int Explor Mer 19:309–328

Sternberg RW (1989) Instrumentation for estuarine research. J Geophys Res 94:C10 14.289–14.301

Sternberg RW, Larsen LH (1975) Threshold of sediment movement by open ocean waves. Deep Sea Res 22:299–309

Sternberg RW, Larsen LH, Miao YT (1985) Tidally driven sediment transport on the East China Sea continental shelf. Cont Shelf Res 4(1/2):105–120

Sternberg RW, Cacchione DA, Drake DE, Kranck K (1986) Suspended sediment transport in an estuarine tidal channel within San Francisco Bay, California. Mar Geol 71:237–258

Stokes GG (1845) On the theories of internal friction of fluids in motion and of the equilibrium and motion of elastic solids. Cambridge Philos Trans 8:287

Strakhov NM (1967) Principles of lithogenesis, vol 1. Oliver & Boyd, London, p 245 (Russian edn 1962)

Stride AH (1988) Indications of long term episodic suspension transport of sand across the Norfolk Banks, North Sea. Mar Geol 79:55–64

Sturm M (1975) Depositional and erosional sedimentary features in a turbidity current controlled basin (Lake Brienz). In: Proc Int Congr Sediment, Nice, pp 1–5

Su J, Wang K (1986) The suspended sediment balance in Chang-jiang estuary. Est Coast Shelf Sci 23:81–98

Su J, Wang K, Li Y (1990) A plume front in Hangzhou Bay and its role in suspended sediment transport. In: Cheng RT (ed) Coastal and estuarine studies 38. Springer, Berlin Heidelberg New York, pp 333–347

Sugimura Y, Suzuki Y (1988) A high temperature catalytic oxydation method of non-volatile dissolved organic carbon in seawater by direct injection of liquid sample. Mar Chem 24:105–131

Sumer BM, Öguz B (1978) Particle motions near the bottom in turbulent flow in an open channel. J Fluid Mech 86:109–127

Sundborg A (1956) The river Klarälven, a study in fluvial processes. Geogr Ann 38:127–316

Sundborg A (1967) Some aspects on fluvial sediments and fluvial morphology. I General views and graphic methods. Geogr Ann 49:333–343

Sündermann J (ed) (1986–1989) Landolt-Börnstein numerical data and functional relationships in science and technology. N S vol 3 (a, b, c) Oceanography. Springer, Berlin Heidelberg New York

Sutcliffe WH, Baylor ER, Menzel DW (1963) Sea surface chemistry and Langmuir circulation. Deep Sea Res 10:233–243

Suzuki N, Kato K (1953) Studies on suspended materials marine snow in the sea. Part 1 Sources of marine snow. Bull Fak Fish Hokkaido Univ 4:132–135

Sverdrup HU, Johnson MW, Fleming RH (1946) The oceans. Prentice Hall, Englewood Cliffs, p 1087

Swart DH (1976) Coastal sediment transport. Computation of longshore transport. Delft Hydraulics Lab Rep R968(1)

Syvitski JPM (1989) On the deposition of sediment within glacier-influenced fjords: oceanographic controls. Mar Geol 85:301–329

Syvitski JPM (ed) (1991) Principles, methods and application of particle size analysis. Cambridge Univ Press, Cambridge, pp 368

Syvitski JPM, Macdonald RD (1982) Sediment character and provenance in a complex fjord; Howe Sound, British Columbia. Can J Earth Sci 19:1025–1044

Syvitski JPM, Murray JW (1981) Particle interaction in fjord suspended sediment. Mar Geol 39:215–242

Syvitski JPM, Silverberg N, Ouellet G, Asprey KW (1983) First observations of benthos and seston from a submersible in the lower St Lawrence estuary. Geogr Phys Quat 37:227–240

Syvitski JPM, Burrell DC, Skei JM (1987) Fjords, processes and products. Springer, Berlin Heidelberg New York, p 379

Takahashi K (1987) Seasonal fluxes of silicoflagellates and Actiniscus in the subarctic Pacific during 1982–1984 J Mar Res 45:397–425

Talbot MMB, Bate GC (1987) Rip current characteristics and their role in the exchange of water and surf diatoms between the surf zone and nearshore. Est Coast Shelf Sci 25:707–720

Tambo N, Hozumi H (1979) Physical characteristics of flocs, II. Strength of flocs. Water Res 13:421–427

Tambo N, Watanabe Y (1979) Physical characteristics of flocs. I: The floc density function and aluminium floc. Water Res 13:409–419

Tan FC, Strain PM (1979) Organic carbon isotope ratios in recent sediments in the St Lawrence estuary and the Gulf of St Lawrence. Est Coast Mar Sci 8:213–225

Teisson Ch, Latteux B (1986) A depth-integrated bidimensional model of suspended sediment transport. In: 3rd Int Symp River Sediment, Univ Mississippi, pp 421–429

Tennekes H, Lumley JL (1972) A first course in turbulence. MIT, Cambridge Mass, p 300

Terwindt JHJ (1977) Deposition, transportation and erosion of mud. In: Golterman H (ed) Interactions between sediments and fresh water. Proc Int Symp, Amsterdam 6–10 Sept 1976, Junk BV, pp 19–24

Terwindt JHJ Breusers HNC (1972) Experiments on the origin of flaser, lenticular and sand-clay alternating bedding. Sedimentology 19:85–98

The Times Atlas of the World (1990) Times Books. London

Thorn MFC, Parsons JG (1980) Erosion and cohesive sediments in estuaries: an engineering guide. In: Proc 3rd Int Symp Dredging Technol, Bordeaux, pp 349–358

Thoulet J (1905) Etude sur la transparence et la couleur des eaux de mer. Res Comp Sci Monaco Mem Oceanogr I fasc XXIX: 113–135

Thoulet J (1912) Etude bathylithologique des côtes du Golfe du Lion. Ann Inst Océanogr IV (6), 67 pp

Thoulet J (1922) L'océanographie. Gauthier-Villar, Paris, p 287

Tingsanchali T, Rodi W (1986) Depth average calculation of suspended sediment transport in rivers. In: Proc 3rd Int Conf River Sediment, Univ Mississippi, pp 1416–1425

Tipping E, Cooke D (1982) The effects of adsorbed humic substances on the surface charge of goethite (–FeOOH) in freshwaters. Geochim Cosmochim Acta 46:75–80

Titley JG, Glegg GA, Glasson DR, Millward GE (1987) Surface areas and porosities of particulate matter in turbid estuaries. Cont Shelf Res 7:1363–1366

Tito de Morais A (1983) Role écologique de la biodéposition d'un organisme benthique filtreur (l'ascide Phallusia mammillata). Thesis, Univ Paris, p 109

Tomi DT, Bagster DF (1978) The behaviour of aggregates in stirred vessels. Trans I Chem E 56:1–18

Trent JD, Shanks AL, Silver MW (1978) In situ and laboratory measurements on macroscopic aggregates in Monterey Bay, California. Limnol Oceanogr 23:626–635

Trimble SW (1977) The fallacy of stream equilibrium in contemporary denudation studies. Am J Sci 277:876–887

Tubman MW, Suhayda JN (1976) Wave action and bottom movements in fine sediments. In: Proc 15th Coast Eng Conf. 2:1168–1183

Twenhofel WH (1933) The physical and chemical characteristics of the sediments of lake Mendota, A freshwater lake of Wisconsin. J Sediment Petrol 3:68–76

Uiterwijk Winkel APB (1975) Microbiologische aspecten en het sedimentatie-gedrag van rivierslib. Rijkswaterstaat, Dir WaWa, District ZW, Rep 44.006.01, p 60

Uncles RJ (1981) A note on tidal asymmetry in the Severn Estuary. Est Coast Shelf Sci 13(4):419–432

Uncles RJ, Elliott RCA, Weston SA (1985) Observed fluxes of water, salt and suspended sediment in a partly mixed estuary. Est Coast Shelf Sci 20:147–167

UNESCO (ed) (1972–1979) Discharge of selected rivers of the world (1969–1972). Stud Rep Hydrol 5, vol 1 (70 pp) vol 3, 2 (124 pp) and vol 3, 3 (104 pp)

Van Andel Tj, Postma H (1954) Recent sediment of the Gulf of Paria. Verh Kon Acad Wetensch Afd Natuurkd 19 XX (5):245

Van Dam H (1987) Verzuring van vennen: een tijdsverschijnsel. Thesis, Univ Wageningen, p 175 (in dutch)

Van der der Gaast SJ, Jansen JHF (1984) Mineralogy, opal and manganese of Middle and Late Quaternary sediments of the Zaire (Congo) deep-sea fan: origin and climatic variation. Neth J Sea Res 17(2–4):313–341

Van der Leeden F (1975) Water resources of the world. Water Inf Center, New York, p 568

Van de Ven TG, Hunter RJ (1977) The energy dissipation in sheared coagulated soils. Rheol Acta 16:534–543

Van der Plas L, Tobi AC (1965) A chart for judging the reliability of point counting results. Am J Sci 263:87–90

Van der Toorn J (1868) Proeven aangaande het slibgehalte van het water der rivieren de Waal en de Maas. Nwe Verh Bat Gen Proefonderv Wijsb Rotterdam, 2e, 1

Van Heerden JL, Wells JT, Roberts HH (1983) River-dominated suspended-sediment deposition in a New Mississippi Delta. Can J Fish Aquat Sci 40 (Suppl 1):60–71

Van Leussen W, Cornelisse J (1992) Direct measurements of sizes and settling velocities of mud flocs in the Ems estuary. Cohesive Sediments Worksh Florida, April 1991 (in press)

Van Leussen W (1988) Aggregation of particles, settling velocity of mud flocs. A review. In: Dronkers J, van Leussen W (eds) Physical processes in estuaries. Springer, Berlin Heidelberg New York, pp 347–403

Van Olphen H (1963) An introduction to clay colloid chemistry. Wiley Interscience, New York, p 301

Vanoni VA (ed) (1975) Sedimentation engineering. ASCE Task Comm, New York, p 745

Vanoni VA, Nomicos GN (1959) Resistance properties of sediment-laden streams. J Hydraul Div ASCE 85 HY5 77–107

Van Rijn LC (1984) Sediment transport, Part II: Suspended load transport. ASCE J Hydraul Eng 110:1613–1641

Van Rijn LC (1985) In situ determination of fall velocity of suspended sediment. In: Proc 21st Congr IAHR, Melbourne 4:144–148

Van Straaten LMJU (1952) Current rips and dip currents in the Dutch Wadden Sea. Proc Kon Ned Akad Wet Ser B 55(3):228–238

Van Straaten LMJU (1954) Composition and structure of recent marine sediments in the Nether-lands. Leidse Geol Meded XIX:1–110

Van Straaten LMJU, Kuenen PhH (1958) Tidal action as a cause of clay accumulation. J. Sediment Petrol 28:406–413

Velikanov MA (1944) The transport of suspended sediment by a turbulent flow. Izv Akad NAUK SSSR OTN 3. (in Russian)

Velikanov MA (1955) Dynamics of alluvial streams. II. Sediment and flow bed. State Publ House Theor Tech Lit, Moscow

Velikanov MA (1958) Alluvial process: fundamental principles. State Publ Phys Math Lit, Moscow

Verdouw H, Gons HJ, Steenbergen CLM (1987) Distribution of particulate matter in relation to the thermal cycle in Lake Vechten (The Netherlands): the significance of transport along the bottom. Water Res 21:345–351

Verger F (1988) Marais et wadden du littoral français, 3rd edn. Paradigm, Caen, p 549

Verreet G, van Goethem J, Viaene W, Berlamont J, Houthuys R, Berleur E (1986) Relations between physico-chemical and rheological properties of fine-grained muds. 3rd Int Symp River Sed Univ Mississippi, March–April 1986, pp 1637–1646

Verweij J (1952) On the ecology of distribution of cockle and mussel in the Dutch Wadden Sea, their role in sedimentation and the source of their food supply. Arch Neerl Zool 10:171–239

Vetter M (1984) Die Anwendung der Gravitationstheorie zur Ermittlung der vertikalen Verteilung der Schwebstoffkonzentration: Mitt Inst Wasserwesen Univ Bundeswehr (München) 13:171–203

Vetter M (1986) Velocity distribution and Von-Karman constant in open channel flows with sediment transport. 3rd Int Symp River Sed Univ Mississippi, March–April 1986, pp 814–823

Villaret C, Paulic M (1986) Experiments on the erosion of deposited and placed cohesive sediments in an annular flume and a rocking flume. Rep UFL/COEL-86/007 Coast Oceanogr Eng Dep Univ Florida, Gainesville

Vincent CE, Green MO (1990) Field measurements of the suspended sand concentration profiles and fluxes of the resuspension coefficient go over a rippled bed. J Geophys Res 95:C7 11.591–11.601

Vincent CE, Young RA, Swift DJP (1983) Sediment transport on the Long Island shoreface, North American Atlantic Shelf: role of waves and currents in shoreface maintenance. Cont Shelf Res 2:163–181

Vital H, do C Faria LE Jr (1990) O sistema hidrodinâmico e a suspensio de Lago-Arare, Ilha de Maraj1 (Pa) In: 36th Congr Brasil Geol 28/10-4/11:1–15

Vittori G (1989) Turbulence simulation in tidal flows. In: Palmer MH (ed) Advances in water modelling and measurement BHRA. Fluid Eng Centre, Crossfield, Bedford, pp 219–229

Vongvisessomjai S, Pongpiridom P (1986) A laterally averaged model for estuarine sedimentation. 3rd Int Symp River Sediments, Univ Mississippi, pp 453–462

Von Helmholtz H (1858) Über Intergrale der hydrodynamischen Gleichungen, welche den Wisbelbewegungen extsprechen. Wiss Abn 1:101

Von Liebig J (1840) Organic chemistry and its applications to agriculture and physiology. Engl edn by Playfair L, Grogory W. London, p 387

Wagner G, Wagner B (1978) Zur Einschichtung von Flusswasser in den Bodensee — Obersee Schweiz. Z Hydrol 40/2:231–248

Wakeham SG, Canuel EA (1988) Organic geochemistry of particulate matter in the eastern tropical North Pacific Ocean: implications for particle dynamics. J Mar Res 46:183–213

Wakeham SG, Lee C, Farrington JW, Gagosian RB (1984) Biogeochemsitry of particulate organic matter in the oceans: results from sediment trap experiments. Deep Sea Res 31:509–528

Walling DE (1974) Suspended sediment and solute yields from a small catchment prior to urbanization. In: Gregory KJ, Walling DE (eds) Fluvial processes in instrumented watersheds. Inst Br Geogr Spec Publ 6:169–192

Walling DE, Webb BW (1981) The reliability of suspended sediment load data. In: Proc IAHS Symp Erosion and sediment transport measurement. IAHS Publ 133:177–194

Walsh J, Dymond J, Collier R (1988) Rates of recycling of biogenic components of settling particles in the ocean derived from sediment trap experiments. Deep Sea Res 35:43–58

Walsh JJ (1989) How much shelf production reaches the deep sea? In: Berger WH, Smetacek VS, Wefer G (eds) Productivity of the ocean: present and past. John Wiley & Sons, New York, pp 175–191

Walsh JN, Buckley F, Barker J (1981) The simultaneous determination of the rare-earth elements in rocks using inductively coupled plasma source spectrometry. Chem Geol 33:141–153

Wang BC, Eisma D (1988) Mud flat deposition along the Wenzhou coastal plain in southern Zhejiang, China. In: De Boer PL, van Gelder A, Nio SD (eds) Tide-influenced sedimentary environments and facies. Reidel, Dortrecht, pp 265–274

Wang BC, Eisma D (1990) Supply and deposition of sediment along the north bank of Hangzhou Bay, China. Neth J Sea Res 25:377–390

Wang H, Liang SS (1975) Mechanics of suspended sediment in random waves. J Geophys Res 80:3488–3494

Wang WC, Evans RL (1967) Variation of silica and diatoms in a stream. Limnol Oceanogr 14:941–944

Ward LG (1981) Suspended-material transport in marsh tidal channels, Kiawah island, South Carolina. Mar Geol 40:139–154

Wassmann P (1985) Sedimentation of particulate material in two shallow, land-locked fjords in western Norway. Sarsia 70:317–331

Wassmann P, Naas KE, Johannessen PJ (1986) Annual supply and loss of particulate organic carbon in Nordåsvannet, a eutrophic, land-locked fjord in western Norway. Rapp P Réun Cons Int Explor Mer 186:423–431

Weatherly GL, Kelley EA Jr (1985) Storms and flow reversals at the HEBBLE site. Mar Geol 66:205–218

Weaver CE (1958) The effects and geologic significance of potassium "fixation" by expandable clay minerals derived from muscovite, biotite, chlorite and volcanic material. Am Mineral 43:389–861

Weilenmann U, O'Melia CR, Stumm W (1989) Particle transport in lakes: models and measurements. Limnol Oceanogr 34:1–18

Weisberg RH (1976) The non-tidal flow in the Providence river of Narragansett Bay: a stochastic approach to estuarine circulation. J Phys Oceanogr 6:721–734

Weisberg RH, Sturges W (1976) Velocity observations in the West Passage of Narragansett Bay: a partially mixed estuary. J Phys Oceanogr 6:345–354

Welch PS (1935) Limnology. McGraw-Hill, New York London

Wells JT (1978) Shallow water waves and fluid-mud dynamics, coast of Surinam, South America. Coast Stud Inst Lanisiama State Univ, Baton Rouge, Tech Rep 257, 56 pp

Wells JT (1988a) Distribution of suspended sediment in the Korea Strait and southeastern Yellow Sea: onset of winter monsoons. Mar Geol 83:273–284

Wells JT (1988b) Accumulation of fine-grained sediments in a periodically energetic clastic environment. Cape Lookout Bight, North Carolina. J Sediment Petrol 58:596–606

Wells JT, Kemp GP (1986) Interaction of surface waves and cohesive sediments: field observations and geologic significance. In: Mehta AJ (ed) Estuarine cohesive sediment dynamics. Proc Worksh Tampa Florida, Nov 1984. Lecture Notes Coast Est Studies 14. Springer, Berlin Heidelberg New York, pp 43–65

Wells JT, Shanks AL (1987) Observations and geologic significance of marine snow in a shallow-water, partially enclosed marine embayment. J Geophys Res 92:C12 13185–13190

Wells JT, Coleman JM, Wiseman WJ (1978) Suspension and transportation of fluid mud by solitary-like waves. In: Proc 16th Coast Eng Conf Hamburg 2:1937–1952

Wells JT, Adams CE Jr, Park Y-A, Frankenberg EW (1990) Morphology, sedimentology and tidal channel processes on a high-tide-range mudflat, west coast of South Korea. Mar Geol 95:111–130

Whitehouse UG, Jeffrey LM, Debbrecht JD (1960) Differential settling tendencies of clay minerals in saline waters. In: Swineford A (ed) Clays and clay minerals. Proc 7th Natl Conf, Pergamon, New York, pp 1–79

Whitney LV (1937) Microstratification of the waters of inland lakes in summer. Science (2200:224–225

Wicker CF (1973) Nature, source and cause of the shoal. US Army Eng Distr Philadelphia Long Range Spoil Disposal Study, pt III, Substudy 2, 95 pp

Williams PJ (1981) Primary productivity and heterotrophic activity in estuaries. In Martin J-M, Burton JD, Eisma D (eds) River inputs to ocean systems. UNESCO, Paris, pp 243–249

Williams JDH, Syers JK, Shukla SS, Harris RF, Armstrong DE (1971) Levels of inorganic and total phosphorus in lake sediments as related to other sediment parameters. Environ Sci Technol 5:1113–1120

Williams JDH, Murphy TP, Mayer T (1976) Rates of accumulation of phosphorus forms in Lake Erie sediments. J Fish Res Board Can 33:430–439

Wilson L (1977) Sediment yield as a function of climate in United States rivers. In: Proc IAHS Symp Erosion and solid matter transport measurement. IAHS Publ 122:82–92

Windom HL (1975) Eolian contributions to marine sediments. J Sediment Petrol 45:520–529

Windom HL (1976) Lithogenous material in marine sediments. In: Riley JP, Chester R (eds) Chemical oceanography, 2nd edn, vol 5. Academic Press, New York London, 103–135

Windom HL, Gross ThF (1989) Flux of particulate aluminium across the southeastern US continental shelf. Est Coast Shelf Sci 89:327–339

Wolanski E, Ridd P (1986) Tidal mixing and trapping in mangrove swamps. Est Coast Shelf Sci 23:759–771

Wolanski E, Ridd P (1990) Mixing and trapping in Australian tropical coastal waters. In: Cheng RT (ed) Coastal and estuarine studies 38. Springer, Berlin Heidelberg New York, pp 165–183

Wolanski E, Chappal J, Ridd P, Vertessy R (1988) Fluidization of mud in estuaries. J Geophys Res 93:C3 2351–2361

Wolaver ThG, Dame FR, Spurrier JD, Miller AB (1988) Sediment exchange between a euhaline salt marsh in South Carolina and the adjacent tidal creek. J Coast Res 4:17–26

Wollast R (1981) Interactions between major biogeochemical cycles in marine ecosystems. In: Likens GE (ed) Some perspectives of the major biogeochemical cycles. John Wiley & Sons, New York, pp 125–142 (SCOPE vol 17)

Wollast R, Mackenzie FT (1983) The global cycle of silica. In: Aston SR (ed) Silicon geochemistry and biogeochemistry. Academic Press, New York London, pp 39–76

Wood JR, Jenkins BS (1973) A numerical study of the suspension of a non-buoyant particle within a turbulent flow. In: IAHR Int Symp River Mechanism, Bangkok, pp 431–442

Woodroffe CD (1985a) Studies of a mangrove basin, Tuff Crater, New Zealand. II. Comparison of volumetric and velocity-area methods of estimating tidal flux. Est Coast Shelf Sci 20:431–445

Woodroffe CD (1985b) Studies of a mangrove basin, Tuff Crater, New Zealand. III The flux of organic and inorganic particulate matter. Est Coast Shelf Sci 20:447–461

Wright LD, Coleman JM (1974) Mississippi River mouth rocesses: effluent dynamics and morphologic development. J Geol 82:751–778

Wright LD, Wiseman WJ, Bornhold BD, Prior DB, Suhayda JN, Keller GH, Yang Z-S, Fan YB (1988) Marine dispersal and deposition of Yellow River silts by gravity-driven underflows. Nature (Lond) 332(6164):629–632

Wright LD, Wiseman WJ, Yang Z-S, Bornhold BD, Keller GH, Prior DB, Suhayda JN (1990) Processes of marine dispersal and deposition of suspended silts off the modern mouth of the Huanghe (Yellow River). Cont Shelf Res 10:1–40

Wright LD, Boon JD, Kim SC, List JH (1991) Modes of cross-shore sediment transport on the shoreface of the Middle Atlantic Bight. Mar Geol 96:19–51

Wu J (1988) Bubbles in the near-surface ocean: a general description. J Geophys Res 93:C1 587–590

Wüst G (1955) Stromgeschwindigkeiten im Tiefen- und Bodenwasser des Atlantischen Ozeans auf Grund dynamischer Berechung der Meteor-Profile der Deutschen Atlantischen Expedition 1925/27. Deep Sea Res 3 (Suppl Bigelow Vol):373–397

Yalin MS (1977) Mechanics of sediment transport. Pergamon, New York, p 298

Yalin MS, Krishnappan BM (1973) A probabilistic method for determining the distribution of suspended solids in open channels. In: IAHR Int Symp River mechanics, Bangkok, pp 603–614

Yamamoto S (1979) Size distribution of detrital mineral grains suspended in surface waters of the Yellow Sea and East China Sea. J Oceanogr Soc Jpn 35:91–99

Young RA, Southard JB (1978) Erosion of fine-grained marine sediments: seafloor and laboratory experiments. Bull Geol Soc Am 89:663–672

Young RA, Clarke ThL, Mann R, Swift DJP (1981) Temporal variability of suspended particulate concentrations in the New York Bight. J Sediment Petrol 51:293–306

Author Index

Subject Index

Index of Geographical Names

Printing: COLOR-DRUCK DORFI GmbH, Berlin
Binding: Buchbinderei Lüderitz & Bauer, Berlin